TRIUMPHS of EXPERIENCE

TRIUMPHS of EXPERIENCE

THE MEN OF THE HARVARD GRANT STUDY

George E. Vaillant

THE BELKNAP PRESS *of*
HARVARD UNIVERSITY PRESS

Cambridge, Massachusetts
London, England

2012

Library of Congress Cataloging-in-Publication Data

Vaillant, George E., 1934–

   Triumphs of experience : the men of the Harvard Grant Study / George E. Vaillant.

     p. cm.

   Includes bibliographical references and index.

   ISBN 978-0-674-05982-5 (alk. paper)

    1. Men—United States—Longitudinal studies.   2. Aging—Social aspects—United States—Longitudinal studies.   3. Aging—Psychological aspects—United States—Longitudinal studies.   I. Title.

   HQ1090.3.V35   2012

    305.310973—dc23   2012028519

For Robin Western, who for twenty years has been
the linchpin holding the Study together

CONTENTS

CAST OF PROTAGONISTS (DECATHLON SCORE)

Chapter 1
Adam Newman (2). Rocket scientist who illustrates repression and growing maturity.

Chapter 2
Dr. Godfrey Camille (5). Lonely physician who spent his life finding love, lifelong personality development, and the ability to let love in.

Chapter 3
Art Miller (0). Elusive high combat veteran and drama professor who found peace in Australia.

Chapter 4
Oliver Holmes (6). Judge from an idyllic childhood who illustrates its gift to old age.
Sam Lovelace (0). Architect with a miserable childhood who illustrates its lasting curse.
Algernon Young (0). Blue-collar worker from an aristocratic family; an illustration of development derailed.

Chapter 5
Charles Boatwright (6). Optimistic boatyard owner; a model of empathy.
Professor Peter Penn (1). College professor. A man who never quite grew up.

Professor George Bancroft (7). College professor who illustrates the sequential steps to maturity.

Dr. Eric Carey (TY★). Polio victim who achieved Eriksonian Integrity in the face of an early death.

Dr. Carlton Tarryton (TY★). A physician who failed to reach Career Consolidation.

Chapter 6

John Adams (6). A lawyer with three divorces and a long happy fourth marriage.

Fredrick Chipp (7). A schoolteacher from a warm childhood who created a lifelong happy family.

Dr. Carlton Tarryton (TY★). A physician with four unhappy marriages.

Eben Frost (7). A mentally healthy lawyer who lacked the gift or desire for intimacy.

Chapter 7

Daniel Garrick (3). An actor, a "late bloomer" who illustrates the importance to old age of good (subjective and objective) physical health.

Alfred Paine (0). A manager with no ability to let love in; he illustrates aging in the absence of good health.

Chapter 8

Dylan Bright (TY★). An English professor gifted in sublimation.

Francis DeMille (2). A businessman. He illustrates personality development from repression and dissociation to maturity.

Chapter 9

James O'Neill (TY*). An economics professor who illustrates the destructive power of alcoholic denial.

Bill Loman, a.k.a. Francis Lowell (2). An aristocratic lawyer who provides simultaneous illustrations of alcoholism as voluntary career and alcoholism as debilitating disease.

Chapter 10

Bert Hoover (TY*). A Study conservative.

Oscar Weil (TY*). A Study liberal.

Chapter 11

Professor Ernest Clovis (3). A French professor who illustrates both sublimation and Guardianship.

TY: Died too young to receive a Decathlon score.

Decathlon Score significance: 0–1 bottom third, 2–3 middle third, 4–9 top third.

TRIUMPHS of EXPERIENCE

1 | MATURATION MAKES LIARS OF US ALL

No man ever steps in the same river twice;
for it is not the same river, and he is not the same man.

—HERACLITUS

THIS BOOK IS ABOUT how a group of men adapted themselves to life and adapted their lives to themselves. It is also about the Grant Study, now seventy-five years old, out of which this story came. In it I will offer tentative answers to some important questions: about adult development in general, about the people who engaged us in this exploratory venture, about the study itself, and, perhaps above all, about the pleasures and perils of very long scientific projects.

Originally the Grant Study was called the Harvard Longitudinal Study. A year later it became the Harvard Grant Study of Social Adjustments. In 1947 it received its now-official name, the Harvard Study of Adult Development. But to its members and its researchers, and in early books, it has always been the Grant Study.

It began in 1938 as an attempt to transcend medicine's usual preoccupation with pathology and learn something instead about optimum health and potential and the conditions that promote them. The first subjects were 64 carefully chosen sophomores from the all-male Harvard College classes of 1939, 1940, and 1941, who took part in an intensive battery of tests and interviews. That first group was joined by sophomores from the next three Harvard classes, resulting in a final *cohort* (as the panel of subjects in a study of this kind is called) of 268 men. The original intention was to follow these healthy and privileged men for fifteen or twenty years, supplementing the intake data from time to time with updates. Thus an abundance of information would accumulate about the men and the lives they constructed for themselves—information that could be analyzed at will over time and across different perspectives. (Interested readers will find much, much more on the history and structure of the Study in Chapter 3.)

That plan was realized, and more. Almost seventy-five years later, the Grant Study still, remarkably, goes on. We're asking different questions now than the founders asked when the Study began, and our investigative tools are different. Of course the participants are no longer the college sophomores they once were; those who are still with us are very old men indeed. Time has called many of the beliefs of those days into question, and some much more recent ones, too. How long our current conclusions will hold up we cannot know.

But whatever the uncertainties, asking questions and trying to answer them is always a fruitful process. We actually have learned some of what they wanted to know back in 1938: who would make it to ninety, physically capable and mentally alert; who would build lasting and happy marriages; who would achieve conventional (or unconventional) career success. Best of all, we have seventy-five years' worth of data that we can refer back to (over and over again, if need be) as we try to understand *why* these things turned out as they did.

We can use those data to try to answer other questions, too. There are old ones still open from the early Study days—about the relative importance of nature and nurture, for instance, or how mental and physical illness can be predicted, or the relationship between personality and health. There are new ones that could never have been dreamed of in 1938, like what the emotions of intimacy look like in the brain. And some old questions have morphed into new ones as we learn to formulate ever more cogently just what we're seeking to accomplish in even making inquiries like these. This last in particular is an abiding concern of good science.

Thus this story of the Grant Study: how it came to be, how it developed, and how those of us who participated in it developed too. It's the story of what we've learned from it and what we haven't (yet). And it's a story about Time—studying it and living in it.

LONGITUDINAL STUDIES: THEIR STRUCTURE AND PURPOSE

The Grant Study is a longitudinal prospective study. Let me say a few words about that as we set out. In a *longitudinal* study, a group of participants (a *cohort*) is observed over time, and data about points of interest *(variables)* are collected at repeated intervals. In contrast, observations in a *cross-sectional* study are made only once. A longitudinal study can be prospective or retrospective. *Retrospective* studies look back over time, seeking to identify the variables that might have contributed to *outcomes* that are already known. *Prospective* studies follow a cohort in real time, tracking target variables as the subjects' lives proceed, and identifying outcomes only as they occur. Accordingly, the Grant Study collected all kinds of potentially (but not necessarily) relevant information about its cohort members over the many years of their lives, looking to learn what it could about health and success. And it has correlated this information periodically with the levels of health and success that each man actually achieved. You'll see many examples of this process as we go on.

Less abstractly, longitudinal studies let us contrast eighty-year-olds with themselves at twenty-five or sixty. Biographies and autobiographies, and the debriefings that elders offer their fascinated grandchildren, do the same thing. But these are all retrospective narratives, inevitably influenced by forgettings, embellishments, and biases. Time is a great deceiver.

Prospective studies are like the baby books compiled by doting parents, or like time-lapse photographs; they document change as it happens, allowing us to visualize the passage of time free from the distortions of memory. While butterflies recalling their youths tend to remember themselves as young butterflies, prospective studies capture the reality (hard to believe, and often avoided!) that butterflies and caterpillars are the same people.

But baby books and even year-long time-lapse videos are a dime a dozen. Prospective studies of entire lives are very rare indeed. None are known to exist before 1995; I'll have more to say on that shortly. In fact, observational data on adult development have been so sparse that when Gail Sheehy and Daniel Levinson produced their groundbreaking books on the subject in the 1970s, the cross-sectional studies they relied on led them to some very erroneous conclusions (such as the inevitability of the so-called midlife crisis).[1]

Another important reason for prospective studies is that they establish contexts for the outcomes we're trying to understand. The most experienced handicapper in the world can't predict for sure which of the well-bred, handsome horses in the Churchill Downs paddocks each May will win the Derby. Certainty comes only after the race is won, and even then we only know who. The hows and whys remain a mystery. It's easy to rate a beautiful woman according to how many people think her so, or how well she conforms to some established ideal, or how long she keeps her looks. But ratings like that can't encompass the difference between being beautiful at one's high school prom and being beautiful at one's great-granddaughter's wedding, nor between the beauty at eighteen that is the luck of the draw, and the beauty at eighty that is the result of a life generously lived. The Grant Study was hoping to learn something about nuances like these in its exploration of success and "optimum" health.

Part of my intent in this book is to show why these nuances matter, and why, therefore, we need longitudinal studies if we want to learn about human lives. I will be focusing primarily on the Grant Study and its cohort, whom I will call the *College* men. But I will occasionally make reference to two other important cohorts. One is the *Inner City* cohort of the Glueck Study of Juvenile Delinquency.[2] The Glueck Study is a second Harvard-based prospective longitudinal lifetime study, begun independently of the Grant Study in 1940; its participants were a group of youths from disadvantaged urban Boston neighborhoods. Since 1970, one cohort of this group has been administered together with the original Grant Study, as part of the Harvard Study of Adult Development. (I will use that term when referring to both studies, but will continue to call the Grant Study per se by its original name.) I will also refer from time to time to the *Terman women,* a cohort from Stanford University's (1920–2011) Terman Study of gifted children, to whose data and participants I have had partial access.[3] It was the coming of age of the Terman cohort (male and female) in 1995 that made prospective lifetime data available for the first time. The Grant Study's access to the Inner City men and the Terman women permitted us to contrast our privileged, intelligent, and all-male College sample with another all-male sample of very different socioeconomic and intellectual profile, and with a group of even more intelligent, though not particularly privileged, women. When these contrasts are enlightening, I will note them. The Glueck and Terman studies (and some of the correlations among them) are described in their own contexts in Appendix B, and are also considered in my previous books *Aging Well* and *The Natural History of Alcoholism Revisited.*[4]

I acknowledge readily that the Grant Study is not the only great prospective longitudinal lifetime study. There are others, three of which are better known than ours. Each has its own strengths and weaknesses. The Berkeley and Oakland Growth Studies (1930–2009) from the University of California at Berkeley include both sexes and began when the participants were younger; they provide more sophisticated childhood psychosocial data but little medical information.[5] These cohorts have been very intensively studied, but they are smaller and have suffered greater attrition than ours. The Framingham Study (1946 to the present) and the Nurses Study at the Harvard School of Public Health (1976 to the present) boast better physical health coverage, but they lack psychosocial data.[6] These are wonderful world-class studies, invaluable in their own ways, and more frequently cited than the Grant Study. But even in this august company the Grant Study is unmistakable and unique. It has been funded continuously for more than seventy years; it has had the highest number of contacts with its members and the lowest attrition rate of all; it has interviewed three generations of relatives; and, most crucial for adult development, it has consistently obtained objective information on both psychosocial and biomedical health.[7] Finally, perhaps alone among the world's significant longitudinal studies, the Grant Study has published, with the men's permission of course, lifespan histories as well as statistical data.

Other studies exist now in Great Britain, Germany, and the United States that are larger and more representative than these older ones, and will join them in length of follow-up in another decade or two. The Wisconsin Longitudinal Study, for example, began in 1957 and included about a third of all of Wisconsin's high school graduates of that year; it has endured for over half a century so far.[8] Eighty-eight percent of its surviving members are still active in the study at age sixty-five. (By way of comparison, 96 percent of the surviving Grant Study members are still active at age ninety!) The Wisconsin Study is more demographically representative than the other studies, and its economic and sociological data are richer and better analyzed. It has a weakness too, however; it lacks face-to-face medical examinations or interviews. We can anticipate a great wealth of prospective life data as these younger studies come into their own. But they will supplement, not supplant, the riches already offered by the Grant Study and its contemporaries.

When I came to the Grant Study in 1966, I was a very young man of thirty-two, and had not yet achieved the pragmatic relativism of Heraclitus. I had been studying stable remissions in schizophrenia and heroin addiction—recovery vs. no recovery, white vs. black. I spent my first ten years at the Grant Study identifying thirty unambiguously good outcomes and thirty unambiguously bad ones out of a randomly selected sample of one hundred middle-aged men from the classes of 1942–44. In 1977 I published a book, *Adaptation to Life,* demonstrating this accomplishment.[9] It made quite a splash. I was forty-three. What did I know?

Today I am seventy-eight. The men of the Grant Study are in their nineties. They're not the same as they were when they joined the Study, and neither am I. I have learned to appreciate how few blacks and whites there are in human lives, and how we and our rivers change from moment to moment. The world we live in is different; science is different; even the technology of documenting difference is different. As the Grant Study becomes one of the longest studies of adult development in the world, it is not only the men of six Harvard classes who are under the microscope. All of us who have worked on the Study—the Study itself, in fact—are now as much observed as observers.

ADULTS, GROWING

The laws of adult development are nowhere near as well known as the laws of the solar system or even the laws of child development, which were only discovered in the last century. It wasn't all that long ago that Jean Piaget and Benjamin Spock were still kids, and the phases of childhood so stunningly elucidated by them still regarded as unpredictable. Now, however, we watch children develop the way our ancestors watched the orderly waxings and wanings of the moon. We worry; we pray; we weep; we heave sighs of relief. But we are no longer particularly surprised. We know what to expect. Our libraries are full of studies of human development—up to the age of twenty-one.

What happens after that remains in many ways a mystery. Even the notion that adults *do* develop, that they don't reach some sort of permanent steady state at voting age, has been slow to gain traction. Some of the reasons for this are obvious. It is much easier to achieve a long perspective on childhood than on an entire life. And as in physics, once Time gets into the picture you can say good-bye to the old Newtonian verities. There are certainly patterns and rhythms to adult life, and when we circumvent the distorting effects of time upon our own vision, we can sometimes discern them.

But even the most carefully designed prospective study in the world can never free us completely from time's confounding influence. Lifetime studies have to last many, many years, and over those years everything will be changing—our questions, our techniques, our subjects, ourselves. I've been studying adult development since I was thirty, and I know now that many of my past conjectures, apparently accurate at the time, were contingent or just plain wrong. Still, the Grant Study is one of the first vantage points the world has ever had on which to stand and look prospectively at a man's life from eighteen to ninety. Having it doesn't save us from surprises, frustrations, and conundrums; quite the contrary. Nevertheless, a continuous view of a lifetime is now possible for human beings. Like that time-lapse film of a flower blooming that Disney made famous in the sixties, it is an awesome gift.

When the Study was undertaken in 1938, what we knew about human development across the whole of life was mostly based on inspiration or intuition. William Shakespeare delineated seven ages of man in *As You Like It* in 1599; Erik Erikson defined eight stages in *Childhood and Society* three hundred and fifty years later.[10] But Shakespeare and Erikson didn't have much by way of real data to go on. Neither did Sheehy and Levinson. Neither did I, in *Adaptation to Life.* *Nobody* had access to prospectively studied whole lifetimes. I hope that this book, written in 2012, will begin to correct that lack.

THE ART OF THE POSSIBLE

A few caveats as we proceed. It is well known that the Grant Study includes only white Harvard men; Arlen V. Bock, the physician who founded it, has frequently been criticized for arrogance and chauvinism on that account. It's less well known that the Grant Study was not an attempt to document average health over time, like the more famous Framingham Study, but to define the best health possible. And we must therefore keep in mind two realities. First, lifetime studies, like politics, are the art of the possible. As Samuel Johnson famously quipped about dogs walking on their hind legs, "It is not done well, but you are surprised to find it done at all." Second, in such huge undertakings, one must optimize one's chance of success. Columbia neuroscientist Eric Kandel did not choose a random sample of the world's population of *homo sapiens* when he did his Nobel-winning work on the biology of memory; he chose the obscure sea snail called *Aplysia.* Why? Because *Aplysia* has unusually large neurons. And it was precisely the gender and privilege of the Grant Study men that made them so useful for a study of human adaptation and development. Men don't change their names in midlife and disappear to follow-up as women do. Well-to-do men don't die early of malnutrition, infection, accident, or bad medical care, as happens much too often to poor ones. These men had a high likelihood of *long* life, a necessity for this sort of study. (A full 30 percent of the Grant Study men have made it to ninety, as opposed to the 3 to 5 percent expected of all white male Americans born around 1920.) Glass ceilings and racial prejudice were unlikely to hold them back from achieving to their fullest potential, or from the careers and lives that they desired. A Harvard diploma wouldn't hurt either. When things went wrong in the lives of the Grant Study men, they would have a good chance of being able to set them right. Last but not least, they were unusually articulate historians. Bock needed all those advantages. You can't study the development of delphiniums in Labrador or the Sahara. The Grant Study's College cohort and *Aplysia* may not be perfectly representative, but they both afford us windows onto landscapes we have never been able to see before.

One further note on Bock's choice of a homogeneous population. If we want to learn what people eat, we have to study many different populations. If we want to learn about gastrointestinal physiology, however, we try to keep variables like cultural habits and preferences uniform. Societies are forever changing, but biology mostly stays the same. This was another reason for the Grant Study's strategy. It was examining healthy digestion, not traditional menus. When possible, however, the Grant Study has checked its findings against other homogeneous studies of different populations, especially the disadvantaged men of the Inner City cohort and the highly educated women of the Terman cohort.

STUDYING THE STUDY

It is reasonable to ask whether this book is necessary. Over its seventy-five years of existence, the Study of Adult Development has so far produced 9 books and 150 articles, including quite a number of my own (see Appendix F). Why another? Well, because it's not always easy to see how reports of any given instant relate to the future, or even to the past. How do we understand a seventy-five-year-old article from a news daily? The short answer is, it depends on what's happened since. Whatever the world's papers were printing in the summer of 1940, England did not in fact fall to the Luftwaffe. Reporters do the best they can with what's available, but momentary glimpses can never capture the totality of history, and things may well look different later on. This in a nutshell is why longitudinal studies are so important. And to reap their full value, we have to take a longitudinal view of the studies themselves. When I sit down to summarize my forty-five years of involvement with the Grant Study, much of what I've written in the past seems to me as ephemeral as the 1948 headlines that trumpeted Dewey's victory over Truman, for only now do I understand how the story has really turned out. So far. As the people's philosopher Yogi Berra observed long ago, "It ain't over till it's over."

So as I see it, there are five reasons for this book. First, the Harvard Study of Adult Development is a unique, unprecedented, and extremely important study of human development. For that reason alone, its history deserves documentation

Second, the Grant Study has been directed by four generations of scientists whose very different approaches beg for integration. The first director focused on physiology, the second on social psychology, the third (myself) on epidemiology and adaptation. The priorities of today's director, the fourth, are relationships and brain imaging. As any methodologist can see, the Grant Study had no overarching design. In 1938, there weren't enough prospective data on adult development even to build solid hypotheses. Like the Lewis and Clark Expedition and Darwin's passage on the *Beagle,* the Grant Study was not a clearly focused experiment but a voyage of discovery (or, as some have less charitably suggested, a fishing trip). The findings I report in this book are largely serendipitous.

This is the first time in forty-five years I have allowed this awareness into consciousness, let alone confessed it in print. In my repeated requests to the National Institutes of Health for funding, I always emphasized the potential of the Study—the power of its telescopic lens, its low attrition rate—rather than any specific hypotheses I planned to test. I analyzed data whenever I had a good idea; like a magpie I'd scope out our huge piles of accumulated material looking for a telltale glint or glimmer. I have often wondered whether there is a Ph.D. program in the country that would have accepted the plans of any of the four Grant Study directors as a thesis proposal. But what a wonderful harvest serendipity has produced. As I will demonstrate with pleasure, unpredictability is an inevitable and sometimes infuriating aspect of large prospective studies, but it gives them a startling richness that more narrowly focused endeavors can never achieve. All the better for magpies. Perhaps it was a good thing that I never took a course in psychology or sociology. At least I had no preconceived ideas ruining my eye for an emperor's new clothes.

Third, this book collects in one place material scattered across seventy years of specialty journals, in which the publications of each decade modified the findings of the previous ones, and were modified again in turn as new contexts cast new light on old data. Until now, for example, there's been no recognition of the importance of alcoholism in studies of development. But it's clear that earlier findings in some very unexpected areas are going to have to give way in the face of the Harvard Study of Adult Development's accumulating evidence about the developmental effects of alcohol abuse. It proved to be the most important predictor of a shortened lifespan, for one thing, and it was a huge factor in the Grant Study divorces, for another. Science is a changing river too.

Fourth, the Grant Study has seen and absorbed many theoretical and technological transitions, especially in the evolving field of psychobiology. It began in a day when blood was still typed as I, II, III, and IV. Race, body build, and (more speculatively) the Rorschach were considered potentially predictive of adult developmental outcomes. Data were tabulated by hand in huge ledgers—even punch cards sorted with ice picks were a yet-undreamed-of technology. For calculation, the slide rule ruled. Today we grapple with DNA analysis, fMRIs, and attachment theory, and 2,000 variables can be stored on my laptop and analyzed instantly as I fly between Cambridge and Los Angeles. Here, too, documentation and integration are called for.

A fundamental paradox is my fifth and final reason for writing this book. Despite all the changes I aim to document, we are all still the same people—the men who joined the Study seventy-plus years ago, and I, who came to direct it in 1966. One of the great lessons to emerge in the last thirty years of research on adult development is that the French adage is right: *Plus ça change, plus c'est la même chose.* People change, but they also stay the same. And the other way around.

A NOTE ON THE LIFE STORIES

I will be making my points in this book not only with numbers, but also with stories. The biographies of all living protagonists have been read and approved by them. All names are pseudonyms, but these narratives are not composites. They accurately depict the lives of real individuals, with the one stipulation that I have carefully altered some identifying details according to strict rules. I may substitute one research university for another, for example, but I do not substitute a research university for a small college, or vice versa—Williams College for Swarthmore, perhaps, but not for Yale. I allow Boston as an acceptable interchange for San Francisco, and Flint for Buffalo, but neither for Dubuque or Scarsdale. I have made similar narrow replacements among types of illnesses and specific careers. In this way I have striven to remain faithful to the spirit of these lives and maintain their distinctive flavor, while altering the letter to ensure privacy. From time to time a prominent member of the Study has made a public comment relating to his participation in it. When I have quoted such comments, I have not disguised their authorship.

THE LIFE OF ADAM NEWMAN

Let me begin my history of the Grant Study with a story that illustrates many of the themes that have intrigued, enlightened, and confounded me over our years together. The protagonist is Adam Newman (a pseudonym), whose life—seemingly different with every telling—confronted us constantly with the realities of time, identity, memory, and change that are the heart of this book.

Newman grew up in a lower-middle-class family. His father was a bank worker who never finished high school. One grandfather had been a physician, the other a saloon owner. There was relatively little mental illness in Newman's family tree. Nevertheless, his childhood was grim. His mother told the Study that her first way of dealing with Adam's tantrums was to tie him to the bed with his father's suspenders. When that didn't work, she took to throwing a pail of cold water in his face. Later she spanked him, sometimes with a switch. Adam became extremely controlled in his behavior. He maintained a strict belief in and observance of Catholic teachings, and concentrated on getting all A's in school. His father was more lenient than his mother, but also more distant. "He only recognized that I was one of his children about once a month," Adam said. There was little show of affection in the family, and in Newman's 600-page record there is not a single happy childhood memory recounted. Rereading that record for this book, I see too that Newman says almost nothing about his father's death when he was seventeen.

In high school, Adam was a leader. He was a class officer in all four years, and also an Eagle Scout. He had many acquaintances, but no close friends. When he was a sophomore at Harvard, a few of his intake interviewers for the Study described him as "attractive," with "a delightful sense of humor." Others, however, described him as aloof, rigid, inflexible, repelling, self-centered, repressed, and selfish—the first of a lifetime of clues that this was a man of contradictions.

Newman's physical self was scrutinized minutely when he entered the Study, because the scientific vogue of the time was that constitutional and racial endowment could predict just about everything important in later life. He was described as a *mesomorph of Nordic race with a masculine body build* (all presumably excellent predictors for later success), but in poor physical condition. He was among the top 10 percent of the Grant Study men in general intelligence, and his grades were superior. As in high school, he had many acquaintances but few friends; he joined only the ornithology club and, eventually, that least social of fraternities, Phi Beta Kappa. One psychologist observed that he was "indifferent to fascism," and Clark Heath, the Study internist and director, noted that he "did not like to be too close to people."

In short, Adam lived mostly inside his own head. His personality was described as *Well Integrated,* but he was also considered a man of *Sensitive Affect, Ideational,* and *Introspective;* you'll hear more about all these traits as we go along. He gained distinction in the psychological testing on two counts: for his intellectual gifts, and for being "the most uncooperative student who ever agreed to cooperate in our experiments." In psychological "soundness" he ended up classified a "C," the worst category. (For much more on the assessment process, see Chapter 3.)

The Study psychiatrists, informed by another theoretical fashion of the period, were more interested in Newman's masturbatory history than in his social life at college. They took pains to classify him as *cerebrotonic,* as opposed to *viscerotonic* or *somatotonic.* (These terms imply constitutional distinctions between people who live by the intellect, by the senses, and by physical vigor, but they were never satisfactorily defined.) It's also worth noting here, and it will come up again, that the early Study designers never put to the test their deep belief that body build is destiny. That had to wait for many years, even though opportunities for an empirical trial soon became available.

As a nineteen–year–old sophomore, Newman took a hard line on sex. He condemned masturbation, and he boasted to the Study psychiatrist that he would drop any friend who engaged in premarital intercourse. The psychiatrist noted in his turn, however, that although Adam disapproved of sexual activity, he was "frankly very much interested in it as a topic of thought." Newman told the psychiatrist a dream, too: two trees grew together, their trunks meeting at the top to form a chest with two drawers side by side, suggestive of a naked woman. He would wake from this dream, which visited him repeatedly, filled with anxiety.

To my mind, there was no adolescent in the Study who better exemplified psychoanalytic ideas about repressed sexuality than Adam Newman. And it was typical of the Grant Study's approach at the time that while Freud's theories were carefully explained to Newman (who vigorously rejected them), no one asked him about dating or friendships.

He was as dogmatic politically as he was sexually, and he regularly tore up the "propaganda" he received from the "sneaky" college Liberal Union. He claimed a commitment to empirical science both as an idea and as a career, but he also remained a practicing Catholic, attending Mass four times a week. When an interviewer wondered how his religious and scientific views fit together, he replied, "Religion is my private refuge. To attack it with my intellect would be to spoil it." More contradiction.

It wasn't until ten years later, when the scientific climate had shifted away from physical anthropology toward social psychology, and relationships had become a matter of interest, that the Study recorded that Newman had had few close friends in college besides his roommate, and that he had dated only rarely. In part this may have been because Adam was working to pay all of his college expenses himself, and was also sending money home to his fatherless family. But it's also true that he met early a Wellesley College math major who would become his wife and "best friend forever."

Adam went to medical school at Penn. He didn't want to minister to the sick, but he did want to study biostatistics and avoid the draft. He married during his second year there, but except for his wife he remained rather isolated. He had no more interest in World War II than he had in patient care. On graduation he fulfilled his military obligations with classified research at Edgewood Arsenal, America's research locus for biological warfare, and he continued in that work after the war. By 1950, Dr. Heath, the Study director, noted that Captain Newman, shaping the Cold War's nuclear deterrents but never seeing patients, was making more money than any of the other forty-five physicians in the Study. One of Newman's unclassified papers was "Burst Heights and Blast Damage from Atomic Bombs."

Despite these oddities, the record shows that by 1952, when he was thirty-two, Newman was steadily maturing. He was settling into a lifetime career in the biostatistics that he loved, using his leadership talents to build a smoothly running department of fifty people at NASA. His ethical concerns were engaged professionally too; in the 1960s his group participated in President Johnson's plan to put the military-industrial complex to work on the economic problems of the third world. His marriage would remain devoted for fifty years, by his wife's testimony as well as his own. It was an eccentric marriage— the two of them acknowledged each other as best friend, but neither had any other intimate friends at all. But many Study men with bleak childhoods sought marriages that could assuage old lonelinesses without imposing intolerable relational burdens, and Adam Newman seems to have found one. Perhaps this is why, unlike many men with childhoods like his, he effortlessly mastered the adult tasks of Intimacy, Career Consolidation, and Generativity (see Chapter 5) so early in his life.

Furthermore, whereas before he had denied intense feelings, perhaps out of fear of being overwhelmed, he was becoming more able to handle them. During the Study intake proceedings when he was nineteen, he described the atmosphere at home as "harmonious, my mother affectionate and my father the same." He was given a Jungian word association test (despite its preoccupation with physique, the Study did try, in an eclectic way, to measure the complete man), and Newman associated the word *mother* with *affectionate, kind, prim, personal, instructive,* and *helpful.* But the seasoned and tolerant social investigator who interviewed his mother experienced her as a "very tense, ungracious, disgruntled person," and Adam's sister commented, "Our mother could make anyone feel small." It wasn't until 1945, six years after he left home and five years after his initial workup, that Newman was able to write to the Study of his childhood, "Relations between me and my mother were miserable." "I don't remember ever being happy," he went on, and he recalled his mother having told him that she was sorry she had brought him into the world.

At age sixty, he took the sentence completion test developed by Jane Loevinger at Washington University to assess ego maturity.[11] The stem phrase was: *When he thought of his mother . . .* and Newman's response was *. . . he vomited.* Yet it would be an oversimplification to say that he went on to become steadily more conscious of difficult feelings, especially the painful relationship with his mother, because at age seventy-two he could not believe he had ever written any such answer. People are complicated; memory, emotion, and reality all have their own vicissitudes, and they interact in unpredictable ways. That's one of the reasons prospective data are so important.

The young Newman was extremely ambitious. "I have a drive—a terrible one," he said of himself in college. "I've always had goals and ambitions that were beyond anything practical." But by thirty-eight he had gained some insight into the "terrible" drive of the college years: "All my life I have had Mother's dominance to battle against." This realization caused a change in his philosophy of life. Now, he said, his goals were "no longer to be great at science, but to enjoy working with people and to be able to answer 'yes' to the question I ask myself each day, 'Have you enjoyed life today?' . . . In fact, I like myself and everyone else much more." He hadn't become a hippie; it was 1958 and that scene was still some years away, and in fact Newman's fierce ambition was still burning in his heart. This was just another manifestation of the complexities of the man.

By the time he was forty-five, the laid-back freedom of his thirties was once more in abeyance as he battened down the hatches to deal with rebellious and sexually liberated daughters. Identifying unwittingly with his mother, he was now insisting that "greatness" be the goal of his intellectually gifted girls. Twenty years later, one daughter had still not recovered from his pressure. She described her father as an "extreme achievement perfectionist," and wrote that her relationship with him was still too painful to think about. She felt that her father had "destroyed" her self-esteem, and asked that we never send her another questionnaire.

I wish I knew what happened between them in Newman's later years, when he had mellowed remarkably. Alas, I do not. But I do know that Time continued to wreak its changes. Occasional backsliding notwithstanding, Newman became more flexible than he had once been. The tectonic shifts of the sixties, and being father to such

adventurous progeny, loosened him up sexually. He stopped disavow-
ing Freud's theories. As his daughters reached adulthood, he (reluctantly) abandoned his prohibitions against premarital sexuality. He
stopped fearing the "sneaky liberals" and began to share the view that
law and order was "a repressive concept." In fact, he came to believe
that "the world's poor are the responsibility of the world's rich." He
quit his job in the military–industrial complex and put his scientific
books under the house "to mildew." At sixty he was solving agricul-
tural problems in the Sudan, using statistical knowledge gleaned in the
planning of retaliatory nuclear strikes. The man who had gone to Mass
four times a week in college proclaimed, "God is dead, and man is
very much alive and has a wonderful future." In fact, as soon as Newman's happy marriage had begun buffering the terrible loneliness of
his childhood, his dependence on religion had diminished greatly.
Eventually he became an atheist. In mid-adulthood his mystical side
was being expressed in meditation, and he began teaching psychology
and sociology at the university level.

This was not a complete changing of spots. To some degree, at
least, we always remain the people that we were. On the one hand,
Newman could write that he had learned from his daughters that
"there was more to life than numbers, thought and logic." But he still
limited himself to only one intimate relationship—that is, with his
wife. And although in his laboratory work he had become an increasingly generative leader, guiding those entrusted to his supervision
with attention and concern, he was still a technician. Even in his
teaching, he approached psychology and sociology not through the
study of feelings, but through explorations of "the linguistic deriva-
tion of words like 'relationship' and 'love.'"

So what had happened? Clearly these weren't drastic personality
changes. They weren't intellectual changes, either. Newman hadn't
taken any riveting new psychology courses or met any charismatic

psychiatrists. This was an evolution of his personality, which could be seen in the increased ability—not uncommon as people get older (see Chapter 5)—to stay aware of uncomfortable feelings without having to control or disown them. As Newman slowly became less anxious about his own sexuality, he had less need to condemn or control the sexuality of others. He was recapitulating a developmental process that is familiar to us in much younger people—the one that accompanies grammar-school children into the tumult of adolescence. As Newman struggled free of parental domination, he achieved a less constricted morality and became more comfortable with himself. In that greater comfort, he moved toward a greater comfort with, and willingness to be responsible for, others. None of the great psychologists of the nineteenth and early twentieth centuries, including Freud and William James, had had anything to say about adult maturational processes like this. But over the decades, we at the Grant Study have watched fascinated as Adam Newman and his fellows changed and grew.

Indeed, one great triumph of the Grant Study has been to make clinical documentation of changes like his available at exactly the moment when psychologists like Erik Erikson were beginning to recognize, schematize, and theorize about them. The loosening of the Sixties played a part in Newman's evolution, but only a part. Not all of the Grant Study men did as he did at that time; some just dug their heels in deeper. This is one important truth of adult development: people grapple with growing up in their own ways, but we all do have to grapple. Obviously the men in the Study didn't all follow the same course of sexual development as Newman, any more than all children handle adolescence identically. But one way or another we all have to come to terms with our own sexuality, and the ways that we do (and don't) will end up shaping our lives.

Newman's story illustrates another endemic issue in long-term studies—the vicissitudes of subjects' memories. (It illustrates the vicis-

situdes of researchers' memories too, as you will see in a moment.)
When I interviewed him at fifty, the only recurrent dream he could
recall was of urinating secretly behind the garage, and he insisted that
he had given up church as soon as he got to Harvard, having come to
doubt the validity of religion. The four Masses a week and the dream
of the naked woman/tree were as unavailable to him at fifty as his
mother's unkindness had been when he was nineteen.

Newman's recollections of his history continued to alter as the
years went on, always in the service of a process of psychological development; he was adjusting his inner self to the real world as he
cautiously admitted more and more of his emotional life into consciousness. For example, when he was fifty-five, I asked Newman for
permission to publish examples of his shifting memory. Repression
again: he never became anxious about the fallibility of his beliefs; he
simply returned my manuscript with the terse comment, "George,
you must have sent this to the wrong person."

At sixty-seven, Newman told me that his approach to the rough
spots in life was, "Forget it; let come what may." But as we might expect by now, that philosophy did not represent pure evolution from
control freak to Zen-like detachment. On the contrary, in many ways
he was returning to an adjustment much more like his earlier one. If
no longer quite so buttoned-down as his mother had tried to make
him, he had always been far more so than his unbuttoned daughters,
and now he was buttoning up again. He abandoned the social intensity of teaching, and returned to the safety of numbers. From age
fifty-five until he retired at sixty-eight he worked in city planning,
managing the complexities of the new Texan megacities. Although
he appeared at first glance to have had three widely varied careers—
ballistic missile engineering, academic sociology, and this last enterprise—his creative expertise in multivariate statistics was the common
denominator. And while he was flexible enough to become the office

computer guru (despite being a generation older than most of the technological hotshots in the department), he never became Mr. Sociable.

"I don't know what the word 'friend' means," he said at seventy. He gave up meditation. "I ruined some of its delights by reading too much neuropsychology," he told me at seventy two, in a comment reminiscent of his early fear that intellectual scrutiny would destroy his Catholicism. Still, his passion at the time was nuclear disarmament—an attempt to undo the aggressive achievements of his military youth. It was also an example of the developmental stage of Guardianship, which we'll explore in Chapter 5. And one of his bulwarks always was his gift for humor, a characteristic not shared by all statistical types. Even as a young man, he could take some distance from the complaint that he had "twice the sex drive" of his wife, and give his grumbling a wry and graceful turn: "We believe that making love should be practiced as an art form."

In our last interview, Adam Newman still maintained that he knew nothing about friendship, yet he loved his wife from his teens to the end of his days. His plan for the dread possibility that she might die before him was to join the Sierra Club and hang out with "tree huggers," yet he was happiest working with numbers. He avoided too much contact with people, but during our interview the birds came to eat seeds literally from his hands, which he lovingly held out for them. He had abandoned "spirituality" years before and had not set foot in a church for years before that, but he showed me proudly the awe-inspiring fractals he was creating on his computer. He was a bundle of superficial contradictions, but he wasn't alone in that. Everyone is consistent in some things and not in others, yet ultimately true to some fundamental essence in themselves. The more things change, the more they stay the same; Newman was part mystic and part engineer, and he remained that way to the end.

He remained true to repression as a defense mechanism, too. When he was seventy-two I asked him for the third time if he could remember any recurrent dreams from childhood. This time he didn't miss a beat: "Do you mean the one when I was on roller skates and ran into the back porch?" Three times over thirty years he is asked for a recurrent dream from his adolescence; each time he comes up with a completely different one with no recollection of the others. He relied on repression at nineteen, and at seventy-two he still did.

But that doesn't mean he hadn't changed. He wasn't living purely inside his head anymore; the narcissism of his youth had given way more and more over the years to interest and to empathy. And his mood was light. A very unhappy college student had become a very contented old man. As I closed our interview, I asked him if he had any questions about the Study to which he had contributed for more than fifty years. He had one. "Are you having a good time?" As I prepared to take my leave I politely shook his hand and he—so uncooperative and self-centered in his college days, so shy and intellectual for the last two hours—exclaimed, "Let me give you a Texan good-bye!" and threw open his arms to give me a huge hug.

Reviewing my write-up of that interview eighteen years later for this book, I found that I had closed with the simple words, "I was entranced." But I also found, to my chagrin, that my own memory was as fallible and as defensive as his. I had been remembering the fifty-five-year-old man who appeared in my first book; I had expected to be relating the tale of an aging hippie. I had completely lost track of his return to city planning (which is, after all, engineering, if of a gentler form than bomb design). One lesson for me in the writing of this book has been that retrospection is as unreliable in investigators as in their subjects. Without prospectively written records it is very easy to repress unpleasant truths, and I was as good at it as Newman was.

Even rereading my own transcript wasn't enough to keep me honest; that took an encounter with his death certificate. For twenty years I had conveniently forgotten that at the time of that interview, when he was seventy-two, he was a man preparing to die of an ugly and debilitating cancer. Out of my own dread of death, I couldn't keep his impending demise in mind. How could I claim any understanding of what was going on in him while failing to take that into account?

Be that as it may, less than a year before his life was consumed by the insatiable greed of his malignancy, Adam Newman wrote his last words to the Study: "I am happy." He had become a model of Erikson's final life stage, the one he calls Integrity (Chapter 5). This was an achievement that was still beyond me; Newman was not nearly as scared of his dying as I was, which probably had a fair amount to do with my failure to recall his cancer. My own father had died too soon to teach me much about adult life, but remembering Newman's life reminded me of how much I have learned from the Grant Study and its members in his stead. The dying can be happy too.

CONCLUSION

As a child I had little interest in microscopes, but I admired the telescope on Mount Palomar—then the most powerful one in the world. I wanted to see forests in their entirety, not just trees. Later on, as a psychiatrist, I longed to see beyond mere snatches of life to the whole picture. In this book I will focus on the rich potential of a telescopic view of life. There will be plenty of surprises, because a long enough view can turn conventional views of causation upside down. For instance, studied prospectively, physical health turns out to be just as important a cause of warm social supports and vigorous exercise as exercise and social supports are causes of physical health. Some readers will surely be outraged at such heresy, but as Galileo discovered, tele-scopes can get people into a lot of trouble. Long-term studies are as unsettling as they are enlightening.

To add to the uncertainty, we don't know how far to trust even our latest findings. Time changes everything, and it makes no excep-tions for longitudinal studies. It transforms the world we live in while we're living in it, and pulls scientific thinking forward even while making it obsolete. None of this can be helped; it's an intrinsic hazard of long endeavors. The more powerful the telescope, the more likely it is that the light we are seeing through it is many thousands of years old. The Grant Study is only seventy-five, but that's more than three-score years and ten, and in the context of a man's life, a very long time. Many of the early findings of the Study are ill-conceived, out-of-date, and parochial; some of our later findings will likely prove to be so too. But some, I hope, will endure. And in the meantime, they give those of us who are curious about our own lives, and the lives of those we cherish, plenty to think about.

It reminds me of my first day of medical school. "Boys," the Dean told us (this was in 1955), "the bad news is that half of what we teach you will in time be proven wrong; and worse yet, we don't know which half." Still, half a century later, our class has done pretty well by its patients. So I maintain hope that the old-fashioned Grant Study, twentieth-century artifact though it be, can offer some fresh wisdom and some real inspiration to twenty-first-century readers.

2 | THE PROOF OF
THE PUDDING

TO FLOURISH FOR THE NEXT SIXTY YEARS

The follow-up is the great exposer of truth,
the rock upon which fine theories are wrecked. —P. D. SCOTT

IN THIS CHAPTER I will get down to particulars, showing exactly
how longitudinal studies work, how they can be used, and why they
matter so much. I will demonstrate in action how different this kind
of information—that is, information derived longitudinally and prospectively—is from the other kinds of data so abundantly compiled
through the "instruments" of social science. And I will illustrate our
findings (as throughout the book) with both nomothetic (statistical)
and ideographic (narrative) data.

In 2009, *The Atlantic* asked me to identify the most important
finding of the Grant Study since its inception.[1] Without any official
evidence to back me up, I answered rashly: "The only thing that really
matters in life are your relations to other people." This impulsive response was quickly challenged by a leading business weekly, which
put it to me straight: What do romantic notions like that have to do
with the real dog-eat-dog world?[2]

Clearly I had dropped myself right into the middle of one of the
oldest and most heated controversies in developmental studies: which
matters more, nature or nurture? Is constitution or environment more
important? Feelings have run high on this subject for a very long time,

and here was an opportunity to bring some real evidence to bear on the matter. After forty years on the Study, did I have enough data to answer a make-or-break question?

At first I hoped to resolve the matter with a straightforward comparison. I would look at the lives of two groups of Grant Study men—one group that had been favored by nature with an enviable constitutional physical endowment, and another that had been favored by nurture with a loving childhood—and compare their lives fifty years later. This comparison had not been done before, and as it turned out I couldn't do it in 2009 either, at least not precisely on those terms. Those two alternatives came out of different temporal universes, at least as far as the Study was concerned.

Remember, I was limited to data provided by the Study itself. The "nature" part was easy. In the 1940s, the Grant Study researchers were betting that a muscular, classically "masculine" body build could predict future success, and they had amassed a huge amount of information about the men's physiques. It was a cinch to tell which men nature had favored, at least with regard to physical constitution. But the "nurture" part was a different story altogether. As I'll detail in the next chapter, the Study originators weren't thinking about tenderness and warmth as shapers of human lives. Personality researchers today give short shrift to physiognomy, and in the 1940s they didn't ask about relationships. Environmental considerations were barely on their radar at all, let alone the nuances of family atmosphere. So it was much harder to discern which of those children had been lucky in love.

That was one problem. Another was obvious from the way the *Australian Financial Review* had responded to my venture: there are more opinions than one on what "matters." Persuading twenty-first-century Wall Street types that love is all you need was going to be a hard sell.

But my appetite was whetted. Now I really did want to see what light the Grant data could shed on the nature/nurture battles, which meant reframing the issues (identifying success and identifying love) in terms that the available data could address.

So I devised a new question: Which best predicts successful old age: physical endowment, childhood social privilege, or early love? Then I set about defining explicitly the terms that would allow me to answer it. The problem raised by the Australians was the first one to tackle: what counts as success? High school football stars are usually considered successful adolescents. But do the traits that make good quarterbacks also make good grandparents?

Eventually it occurred to me that while people can wrangle end-lessly over whether the 400 meters or the high jump is the more chal-lenging event, almost no one will deny that a person who scores high in a decathlon is a fine athlete. So instead of incurring the wrath of the *Financial Review* and who knows who else by settling on any sin gle criterion, I established a Decathlon of Flourishing—a set of ten accomplishments in late life that covered many different facets of suc-cess. Then I set out to see how these accomplishments correlated, or didn't, with my three gifts of nature and nurture—physical constitu-tion, social advantage, and a loving childhood.

This exercise, as I pursued it, proved an elegant demonstration of the power of prospective longitudinal material in general, and of the long-term perspective that gives it meaning. It also brought to the fore, and put into a larger context, many of the specific individual is-sues that the Study had been dealing with, and that I'll be addressing in detail in coming chapters. It served as a giant telescope, quickly bringing the big picture into focus, and clarifying decades worth of theoretical assumptions and disagreements. It's an ideal introduction, therefore, to this biography of the Grant Study.

THE DECATHLON

We all have our own definitions of what "flourishing" means, and of what makes life well lived and worth living. I covered as many of these as I could in my Decathlon of ten rewarding late-life outcomes—in accordance with my Olympic metaphor I'll call them *events*—that could be measured with some objectivity. I avoided abstractions like virtue, happiness, and self-actualization, and focused instead on concrete behaviors and achievements. Many readers, I'm sure, will object to at least some of the criteria I chose. But they were pragmatic choices, not dictated by taste, political correctness, or even principle. I had to be able to score the men's capacity to survive, to work, to love, and to play, and I had to be able to do it with data that were already on hand, so as to assure that I was exerting no control over the *predictor variables* of my three childhood conditions. I picked my ten events (that is, measurable achievements) as *outcome variables* because they were available. I had no advance knowledge of how they would correlate with the predictor variables. Some overlaps among the events, and some of the age limits that defined them, may seem arbitrary, but these too were enforced by the way the available data were collected and assessed, sometimes many years ago. As we go along, you'll become familiar with the kinds of questions we asked and the way we handled the answers. In the meantime, examples of our interview protocols are available for reference in Appendix A, and of our adjustment assessment schedules in Appendix D. The scales we used to assess childhood temperament and environment are in Appendix C. The background on how all this developed is in Chapter 3, which recounts the history of the Study.

Table 2.1 defines the Decathlon—the set of ten outcomes that I used to denote "flourishing." The first two Decathlon events reflected occupational success: inclusion by age sixty in *Who's Who in America*

Table 2.1   A Decathlon of Flourishing, from Age 60 to Age 80*

1.	Included in *Who's Who in America*
2.	Earning income in the Study's top quartile
3.	Low in psychological distress
4.	Success and enjoyment in work, love, and play since age 65 (Appendix D)
5.	Good subjective health at age 75 (that is, physically active at age 75)
6.	Good subjective and objective physical and mental health at age 80
7.	Mastery of the Eriksonian task of Generativity**
8.	Availability of social supports other than wife and kids between ages 60 and 75
9.	In a good marriage between ages 60 and 85
10.	Close to kids between ages 60 and 75

* The Decathlon was conceived to assess the men's success from 65 to 80, but all but one of the events could be estimated for the 14 men who died between 58 and 64 years of age, and therefore they were included. Men who died before their 58th birthdays, however, were excluded.

** The men were coded for the highest Eriksonian psychosocial task they had mastered (see Chapter 5).

and earned income in the top quartile of the Grant Study sample. I chose the former admittedly elitist criterion because almost all the men in the Study enjoyed high-prestige professions, and I needed a documented way to distinguish levels of achievement among them. Only 21 percent of the men were included in *Who's Who*. I recognized immediately that this distinction would favor writers, educators, politicians, and businessmen over doctors and lawyers. It took me longer to realize that the Decathlon did not adequately reflect the achievements of creative artists, however successful.

The next four events reflected mental and physical health. Men who didn't need psychotherapy to cope with life's problems, or psychopharmaceuticals to relieve life's pains, were rated low for psychological distress. Event number four measured the men's enjoyment of work, love, and play between the ages of sixty-five and eighty (see Appendix D for how we measured this).

Events five and six reflected different aspects of biological success over time: being physically active (subjectively) at age seventy-five, and being physically and mentally healthy (both subjectively and objectively) at eighty. To meet the criteria for event number six required: that a man survive in good subjective and objective physical health to his eighty-first birthday; that at eighty he be without alcoholism, depression, chronic anxiety, or social isolation; and, finally, that he find subjective joy in many facets of his life.

Some readers may question my use of physical health as a Decathlon benchmark. Certainly longevity is not in itself a guarantee of flourishing. And while most of us would consider a happy and productive long life to be desirable, it is debatable whether an unhappy long life is better than a happy and productive short one. It's also true that a few vibrant and loving men received low Decathlon scores because they died young from genetic illnesses or chance events, while several inhibited men scored higher because they took few chances and lived a long, long time. Nevertheless, mental and physical well-being are integral to each other, and the challenge I had set for myself had to do with predicting success in old age.

The next four events reflected good relationships. The first was the achievement of Eriksonian Generativity (that is, the capacity for empathic nurturing of adolescents and adults other than one's own children; more on this in Chapter 5). The others were a happy marriage in late life, close father-child relationships in late life, and good social supports (friends, confidants, people to play tennis and bridge with, and so on) between the ages of sixty and seventy-five.

Some definition was necessary here, too. A man who had been happily married for most of the years between sixty to eighty-five was scored as having achieved this event even if there were five unhappy years and a divorce in the middle. In contrast, an intact thirty-five-year marriage was called "unhappy" if most of the time was spent in discord. (On these issues wives were consulted as well, separate from their husbands.)

Construing a variable as complex as good marriage as "present" or "absent" sounds simplistic, but in fact it was a pragmatic and useful device. It's very difficult to rank the relative happiness of marriages—your siblings', for example, or your three best friends'. But if both partners maintain over twenty years that their marriage is happy (which we scored as a 1), or if the state of their happiness is unclear to them during that time (scored as 2), or if their marriage is consistently rocky or they get divorced and never remarry (scored as 3), a three-point scale of marital happiness becomes plausible.

Throughout this book you'll see us transforming value judgments into science this way. What makes it possible is the availability of repeated measurements over time, one of the great strengths of longitudinal studies. And if these assessments can be defined and systematically derived with rater reliability—that is, if independent raters come to virtually identical scores—then it becomes possible to codify value judgments in a way that can be examined and tested statistically years or decades later, without the risk of their being altered or otherwise reinterpreted to suit an investigator's whim.

However annoying my sins of omission and commission, therefore, I hope my readers will agree that taken as a whole, doing well in the Decathlon is preferable to the alternative.

Although the ten variables that I employed to reflect flourishing appear disparate, they turned out to be, in fact, highly correlated with each other that is, they tended to appear together—and up to a point this makes intuitive sense. Maintaining a high income and good social supports requires a certain amount of empathy and social intelligence; so does maintaining close family relations. Of forty-five possible correlations among the ten variables (that is, the forty-five different ways that ten variables can be paired) in Table 2.1, twenty-four pairs were very significantly correlated. Twelve more pairs were significantly correlated, and only nine pairs of correlations were not significant.

Please note, by the way, that throughout this book the word *significant* is used only in its statistical sense. *Very significant* (VS) refers to correlations for which the probability of their occurring by chance is less than one in a thousand; *significant* (S) refers to correlations for which the probability of occurring by chance is less than one in a hundred. Correlations significant at a level of p <.05—that is, occurring by chance fewer than one time in twenty—are conservatively called *not significant* (NS).

SCORING

It's important to keep in mind that even though the Decathlon itself was not conceived until 2009, the data pertaining to the ten events I targeted for study were collected long before then, and usually during the time when the events were actually taking place. This is a prospective study, remember. The information that went into the Decathlon assessments came from the men's initial intake processing in 1938–1942, and from the extensive follow-up that succeeded it, up through the year 2009. All of these records were available to the scoring process, so judgments were made about a man's status based not on the fluctuating conditions of a single moment, but on information amassed over decades.

For every Decathlon event in which a man was in the top quartile of the Study sample, he received a point. If he had died before the assessment could be made, he received a zero in that event. Total scores, therefore, ranged from 0 to 10. A full third of the men scored 2 or 3; they were considered average on the flourishing scale. If we accept that this Decathlon does address, however imperfectly, several vital aspects of flourishing in late life, then the one-third of the men who received fewer than 2 points from most raters were living less desirable lives than the one-third of the men who scored 4 points or more. A cast of protagonists and their Decathlon scores can be found at the front of the book. Adam Newman received a midlevel Decathlon score of 2; Godfrey Camille, whom I will introduce shortly, received a 5. Of course, judgments about the "good life" can be very annoying. I had an academic partner once who challenged me for saying that Jack Kennedy was mentally healthier than Lee Harvey Oswald. Tastes differ.

ANTECEDENTS OF SUCCESS

Having established my criteria for success and scored the men on their achievement of them, I could now test statistically which of the myriad variables and traits that the Study had documented when the men were young—variables of nature and variables of nurture—could predict a good life in their later years.

In fact, to uncover the antecedents of flourishing had been one of the avowed purposes of the Study at its inception. Furthermore, the early years of data collection coincided with the early years of World War II, when efficient officer selection was very much on the minds of the investigators. Typical of the time, the original Study staff expected that the gift of leadership could be predicted by a particular constellation of constitutional endowments; this meant a mesomorphic (that is, athletic) and "masculine" (that is, narrow-hipped and broad-shouldered) body build. The staff used a study of ROTC recruits to justify this expectation retrospectively in 1945; in that study, 41 percent of the men with a strongly masculine physique were considered "excellent officer material," while not a single man who was considered weak in the so-called masculine component was equally valued.[3] However, there was no follow-up to document that these

men actually made good officers, and as P. D. Scott implies in the epigraph of this chapter, that left this fine theory very much an open
question. My intention was to pit the hard-nosed, body-build, nature
values of the early Study investigators and the money-centered preoccupations of the modern business world against my own sentimental
twenty-first-century values about the power of love. Which—constitution, money, or tender care—had the greatest power to predict success in the Decathlon of Flourishing? It's better to know than to believe, at least in matters of science, and it was time for conflicting
beliefs to be resolved by facts.

Section A of Table 2.2 depicts ten characteristics, some physical
and some not, that I chose to test as potential constitutional predictors
of a rewarding late life. The first six had been thought likely suspects
by the original Study staff, especially for predicting future officers and
store managers. (This was a high priority of William Grant, the chain-
store impresario who funded the Study and gave it his name; more
about Grant and the first four characteristics in Chapter 3.) They were:
vital affect (a rich, exuberant social presence), *sociability, masculine body
build* (narrow hips, broad shoulders), *mesomorphy* (muscularity), *treadmill endurance,* and *athletic prowess.* To these I added four other constitutional or biological factors: *childhood temperament* (see Chapter 4),
hereditary loading for (that is, family history of) *alcoholism and major depression,* and *longevity of parents and grandparents.*

I also included three variables to test the effect of socioeconomic
status on the men's future. Parental *social class* was assessed by paternal
income, occupational status, neighborhood, and a home visit. The *education* of each parent was also assessed.

That done, the Australian challenge could no longer be avoided.
And this brought me straight back to my initial problem, the one that
had foiled my simple group-to-group comparison and driven me
to Olympic elaborations. Namely, there were no data available—*the*

Table 2.2 What Variables Predict a High Decathlon Score at Age 60–80?

	Association with Total Decathlon Score
A. Variables Reflecting Constitutional Factors	
1. Mesomorphy (0 items*)	NS
2. Masculine body build (0 items)	NS
3. Vital affect (1 item)	Significant
4. Athletic prowess (2 items)	Very significant
5. Treadmill run (2 items)	Significant
6. Sociable/Extroverted (0 items)	NS
7. Mean ancestral longevity (1 item)	NS
8. Alcoholic relatives (2 items)	NS
9. Depressed relatives (0 items)	NS
10. Good childhood temperament (2 items)	Significant
B. Variables Reflecting Social Class	
1. Social class (upper-upper to blue collar) (0 items)	NS
2. Maternal education (6–20 years) (0 items)	NS
3. Paternal education (6–20 years) (0 items)	NS
C. Variables Reflecting Attachment	
1. Warm childhood (6 items)	Very significant
2. Overall college soundness (8 items)	Very significant
3. Empathic coping (defense)** style (age 20–35) (7 items)	Very significant
4. Warm adult relationships 30–45 (10 items)	Very significant

Very significant = p<.001; Significant = p<.01; NS = Not significant.

* "Items" refers to the number of Decathlon events predicted significantly by the antecedent variable.

** See Chapter 8.

Study had collected no data—that directly addressed the amount or quality of the love that the College men had received as children.

I had to find a way to test definitively my impulsive assertion that warm intimate relationships are the most important contributing factor in the establishment of a good life. Relational warmth is tough to measure even in the twenty-first century, and we've been working on it for a long time. In 1940 no one had even put forth the concept of attachment, certainly not the biometric psychologists or the Freudian psychoanalysts to whom the Study first gave ear. But more on that later. What matters here is that only four objective measures of relationship skills had been gathered before the measures of Decathlon success were obtained, so I had to content myself with those. They became my third set of predictors.

The men and their parents were not accessible to the Study until the men were in college, at which time both men and parents were interviewed in depth. These interviews were the first indicators the Study had of the men's early home life, and of their relationships with their parents and siblings—obviously important predictive variables. Yet the men's childhood environments were not assigned nuanced numerical scores at the time they were assessed; that was done only after I joined the Study. (I will describe our after-the-fact scoring of the men's childhoods in Chapter 4.)

The second relational predictor was the Study staff's consensus rating (as the men finished their senior years at about age twenty-one) of the men's overall "soundness." That rating had been based on a 1-to-3 scale as follows:

1 = A man seen as without "serious problems in handling problems that might confront them."

2 = A man seen as "lacking in warmth in his touch with people," or too "sensitive."

3 = A man seen as "markedly asocial," or manifesting "marked mood variations."

The third predictor was the maturity or immaturity of the men's *involuntary coping style* (perhaps better known as *defensive style*) between ages twenty and thirty-five. This was assessed by me retrospectively when the men were forty-seven from information provided earlier (retrospective data *analyses* are permissible in a prospective study; retrospective *data* are not). This too will be explained more fully (in Chapter 8); keep in mind that this chapter is a sketch, a roadmap of terrain that we will cover in more detail later on. For now, what matters is that our defensive styles influence our relationships. Mature coping (sometimes called *defense*) mechanisms like humor and patience tend to attract other people; immature coping mechanisms like projection and hypochondriasis, while temporarily soothing to their users, appear self-centered to other people and tend to alienate them.

The final predictor was warm adult relations between ages thirty and forty-seven. Although this variable was not assessed until 1975, when the men were in midlife, we included it because there really were no earlier objective variables reflecting capacity for close attachment. This crude scale asked six objective, if simplistic, yes-or-no questions:

- Had the man been married for more than ten years?

- Was he close to his children?

- Did he have close friends?

- Did he maintain pleasant contact with his family of origin?

- Did he belong to social organizations?

- Did he play games with others?

WHAT VARIABLES PREDICT SUCCESS?

Table 2.2 sets out the seventeen variables that we tested for their power to predict flourishing late in life—ten constitutional variables that recreated the worldview of the early investigators, three socioeconomic variables that recreated the worldview of modern social psychology (and possibly the *Australian Business Review*), and finally the four available relational variables that reflected the view of attachment theorists and ethologists. I admit readily that these were sketchy, but they were the only ones available to us; any measures of relationship that post-dated the Decathlon would of course not be predictive. The table also indicates how each variable associated with the total Decathlon score.

As the table indicates, the ten constitutional variables and the three social variables were of limited predictive use. The two manifestations of the body-build hypothesis so beloved of the original researchers had no significant correlation at all with later success. Nor did the measures of socioeconomic status. Alcoholism and depression in family histories proved irrelevant to Decathlon scores at eighty, as did longevity. The sociable and extraverted personality type that was so esteemed in the Study's selection process (see Chapter 3) did not correlate with later flourishing either. (We'll see in the last chapter, though, that *Extraversion* when measured by a sophisticated psychological "instrument" did prove very significant.) In fact, of the thirteen constitutional/socioeconomic factors that we tested for their predictive strength, only four correlated significantly with the total Decathlon scores, and those only with one or two events apiece. The correlations between the constitutional and socioeconomic variables and Decathlon success were weak and scattershot.

The four measures of relationship, however, were all highly predictive of Decathlon success. Each of these four variables predicted at least six events apiece, and collectively they predicted all ten. Furthermore, each relationship predictor correlated highly significantly with the other three, suggesting that they had something in common. In short, it was the capacity for intimate relationships that predicted flourishing in all aspects of these men's lives, as can be seen in Table 2.3.

Table 2.3 Childhood Predictors of Flourishing in Old Age

Decathlon of Flourishing	Environmental Childhood Strengths	Social Health Rating, Age 47	College Soundness, Age 21	Maturity of Defenses, Age 20–35
Total Decathlon score	Very Significant	Very Significant	Very Significant	Very Significant
Who's Who in America	NS	Significant	NS	Very Significant
Highest earned income	Significant	Very Significant	Significant	NS
Low lifetime distress	NS	Very Significant	Very Significant	Significant
Access and enjoyment in work, love, & play (age 65–80)	Very Significant	Very Significant	NS	NS
Subjective health (age 75)	NS	Very Significant	Significant	Significant
Successful aging (age 80)	NS	Very Significant	Significant	NS
Achieved Generativity	Very Significant	Very Significant	Very Significant	Very Significant
Social supports (other than wife and kids) (age 60–75)	Very Significant	Very Significant	Very Significant	Very Significant
Good marriage (age 60–85)	Significant	Very Significant	Significant	Significant
Close to kids (age 60–75)	Significant	Very Significant	NS	Significant

Very Significant = p<.001; Significant = p<.01; NS = Not Significant.

Here's a concrete example, to make these abstractions a little more vivid. We found, for instance, that there was no significant difference between the maximum earned incomes of the men with IQs of 110–115 and the incomes of the men with IQs of 150-plus. Nor was there a significant difference between the maximum earned incomes of men with mesomorphic (muscular) physiques and men with ectomorphic (skinny) or endomorphic (plump) physiques, or between men with blue-collar fathers and men with fathers from the upper class (Table 2.2).

On the other hand, the men who had good sibling relationships (one factor of a warm childhood; see Table 2.2 and Appendix C) when young were making an average of $51,000 (in 2009 dollars) more a year than the men who had poor relationships with their siblings, or no siblings at all. The men from cohesive homes made $66,000 per year more than men from unstable ones. Men with warm mothers took home $87,000 more than those men whose mothers were un-caring. The fifty-eight men with the best scores for warm relation-ships were three times more likely to be in *Who's Who,* and their max-imum income—between the ages of fifty-five and sixty, and in 2009 dollars—was an average of $243,000 a year. In contrast, the thirty-one men with the worst scores for relationships earned an average maxi-mum salary of $102,000 a year. The twelve men with the most mature coping styles reported a whopping $369,000 a year; the sixteen men with the most immature styles a much more modest $159,000. These same variables were equally able to predict warm relationships at the end of life.

So my rash contention held up. Nurture trumps nature—at least for predicting late-life success as the Decathlon embodied it. And the aspect of nurture that is most important of all is a warm and intimate relational surround (environmental childhood strengths, Appendix C). There is another piece to be reckoned with too. In a moment I will offer a life story to illustrate it, and what these actualities look like in the real world. But first let me offer one caveat, and then an illuminat-ing historical aside.

The caveat: Throughout this book I will seem to be saying that the mentally healthy have better characters than the mentally un-healthy. This sounds like victim-blaming, but it is not a moral judg-ment. It is a reflection of a tough and pragmatic reality: that empathy is easier for those with full (literally and figuratively) stomachs. Someone who is already suffering the stresses of (literal and figurative) hunger is more likely to respond self-protectively or to lash out when punched in the gut.

The historical aside: I was chary from the beginning of the body-build approach to prediction, and of some of the other cherished hypotheses from the early days of the Study. Remember that in the thirties and forties constitutional medicine and physical anthropology were very much in the theoretical driver's seat; it wasn't only in Germany that people were preoccupied with the fine points of putative racial superiorities. But since the data were so extensive, and since I was in the neighborhood anyway while constructing the Decathlon, I decided to test out the relationship, so confidently assumed at the time, between officer potential and body build.

At the end of World War II, some of the Grant Study men were majors; others were still privates. What made the difference? It turned out that the men's attained military rank at discharge bore no relation to their body build, their parents' social class, their endurance on the treadmill, or even their intelligence. What did correlate significantly with attained military rank was a generally cohesive home atmosphere in childhood and warm relationships with mother and siblings. Twenty-four of the twenty-seven men with the warmest childhoods made at least first lieutenant, and four became majors. In contrast, of the thirty men with the worst childhoods, thirteen failed to make first lieutenant, and none became majors. We don't breed good officers; we don't even build them on the playing fields of Eton; we raise them in loving homes. This result would undoubtedly have astonished physical anthropologist Earnest Hooton (see Chapter 3), whom the Study asked to write its first book.[4]

I offer this story for its morals. One is that belief isn't enough—however impassioned our convictions, they need to be tested. Another is that information does nothing for us if we don't make use of it. My brief excursion here answered a question that the Study had been entertaining from its very beginnings; the data that finally answered it had been available for almost seventy years. And a third: that longitudinal studies protect us from exactly such pitfalls, and from our other shortcomings of foresight and method. They give us the flexibility to re-ask old questions in new contexts, and to ask new questions of old data. That is a very important point of this book, and one I'll keep returning to.

THE LIFE OF GODFREY MINOT CAMILLE

Now let me abandon statistics and present some idiographic evidence for the power of love; a life is worth a thousand numbers. In 1938, when Godfrey Minot Camille first presented himself to the Grant Study, he was a tall redheaded boy with a charming manner who planned to enter medicine or the ministry. Only gradually did the staff discover that the allegedly "normal" Godfrey was an intractable and unhappy hypochondriac. On the tenth anniversary of his joining the Study, each man was given an A through E rating anticipating future personality stability, and when it was Godfrey's turn, he was assigned an "E." But if Godfrey Camille was a disaster as a young man, by the time he was an old one he had become a star. His 5 on the Decathlon put him in the top quartile of the surviving men of the Study. What made the difference? How did this sorry lad develop such an abundant capacity for flourishing? He spent his life searching for love.

Camille's parents were upper class, but they were also socially isolated and pathologically suspicious—both of them; social privilege did not predict the warmth, or the chill, of a man's childhood. "Mother hasn't exactly made up for Dad's shortcomings," Camille put it at nineteen. At forty-six, he sadly reaffirmed a truth he had recognized earlier: "I neither liked nor respected my parents." Lewise Gregory, the Study's family interviewer, described Mrs. Camille as "one of the most nervous people I have ever met . . . a past mistress in self-deception." A child psychiatrist who reviewed Camille's record thirty years later thought his childhood one of the bleakest in the Study. (Evidence used in these assessments rarely came from just one source.)

Unloved and not yet grown into a sense of autonomy, Camille as a student adopted the unconscious survival strategy of frequent visits to the college infirmary. No evidence of tangible illness was found at most of his visits, and in his junior year a usually sympathetic college physician dismissed him with the disgusted comment, "This boy is turning into a regular psychoneurotic." Camille's constant complaining was an immature coping style. It didn't connect with other people, and it kept them from connecting with him; they didn't see his real suffering and just got angry at his apparent manipulations.

When the war began, Camille was destined for failure, at least by Study criteria. He was an ectomorph, skinny and without the broad shoulders and narrow hips that might have redeemed his scrawniness as "masculine." His athletic conditioning was poor, and he hadn't done particularly well in college academically. He would have been doomed by my criteria, too. He had had the worst possible childhood, he used unempathic coping styles, and he seemed clumsy at human relations. To no one's surprise, he emerged from World War II still a private—everyone had predicted he would be a loser. So, a little later, did I.

After graduation from medical school, the newly minted Dr. Camille made a suicide attempt. The Study consensus at the time of his ten-year personality assessment was that he was "not fitted for the practice of medicine," and, unloved as he was, he found taking care of other people's needs overwhelming. But several sessions with a psychiatrist gave him a different view of himself. He wrote to the Study,

"My hypochondriasis has been mainly dissipated. It was an apology, a self-inflicted punishment for aggressive impulses."

Having grasped that he was paying for anger with depression, Camille abandoned his hypochondriacal defense and the unconscious purpose it served for a more mature coping mechanism—displacement. He learned to shift his attention away from uncomfortably intense emotional issues toward more neutral ones. When his sister died, he sent the Study the autopsy report with the simple comment, "Enclosed is a copy of an autopsy protocol, which I expect is also an item of news." He never mentioned his feelings about her death, or even the fact of it. He did not tell the Study directly about his mother's death either, but—displacing his attention from heart to brain—wrote cryptically, "I received an inheritance from my mother." Whatever the limitations of this approach, it was a more empathic one than his hypochondriasis had been; whereas people had once felt overwhelmed, helpless, and angry in the face of his constant physical complaints, they now found him much more comfortable to deal with.

For all his adolescent hypochondria, Camille hadn't really been very aware either of his body or of his feelings. He felt things, yes. But what he was feeling—a symptom of illness? an upsurge of anxiety? a chimerical fantasy?—was not a distinction he knew how to make. Under stress he had sensations, and he felt that they required immediate attention. After his post-suicidal insight at thirty-two, however, he began to distinguish between physical and emotional sensations, and to pay attention to their causes. Emotional stress still led to indigestion, abdominal pains, cold hands, and gastric distress, but Camille no longer insisted to physicians that he was ill, or used his physical complaints to convey his outraged sense of need or hold someone else responsible for it. Instead, after psychotherapy, he used his physical symptoms as a clue to the presence of emotional distress.

Then, at age thirty-five, he had a life-changing experience. He was hospitalized for fourteen months in a veterans' hospital with pulmonary tuberculosis. Ten years later he recalled his first thought on being admitted: "It's neat; I can go to bed for a year, do what I want, and get away with it." "I was glad to be sick," he confessed. And in fact that illness, a real one, ended up giving him the emotional security that his childhood, his hypochondriacal symptoms, and his subsequent careful neutrality never had. Camille felt his time in the hospital almost like a religious rebirth. "Someone with a capital 'S' cared about me," he wrote. "Nothing has been so tough since that year in the sack."

Released from the hospital, Dr. Camille became an independent physician, married, and grew into a responsible father and clinic leader. Over the next five years he mastered the adult developmental tasks of *Intimacy, Career Consolidation,* and *Generativity* (see Chapter 5) in rapid succession. His marriage, which lasted for more than ten years but was not particularly happy, finally ended in divorce. But one of his daughters at fifty told me in an interview that his children remembered him as an exemplary father.

Camille's coping style was changing as the decades passed, too. His transitional reliance on displacement (the unconscious avoidance of emotional intensity) was replaced by the still more empathic involuntary coping mechanism of altruism, including the generative wish to nurture others' development. He was now functioning as a giving adult. Whereas at thirty he had hated his dependent patients, by forty his adolescent fantasy of caring for others had become a reality. He started and directed a large Boston clinic for allergic disorders—the first time he had ever been in charge of anything. He wrote papers to help other clinicians understand and deal with the emotional needs of asthmatic patients with deprived childhoods. In vivid contrast with his post-graduation panic, he now reported that what he liked most about medicine was that "I had problems and went to others, and now I enjoy people coming to me." His daughter told me when I met her that "Dad had the innate ability to just give. He could play like five-year-olds do."

When I was fifty-five and Camille was almost seventy, I asked him what he had learned from his children. "I haven't stopped learning from them," he said. He seemed to appreciate that that was a facile answer, and added thoughtfully, "That's a tough question. . . . Isn't that a whopper!" I was disappointed; I had felt sure that this sensitive man would come up with a more profound response. But two days later he came up to me in Harvard Yard as I was about to give a talk to his reunion class. With tears in his eyes he blurted out, "You know what I learned from my children? I learned love!" Many years later, having seized a serendipitous opportunity to interview his daughter, I believed him. I have interviewed many Grant Study children, but this woman's love for her father remains the most stunning that I have encountered among them.

When I first wrote about the life of Godfrey Camille, I had no idea what had led to his recovery. Clearly it had been catalyzed by his year of enforced invalidism, but how? At age fifty-five, he attributed everything to a visit from Jesus at the hospital; I myself (at forty) was inclined to think that it was all the loving nursing care he had received over those many months. But neither was a very satisfactory explanation. What I know now is that it doesn't really matter. It took me many more years of prospective follow-up, and many more years of emotional growth, to learn to take love seriously. What it looks like—God, a nurse, a child, a good Samaritan, or any of its other guises—is different for everybody. But love is love.

At age seventy-five, Camille took the opportunity to describe in

greater detail how love had healed him. This time he needed no recourse to Freud or Jesus.

Before there were dysfunctional families, I came from one.
My professional life hasn't been disappointing—far from it—
but the truly gratifying unfolding has been into the person
I've slowly become: comfortable, joyful, connected and effective. Since it wasn't widely available then, I hadn't read
that children's classic, *The Velveteen Rabbit,* which tells how
connectedness is something we must let happen to us, and
then we become solid and whole.

As that tale recounts tenderly, only love can make us real.
Denied this in boyhood for reasons I now understand, it
took me years to tap substitute sources. What seems marvelous is how many there are and how restorative they prove.
What durable and pliable creatures we are, and what a store-
house of goodwill lurks in the social fabric. . . . I never
dreamed my later years would be so stimulating and rewarding.

That convalescent year, transformative though it was, was not the
end of Camille's story. Once he grasped what had happened, he seized
the ball and ran with it, straight into a developmental explosion that
went on for thirty years. A professional awakening and a spiritual one;
a wife and two children of his own; two psychoanalyses; a return to
the church of his early years—all these allowed him to build for himself the loving surround that he had so missed as a child, and to give to
others out of its riches.

More years passed. At seventy-seven, Camille was viewing his past
five years as the happiest in his life. He had a new love; he was whip-

ping men thirty years his junior at squash; he was nurturing a beautiful garden. He was also deeply involved in the community at Trinity Church. Perhaps I had been wrong to dismiss his hospital trust in a loving Christ so offhandedly.

At eighty, Camille threw himself a potluck supper birthday party. Three hundred people from his church came. He provided the jazz band.

At eighty-two, Godfrey Minot Camille had a fatal heart attack while climbing in the Alps, which he dearly loved. His church was packed for the memorial service. "There was a deep and holy authenticity about the man," said the bishop in his eulogy. His son said, "He lived a very simple life, but it was very rich in relationships." Yet prior to age thirty, Camille's life had been essentially barren of relationship. Folks change. But they stay the same, too. Camille had also spent his years before the hospital looking for love. It just took him a while to let himself find it.

By the time Godfrey Minot Camille was eighty, even Aristotle would have conceded that he was leading a good life. But who could have foreseen, when he was twenty-nine and the Study staff ranked him in the bottom 3 percent of the cohort in personality stability, that he would die a happy, giving, and beloved man? Only those who understand that happiness is only the cart; love is the horse. And perhaps those who recognize that our so-called defense mechanisms, our involuntary ways of coping with life, are very important indeed—the missing piece I mentioned above. Before age thirty, Camille depended on narcissistic hypochondriasis to cope with his life and his feelings; after fifty he used empathic altruism and a pragmatic stoicism about taking what comes. There are two pillars of happiness revealed by the seventy-five-year-old Grant Study (and exemplified by Dr. Godfrey Minot Camille). One is love. The other is finding a way of coping with life that does not push love away. And that is why I offer Dr.

Camille's story as a sort of outline of the terrain we'll be covering through the rest of this book.

CONCLUSION

Pictures are worth a thousand words, but sometimes they speak more to the passions than to reason. Numbers don't lie, but sometimes they do. Together, however, pictures and numbers can keep each other honest, and that is why I offer in this book a bifurcated exposition of what appear to me (so far) to be the ultimate (so far) lessons of the Grant Study.

One is that positive mental health does exist, and to some degree can be understood independent of moral and cultural biases. But for this to happen, we must acknowledge our value judgments and define them operationally, and we must prove the validity of our definitions not by persuasive argument, but by documented outcomes. This is important. Eighty years ago it was common to hear Hitler and Stalin called great leaders, and Churchill dismissed as a disastrous failure of a statesman. It will be another eighty years before we'll be able to rank realistically the leadership skills of Presidents Nixon, Reagan, Clinton, Bush father and son, and Obama. In presidencies, in studies, and in lives, the proof of the pudding is in prolonged follow-up—not for years, but for decades. More on this throughout the book, but especially in Chapter 7.

The second lesson is that once we leave the study of psychopathology for positive mental health, an understanding of adaptive coping (explored in Chapter 8) is crucial. As in the inflammations and fevers of physical illness, what looks like trouble may be the very process by which healing takes place. As we become better able to endure life's slings and arrows, our coping mechanisms mature, and vice versa.

The third lesson is that the most important influence by far on a flourishing life is love. Not early love exclusively, and not necessarily romantic love. But love early in life facilitates not only love later on, but also the other trappings of success, such as prestige and even high income. It also encourages the development of coping styles that facilitate intimacy, as opposed to ones that discourage it. The majority of the men who flourished found love before thirty, and that was why they flourished. I will discuss some of the vicissitudes of intimacy in Chapter 6.

But—this is the fourth lesson—people really can change, and people really can grow. Childhood need be neither destiny nor doom, as Chapter 5 will make clear. So does the story of Dr. Camille, a model of posttraumatic growth.

Lessons Two, Three, and Four are closely related. Throughout our lives we are shaped and enriched by the sustaining surround of our relationships. The seventy-five years and twenty million dollars expended on the Grant Study points, at least to me, to a straightforward five-word conclusion: "Happiness is love. Full stop." Virgil, of course, needed only three words to say the same thing, and he said them a very long time ago—*Omnia vincit amor,* love conquers all—but unfortunately he had no data to back them up.[5]

A fifth lesson, to be expanded in Chapter 4, is that what goes right is more important than what goes wrong, and that it is the quality of a child's total experience, not any particular trauma or any particular relationship, that exerts the clearest influence on adult psychopathology. Consider Camille's "year in the sack." Let me repeat myself: what goes right is more important than what goes wrong.

A sixth lesson is that if you follow lives long enough, they change, and so do the factors that affect healthy adjustment. Our journeys through this world are filled with discontinuities. Nobody in the Study was doomed at the outset, but nobody had it made, either. Inheriting the genes for alcoholism can turn the most otherwise blessed golden boy into a trainwreck (Chapter 9). Conversely, an encounter with a very dangerous disease liberated the pitiful young Dr. Camille from a life of dependency and loneliness.

The final lesson is that prospective studies really do elucidate life's mysteries. I've given only a tantalizing taste of this lesson in Chapter 10, because it really does involve mysteries, and we are a long way from solving them in full. Yet it is likely that when they *are* solved, it will be with the help of lifetime studies like this one. The Grant Study has given us seventy-five years of real behaviors—not just trues and falses, not just A, B, C, D, or none-of-the-aboves—on which notions old and new about a rewarding life can be tested, retested, and refined.

Even as they first set out, the originators of the Grant Study were beginning to gather exactly the evidence that contrarians like me would need to disagree with them. I inherited an orchard planted and nurtured with great care by a group of dedicated gardeners. For forty years I harvested its yield and took it to market. And since it is on follow-up that our theories stand or fall, my successors will continue the harvest when I no longer can. But credit for the fruit goes back to Arlie Bock, Clark Heath, and Lewise Gregory Davies who gave it the place and the time to grow, and to Charles McArthur who fertilized and pruned it once they had hung up their trowels. The creation of the Study is a life story all its own, and in the next chapter I will tell it.

3 | A SHORT HISTORY OF THE GRANT STUDY

> The sign on the door read "The Grant Study of Adult Development."
> Financed by W. T. Grant, the department store magnate, and run by
> Harvard's Health Services Department, the study proposed to investi-
> gate "normal" young men, whatever that might mean.
>
> On that particular afternoon, I was a sophomore, just turned nine-
> teen. The Depression and a six-month siege of polio had been the sole
> departures from an otherwise contented, if not stimulating, life.
>
> —BENJAMIN BRADLEE, *A Good Life,* 1995

IN THIS CHAPTER, I RECORD the Grant Study's seventy-five-year history for posterity—to be pondered, skimmed, or skipped at the reader's pleasure. It's the story not only of the Study, but also of seventy-
five years of social sciences in America and the worldviews that came
and went over that period. The research program of the Grant Study
was directed by Clark Heath, M.D., from 1938 until 1954, by Charles
McArthur, Ph.D., from 1954 to 1972, and from 1972 until 2004 by
me. Since 2005, Robert Waldinger, M.D., has been the director of the
Study.

In the academic year 1936–1937, Arlen V. Bock, M.D., became Oliver Professor of Hygiene at Harvard and chief of the student health
services. In a report to President James Conant, he proposed broadening the scope of the Department of Hygiene and the role of the college physician. As a first step, he suggested a scientific study of healthy
young men. Bock believed that health should be as much a central

focus of medicine as pathology. As his proposal gained ground, he became more explicit about the kinds of issues he had in mind: the
problem of nature vs. nurture; connections between personality and
health; whether mental and physical illnesses can be predicted; how
constitutional considerations might influence career choice. But his
primary interest was: What is health? That question (and others that
grew out of it) inspired the Grant Study for seventy-five years, and in
this book I will start to answer it.

In his report to Conant, Bock cited "the stress of modern pressures" which, he felt, "the current generation of students had to face
largely unprepared."[1] And he thought that Harvard should be addressing these pressures.

> As time passes, it seems logical to expect a different emphasis
> on the work of the Department of Hygiene than has been
> customary because of the growing complexity of human re-
> lations and the need to turn men out of the University bet-
> ter qualified to take their places in the affairs of life [2]

To further this goal, Bock enlisted the support of his friend and
patient William T. Grant, owner of the chain of stores bearing his
name. Funding began on November 1, 1937, with the arrival of the
first check from what would soon become the Grant Foundation, in
the amount of $60,000 ($900,000 in 2009 dollars), and President Conant and the Harvard faculty approved Bock's project.

Originally called the Harvard Longitudinal Study, it soon became
known as the Harvard Grant Study of Social Adjustments. (This name
reflected a major business preoccupation of Grant's—namely, what
makes a good store manager.) After Grant withdrew funding in 1947
the name was changed again, this time to the Harvard Study of Adult
Development. But colloquially it has always been the Grant Study,

and I will follow that convention here. (In 1967, when I was new to the Study, I naively asked why it was called this. A more senior investigator replied with a straight face, "Because it took an awful lot of grants to keep it going.")

The Grant Study got under way in the fall of 1938, in a squat red-brick building on Holyoke Street in Cambridge, next to the Department of Hygiene. Its multidisciplinary aims were reflected in the composition of the original staff: an internist, a psychologist, a physical anthropologist, a psychiatrist, a physiologist, a caseworker, and two secretaries. Dan Fenn, Jr., editor of the Harvard *Crimson* at the time, said of the eight pioneers that they were "working on what might one day be one of Harvard's important contributions to society, the analysis of the 'normal' person. . . . They may be able to draw up a formula which will easily and correctly guide a man to his proper place in the world's society."[3]

In 1939, Harvard celebrated the opening of the Grant Study with a conference. Alas, I can find no record of what was discussed under its aegis, but it brought together a group of scientists of international distinction, who in their own ways shaped the Study profoundly. One was Adolf Meyer, who was the founding chairman of the department of psychiatry at Johns Hopkins and the new Study's patron saint. Meyer was perhaps the greatest advocate in America of a long-range view in psychiatry. He had come to this country in 1892 to look at the way the brain changes after death. Ten years later his interest had shifted from the neuropathology of the dead to the adaptive neurophysiology of the living. Meyer insisted that the study of psychiatry was the study of lives, and published a famous, if rarely read, paper on the value of the "life chart."[4] In this he pleaded with his fellow psychiatrists for "a conscientious study of the mental life of patients," insisting that "we need less discussion of generalities and more records of well observed cases—especially records of lifetimes—not merely snatches of picturesque symptoms or transcriptions of the meaning in traditional terms." As the sagas of Adam Newman and Godfrey Camille have already made clear, the Grant Study made Meyer's dream a reality. Originally planned to last fifteen to twenty years—a mightily ambitious goal even today—it has now been making "records of lifetimes" for seventy-five years.

America's greatest physiologist, Walter Cannon, was at the conference too. Like Meyer, Cannon has been a role model for the Study throughout its existence. It was he who formulated the concept of the fight-or-flight response, and as a Harvard professor he wrote a classic monograph on physiological homeostasis, *The Wisdom of the Body*.[5] Psychological homeostasis has been an enduring concern of the Grant Study, and I chose the title of my book *The Wisdom of the Ego* in homage to Cannon.[6]

President James Conant was present; he led Harvard through World War II, and served—importantly if less publicly—as civilian administrator of the Manhattan Project, where he opposed the development of the hydrogen bomb. Arlie Bock was there too.

Arlen Vernon Bock, once described by the Harvard Gazette as "blond, brisk, brusque, benign, belligerent, and always busy," was a nononsense physician.[7] He had grown up one of eleven children on a farm in Iowa, and Harvard Medical School accepted him even though no one on the admissions committee had ever heard of his college. Bock began his career in the 1920s with a Moseley Traveling Fellowship for the study of medicine in Europe, following which he undertook a study of the physiological adaptation of men living in the high Andes. This experience led to his interest in physical fitness and positive health.

Bock and his colleague John W. Thompson (also at the conference) were pioneers in the study of normal human physiology, and in 1926 had helped found the Harvard Fatigue Laboratory, where physiologists, biologists, and chemists studied men's ability to adapt to physical stress. There they developed a technique of exercising subjects by having them step on and off locker-room benches, which is still part of modern cardiac testing.[8] The Laboratory was officially located on the campus of the Graduate School of Business Administration, but it sent teams traveling around the world, from the tropics of the Canal Zone to the Andean peaks. Its work eventually led to the decision of the United States Air Force to equip its new high-altitude bombers with supplementary oxygen.

Bock had an expansive vision, and he never stopped inveighing against medicine's tendency to think small and specialized. Writing to accept the directorship of the Harvard Department of Hygiene, he asserted that medical research paid too much attention to sick people, and that dividing the body up into symptoms and diseases could never shed light on the urgent question of how to live well. It was he who first conceptualized the concept of positive health; sixty years later, University of Pennsylvania psychologist Martin Seligman would take Bock's challenge into the new field of study we call positive psychology.

Even after the Grant Study got under way Bock's close affiliation with the Fatigue Lab continued, and it's worth noting that he himself walked two miles a day until his death at 96. He never forgot that "normal" and "average" are not the same thing—20/20 eyesight is normal, but unfortunately not average—and his interest was not in elucidating average fitness, but the best fitness possible. To accomplish this, it seemed sensible at the time to study an elite sample of men. And that's what the Grant Study did.

For its first seventeen years (from its beginnings in 1937–1938 until Charles McArthur assumed the directorship in 1955), the Study was dominated by its founder Arlie Bock, its first director, Clark Heath, M.D., and its social investigator, Lewise Gregory. It was the dedication and kindness of these three people that created the loyalty and gratitude that bonded the men to the Study until death did them part.

THE PIONEERS

Clark Heath, M.D., who served as director from 1938 to 1954, was a promising research scientist. He had once worked with Professor William Castle, a co-discoverer of vitamin B-12. He was also the staff internist, but his job description covered far more than that, and it expanded steadily over the years. He was responsible for the budget, for all necessary reports, for planning for the future, and for putting together case summaries of the Grant men. Still, it was not his administrative skills that made him so important to the young Study, but his clinical gifts.

The files of those early years make clear just how completely Heath epitomized the warm and caring physician. As each man joined the Study, Heath gave him an unusually complete two-hour physical exam. That was part of the Study routine. But until he left Harvard, returning Grant Study alumni came back to him voluntarily, seeking consultation for themselves and their families. When they needed more than mere medical advice, Heath saw a number of these ex-students for psychological counseling.

Lewise Gregory (later Davies), a perceptive Virginian, was the third member of this key trio. Bock needed a social investigator to interview all the Study members and their families. Gregory's only professional education was secretarial school, but Bock chose her for her dazzling interpersonal skills and her innate talent as an interviewer. After her death, Charles McArthur, the second director, described her technique for a staff member who was writing a brief (unpublished) history of the Study. "When she would visit a student's family, she would sit daintily with her legs crossed, affix her large blue eyes on the conversationalist, and the smitten parents would bare their lives to her." She was an attentive and sympathetic listener, and the loyalty she engendered among the participants and their families would last a lifetime. She also endeared herself to the Study men as a big sister. (A very pretty one. And her sister was Margaret Sullavan, the movie star.) Miss Gregory was invaluable in bringing lost sheep back into the fold, and the Grant Study's attrition rate—the lowest of any similar study in history—is a tribute to her diligence and her diplomacy.

In the beginning, Gregory's job was to take a careful social history from each subject as he entered the Study, and then interview his parents. In those early years she traveled the length and breadth of the country to meet the men's families in their homes. She obtained from the mother a detailed history of the early development of each Study man. She also took a family history that included descriptions of grandparents, aunts, uncles, and first cousins, and of any mental illnesses in the extended family. I encountered some of these family members many years later, and they still recalled her visit with great pleasure.

In her family interviews, Gregory accentuated the positive. The usual psychiatric social history with its emphasis upon pathology tends to make all of us look like fugitives from a Tennessee Williams play. Gregory asked about problems in the men's growing up, but she also wanted to hear about what had gone well. Not only did this foster the alliance between the Study and its members, but it also meant that when pathological details did come up in the social histories, they were usually significant.

After her great work of 268 family interviews was complete, Gregory remained with the Study part time. Even in the 1980s, at the end of her tenure, she was still the one who could charm nonresponders back into continued participation—which she did with astonishing success.

William T. Grant, of course, was an important figure in the founding of the Grant Study. Billy Grant, as his friends called him, was a tenth-grade dropout who founded a store for low-priced household wares in 1906. Nothing in his first store cost more than twenty-five cents—he prided himself on that—but it grew into the Walmart of the 1930s. Grant's foundation has become world-famous, too, but it got its start in that first gift to Arlie Bock in 1937. The two men had different visions. Bock hoped that his research into an optimally healthy population would help the United States military select better officer candidates; Grant hoped that his new foundation's first project would help him identify effective managers for his multitude of chain stores. They both hoped to identify superachievers. But Grant was more interested in social and emotional intelligence than the other Grant Study investigators, who shared the contemporary preoccupation with constitutional medicine. By 1945 this would lead to friction. But until the end of the war Bock maintained a close relationship with Grant, visiting him at his houses in Florida and Connecticut.

Frederic Lyman Wells, Ph.D., the Study's chief psychologist, was a psychometrician and one of the developers of the Army Alpha Tests, the major test for intelligence screening in World War I. Wells came from a distinguished New England academic family. He started college at fifteen, and by twenty had earned his M.A. From 1925 to 1928 he served on government advisory boards, such as the National Research Council and the National Committee for Mental Hygiene. His Grant Study brief was to determine the personality organizations, interests and aptitudes, and intelligence levels of the Grant Study subjects.[9] Wells was probably the most distinguished scientist on the staff, and he gained further eminence as a consultant to the War Department between 1941 and 1946, when he helped to develop the Army General Classification Test (a measure of intelligence and vocational skills). He was a hard worker and a methodical and systematic analyst, but given to dry and lengthy statistical expositions. His reports reveal little of the rich humanity of the Study men.

Carl Seltzer, Ph.D., a young physical anthropologist who had worked closely with Earnest Hooton and William Sheldon (on whom more below), was another exponent of constitutional medicine, specifically the relationship between body type and personality. John W. Thompson, Ph.D., a Scot and co-founder of the Fatigue Lab with Bock, and then Lucien Brouha, Ph.D., a Belgian refugee from wartorn Europe, served as early physiologists to the Grant Study staff. Both were gone by 1943, however, and funding for the Fatigue Lab seems to have halted by 1944. Vicissitudes in funding are one of the ways we know when scientific fashions begin to change. Constitutional medicine was on the way out.

The Study's staff psychiatrists were responsible for extensive interviews of the Study members—about ten hours each. None of the five men who filled the position in the Study's early days stayed longer than three years. Two of them went on to distinguished academic careers: Donald Hastings (1938) as chairman of the Department of Psychiatry at the University of Minnesota Medical School, and Douglas Bond (1942) as dean of Case Western Reserve Medical School. Psychiatrists William Woods (1942–45), John Flumerfelt (1940–41), and Thomas Wright (1939–40) also served as interviewers. Woods was also responsible for an assessment scheme of twenty-six personality traits that underpinned much of the early research.[10]

Unfortunately, however, the early staffing and implementation of the Grant Study did not reflect the early work of four important investigators of personality. Their work profoundly affected my own later interpretation of the Study's data, but in 1937–42 it was still too new to inform the Study itself. Heinz Hartmann, Anna Freud, Erik Erikson, and Harry Stack Sullivan all significantly shaped the modern understanding of healthy personality. The first three were instrumental in replacing Sigmund Freud's view of personality as an often-pathological compromise between cognitive morality *(superego)* and irrational passion *(id)*. Hartmann, Anna Freud, and her student Erikson offered an alternative conception of personality as a product of involuntary, but usually healthy and creative, adaptation. Anna Freud's *The Ego and the Mechanisms of Defense* was first published in 1937; two years later, Hartmann published his own classic work on ego psychology, which would appear in English as *Ego Psychology and the Problem of Adaptation*. However, it wasn't until 1967 that the Grant Study began to focus on the men's styles of psychological coping (see Chapter 8).

Harry Stack Sullivan was another pioneer, who opened psychiatry up to the science of relationships that John Bowlby and his student Mary Ainsworth would make famous—but not until the second half of the twentieth century.

FROM BIOLOGY TO PSYCHOLOGY

Let me expand on these omissions for a bit, to show how time changes both science and scientists. It is easy to forget how very recent is our current interest in intimate relationships. For the first ten years of the Study, biological theorizing held undisputed pride of place. In 1938, constitution and eugenics (as in breeding) were considered far more potent forces than environment in how people turned out. Biological indicators were tracked in minute detail, but it was the rare social scientist who paid any attention to what would become known as "emotional intelligence," particularly the capacity for love and close friendship. Harry Harlow was one of these, a psychologist/ethologist who earned renown for groundbreaking studies of relationship deprivation in monkeys. In his 1958 presidential address to the American Psychological Association, he felt driven to lament, "Psychologists not only show no interest in the origin and development of love and affection, but they seem to be unaware of its very existence!"[11]

At the time of Harlow's speech, behaviorists like B. F. Skinner and John Watson were assuming that babies became attached to their mothers because their mothers fed them. Psychoanalysts Sigmund and Anna Freud believed essentially the same thing. However different the behaviorist and psychoanalytic psychologies, they shared a rather concrete view of the interplay between biology and emotion. Lust, hunger, and the vicissitudes of power ruled the psychological universe. Love was conceptualized as *Eros*—a matter of individual hedonistic instinct, not a process of reciprocal pair bonding,

It was only in 1950 that John Bowlby, who was both a psychoanalyst and an ethologist, began to establish awareness that attachment experiences are fundamental shapers of personality, and that babies "imprint" on their mothers not because their mothers fill their bellies, but because they cuddle them, sing to them, and gaze into their eyes. Experimental evidence soon followed. But I can attest that years after the Grant Study began, English teachers were still drilling the youth of the 1940s on Rudyard Kipling's Victorian mantra "He travels the fastest who travels alone." A relational world governed by oxytocin, mirror neurons, and limbic maternal attachment (a.k.a. love) was inconceivable in the psychology of the Study's first decade.

There is an interesting parallel here with infantile autism. This fairly common disorder, which is due to a congenital absence of empathy, was not spotted until 1943, when a child psychiatrist finally noticed it in his own son. Its close relative, Asperger's syndrome, was identified in 1944. But it was fifty years more before those genetic disorders were absorbed into psychiatry's diagnostic framework. In other words, in the 1930s, the congenital impairment of attachment reflected in childhood autism was harder for scientists to grasp than quantum physics. The functional reality of relatedness had not been incorporated into the consciousness of the social sciences.

Cultural anthropology, pioneered by Franz Boas, captured the hearts and minds of college students in the sixties, but it was a relatively esoteric discipline in 1940, and physical anthropology still dominated. Ernst Krestchmer, a German psychiatrist, was nominated in 1929 for the Nobel Prize in medicine for his work on body build and character.[12] Investigators inspired by him believed that personality was determined by *somatotype*—three physical bodily conformations (thin, frail *ectomorphs,* robust muscular *mesomorphs,* and soft, plump *endomorphs*). Social scientists still believed that the British Empire was the result of genetic racial superiority, not the environmental luck of owning the "guns, germs and steel" that Jared Diamond made famous in the brilliant book of that name, in which he proved once and for all that racial dominance was a product of culture and geography, and not of biological heredity.

So while the biological determinism of the early Grant investigators sometimes sounds like racism to twenty-first-century ears, it had less to do with cryptofascism than with the absence of an alternative paradigm. Many hours were spent assessing the physiology, somatotype, and "racial typology" of the Grant Study sophomores. During their ten hours of psychiatric interviews they were asked about masturbation and their opinions on premarital sex. But nobody asked them about their friends or their girlfriends. Until the 1970s at least, attachment and empathy were the province of sentimental novels, not of science. A pity.

THE MEN OF THE GRANT STUDY

As of this writing, all but seven of the surviving members of the Grant Study have reached their ninetieth birthdays. When the Study began, however, they were college sophomores, most of them nineteen years old. The original College cohort—the group of Grant Study men, who liked to refer to themselves as "the guinea pigs"—numbered 268. Sixty-four men were drawn from the Harvard classes of 1939, 1940, and 1941, and 204 came from a more systematic 7–8 percent sample drawn from the sophomore classes of 1942–1944.

About 10 percent of the men ended up in the Study by chance: a few volunteers; some younger brothers of already participating men; the occasional student referred by an advisor. The other nine-tenths of the sample were selected according to the following rough protocol, of which the details varied slightly from year to year.

First the Study investigators went through the new class, screening out any students who might not graduate. On the advice of the dean of students, they used as criteria the students' SAT scores in combination with grade point average and observed measures of natural ability. A high school valedictorian with modest SATs would be preferred to a boy who tested well but whose class standing in high school was low. These criteria eliminated 40 percent of each Harvard entering class.

Known medical or psychological difficulties led to the exclusion of another 30 percent. The (roughly 300) names remaining were submitted to the college deans, who picked from among them about 100 students each year whom they perceived as "sound." In essence, this meant students the deans were glad they had admitted, especially the ones involved in extracurricular activities and freshman athletics. For a single class year, the Grant Study selected the men who would become editor-in-chief of the *Crimson,* president of the *Advocate* (the college literary magazine), and president of the Harvard *Lampoon.* Four times as many of the Grant Study men as chance would have predicted held class offices, both in college and, as it turned out over the next half-century, at college reunions. However, the deans also chose a disproportionate number of "national scholars," gifted men from poor families for whom all expenses, including transportation, were provided by Harvard. These youths, often relatively inept socially, were chosen solely for potential academic brilliance.

The freshman physicals of the classes as a whole tell us—and it's no surprise—that the Grant men were twice as likely as their class-mates to be muscular mesomorphs, rather than skinny ectomorphs or pudgy endomorphs.

Out of the 90 (10 percent of each class) sophomores chosen, roughly 1 in 5 ended up not joining the Study for reasons of his own (schedule conflict, disinclination, or failure to show up for intake procedures). About 70 men a year were added to the Study from the classes of 1942 through 1944, bringing the total College cohort to 268.

The Grant Study deliberately cast its net for men who were likely to lead "successful" lives. All of them had already been selected for admission to a competitive and demanding college, and then they were selected further for their capacity to master college life and, in the words of Arlie Bock, "paddle their own canoe." Many of them were firstborn sons, and independent men were preferred over less autonomous ones. The Study selectors were looking for men with the capacity to live up to or exceed an already high level of natural ability.

The men of the Grant study were homogeneous in many ways. They were well matched in physical and mental health, skin color, education, intellect and academic achievement, and culture and his-torical epoch (see Table 3.1). They all lived through the Great Depres-sion, and they all shared the likelihood of active participation in World War II in the immediate future.

The Study administered the Army Alpha Intelligence Tests, which put the IQs of most of the men in the top 3 percent of the general

Table 3.1   The Historical Setting of the Grant Study Birth Cohort

	Personal Events	Historical/Public Events
1919–1922	Birth	End of World War I
		Women's suffrage in place
1923–1929	Childhood	Roaring Twenties
		Fathers getting richer
		Era of strict toilet training
		Fear that kissing children spreads disease
1930–1940	Adolescence	The Great Depression
		Fathers getting poorer
		Isolationism and pacifism
1941–1945	College	
	Military service	World War II
1946–1960	Marriage	G.I. Bill
	Young families	Strong U.S. dollar
	Career uncertainty	Rapid growth
		Eisenhower years
		"Man in Grey Flannel Suit"
1961–1975	Consolidated careers	Vietnam
	Empty nests	Surgeon general's report on smoking
	Leading and caring for adults	Civil rights and Medicare
		U.S. dollar weaker
		End of Cold War
1990–2010	Retirement	Clintons and Bushes
	Poorer health	Iraq War
	Preserving culture	

population. Their College Board SAT scores (average 584) put them among the top 5 to 10 percent of college-bound high school graduates, but not beyond the range of many other able college students. Of course, in 1940 the population taking the SAT was more select than today; still, many readers of this book will be entitled to sneer, "Well, I scored far better than that."

The average Grant Study man was fifth-generation American. A few were immigrants, but more than a few had families that had been here for ten generations. There were no African Americans. Ten percent of the men were Catholics and 10 percent Jews. The remaining 80 percent, a higher percentage than among their classmates, were Protestants. Eighty-nine percent of the men came from north of the Mason-Dixon Line and east of the Missouri. Twenty-five years later, 75 percent of the sample remained within these boundaries, and 60 percent of the total sample had migrated to the five urban meccas of San Francisco, New York, Washington, Boston, and Chicago.

They were mostly a privileged group, but here they were less perfectly matched. Half of the men had had some private education, but often on scholarship. In college, 40 percent received financial aid (at that time, a year at Harvard cost about $22,500 in 2009 dollars), and half worked during the academic year, paying college costs not covered by their scholarships themselves. The men's families were classified on the basis of paternal education, occupation, and selection for *Who's Who in America*. One-third of the men's fathers had had some professional training, but half of the parents had no college degree. Only 11 percent of the men's mothers had ever worked, and of those who had, most had been single parents. Of the thirty-two working mothers, two were writers, five schoolteachers, one an artist, and one a lawyer; the rest were secretaries and waitresses.

Once Lewise Gregory began her home interviews, the families were also classified more subjectively by reference to such class- and status-related markers as household furnishings, books, art, and size of house. There was significant variation there. Sixteen percent of the families were categorized as upper class. Even during the Depression they enjoyed multiple houses, motorcars, and servants. Their mean yearly income was $225,000 per capita—yes, I mean per family member—in 2009 dollars. Four percent of the families were classified as lower class. Their annual mean per capita income was $5,200 in 2009 dollars.

So the men did not all come to college with silver spoons in their mouths; and even when they did, their parents or grandparents might have had more humble beginnings. There was Alfred Paine, for example, whom we will get to know more extensively in Chapter 7. He had a trust fund from birth; his father had been a head of the New York Stock Exchange; his grandfather was a successful merchant banker. But that grandfather had made his first thousand dollars as an itinerant pioneer, picking up buffalo horns on the Great Plains at night and shipping them back to New England for resale.

The father of another study member, Brian Farmer, was a painter and paperhanger. Soon after Brian's birth, work grew so scarce that Mr. Farmer moved his family to South Dakota, where he and his wife and older children worked in the sugar beet fields as laborers. For their combined efforts, the family received eleven dollars for every acre they cleared. They had barely enough to eat until a kindly neighbor told them that they could have all the beans and potatoes they wanted from his fields. Together they gathered enough beans and potatoes to last them through the winter, plus a surplus that they traded for staples such as sugar, salt, and other groceries. During those years the Farmers did not know what it was to taste fresh vegetables or fruits. Mr. Farmer picked up odd jobs here and there, but his neighbors were too poor to pay him in cash. When Brian entered Harvard, his father was still earning only five dollars a day.

Some of the men had other kinds of difficulties. One Study member, whom his adult friends knew as "a very happy guy," talked about growing up with an alcoholic mother. "I had a terrible time in my second, fourth and sixth grades. I got trial promotions probably because the teacher wanted to get rid of me. Throughout this period, also, there was hardly any income in the family. My folks had lost the variety store and gas station and during the winter, warmth at night consisted of getting under a pile of blankets. The winter days of my early years were spent curled up on a bench behind a big potbelly stove in a variety store listening to men talk. I would stay there because it was warmer than it was in the house."

Whatever a man's origins, his Harvard degree was a ticket of entry to the upper middle class. When the men returned from their service in World War II, they also benefited from high employment, a strong dollar, and the G.I. Bill, which virtually guaranteed an affordable graduate school education. They were just young enough to participate in the physical fitness and anti-smoking trends of the sixties, seventies, and eighties. Most of the 1940 generation of Harvard men were upwardly mobile, and most ended up more successful than their fathers. (There were a few exceptions, like the man whose Wall Street father made two million 1935 dollars a year in the midst of the Depression. No, this father was not Joseph P. Kennedy.)

HOW THE MEN WERE STUDIED: 1939–1946

Three investigators interviewed each sophomore accepted into the Study: Clark Heath, a staff psychiatrist, and Lewise Gregory. The meeting with Dr. Heath included a comprehensive two-hour physical and a history of the young man's dietary habits, medical history, and physical responses to stress. Each man was also studied by Carl Seltzer, the physical anthropologist, who recorded his racial type (Nordic, Mediterranean, etc.) and his somatotype (mesomorph, endomorph, ectomorph), determined whether his body build was predominately "masculine" or "feminine," and made exhaustive anthropometric measurements.[13] The Study investigators noted his every physical detail, from the functioning of his major organs to his brow ridges and moles to the hanging length of his scrotum. They took a careful dietary history, too, including the number of teaspoons of sugar in his daily coffee or tea (the range was 0 to 7!).

As I've noted, the emphasis on somatotyping was an ill-fated effort to advance the then-fashionable science of physical anthropology. Carl Seltzer's mentor was the Harvard anthropologist William H. Sheldon, who had been influenced by Kretschmer's work and believed that human personality was significantly correlated with body build. The ectomorphic build, that is, was thought to be correlated with a schizoid temperament, the mesomorphic build with a sanguine temperament, and the endomorphic build with a manic-depressive temperament.[14] This is a cat that I let out of the bag in the last chapter; a third of a century later, Grant Study follow-up revealed that none of the ten outcomes in the Decathlon of Flourishing (or the men's officer potential) correlated significantly with body type.

Eight to ten one-hour psychiatric interviews focused on the man's family, his values, his religious experience, and his career plans. The psychiatrists were trying to get to know the men as people, not as patients. No attempt was made to look for pathology or to interpret the men's lives psychodynamically. The interviews included a history of early sexual development, but unfortunately the psychiatrists did not inquire into the men's close relationships.

Lewise Gregory integrated individual interviews with the men with the careful social histories that she gathered in her home visits with the men's parents; I've described these above. In keeping with the research methodology of the 1930s, those histories were sometimes more anecdotal than systematic. Her chief informants were usually the men's mothers, although fathers and siblings at times contributed information as well.

Frederic Wells, the Study psychologist, gave each man tests designed to reflect native intelligence (the Army Alpha Verbal and Alpha

Numerical). In many cases he also administered two projective tests: a word association test and a shortened version of the Rorschach. The intention here was to test imagination, not to explore the unconscious (for which purpose the Rorschach is usually employed). Also included was the Harvard Block Assembly Test, an assessment of manipulative dexterity and comprehension of spatial relationships.[15]

Physiologist Lucien Brouha studied each man in the Fatigue Lab. Brouha measured each man's respiratory functions and the physiologic effects of running on an 8.6 percent incline treadmill at seven miles per hour for five minutes or until exhausted, whichever came first. Examiners took measures of pulse rate, blood lactate levels, exercise tolerance, and so on as a way of sorting the students by physical fitness. Surprisingly, I noted in 2000—more than fifty years later—that treadmill endurance correlated better with successful relationships than with physical health. (As it turned out, endurance and stoicism turned out to be better predictors of love than health in other areas, too.)

In 1940, the Study received a one-time grant of $2,400 ($35,000 in 2009 dollars) from the Macy Foundation, a philanthropic organization sympathetic to psychosomatic medicine. This windfall enabled the recording of primitive single-channel electroencephalograms, which had just begun to come into use. The neophyte encephalographer's interpretations of these EEGs sometimes sound more like Tarot card readings than physiological analysis; in some cases the tracings were thought to reveal "latent homosexuality." The Study also hired an experienced forensic graphologist to interpret the men's handwriting. It soon became clear that neither handwriting nor EEGs were useful predictors of personality. But however naive or even comical some of these efforts sound now, they reflect a serious and important truth—that the Grant Study was collecting information even before there was science available to capitalize on its investment, in hopes

Table 3.2   The Sequence of Contacts

1938–1945
   8–10 psychiatric interviews
   Complete physical exam by Dr. Heath
   Interviews by Dr. Heath and Miss Gregory; she also made home visits
   Anthropological and physiological testing
   EEG, Rorschach, and handwriting analysis (on many of the men)
   Complete psychometric testing and some projective testing by Dr. Wells
   The 26 personality traits assigned by Dr. Woods (Appendix E)
   Brief childhood assessment (1–3) by Miss Gregory
   At age 21 the ABC College adjustment (soundness) ratings assigned

1946–1950
   Debriefing of combat experience by Dr. Monks
   Annual questionnaires begun
   Interviews with wives by social anthropologist Dr. Lantis
   Thematic Apperception Test by Dr. McArthur
   Full staff conferences on each man at age 29; ABCDE ratings of personality
       soundness for future adjustment made

1950–1967
   Mostly biennial questionnaires; little other contact

1967–1985
   All men re-interviewed, mostly by Eva Milofsky or Dr. Vaillant (Appendix A)
   Beginning age 45, complete physical exams every five years until the present
   Objective health scored 1–5 (Well, minor illness, chronic illness, disabling illness,
       dead)
   Childhood environment assessed, blind to events after age 19
   Wives twice sent questionnaires
   Adjustment at work, love, and play assessed from age 30–47 on all men (Appen-
       dix D)
   Adaptive coping style (narcissistic, neurotic, empathic) assessed
   NEO (Costa and McCrae) administered by mail
   Lazare Personality Inventory administered by mail

1985–2002
   All men re-interviewed (Appendix A)
   Adjustment at work, love, and play assessed from age 49–65 and from age 65–80
       (Appendix D)
   Wives and children sent questionnaires
   Physical exams every five years and biennial questionnaires continued

Gallup Organization's Wellsprings of a Positive Life administered by mail
Aging at 80 computed (subjective and objective mental and physical health)

2002–2010
Physical exams every five years and biennial questionnaires continued
Joint marital interviews and "daily diaries"
Telephone Interview for Cognitive Status (TICS) at ages 80, 85, and 90
Re-interviewed at retirement, 1985–2005

2010
Decathlon scores compiled

---

that it would prove meaningful later. And in many cases it did. Prewar
psychological and medical science were very different from today's
counterparts. It wasn't only attachment theory that had to wait until
the nineteen-fifties; so did double-blind placebo-controlled drug tri-
als. Good science is always reaching a little ahead of itself.

Case in point: the scientific pendulum has swung once more, and
in the early twenty-first century genetic research once more dominates studies of environment. Instead of handwriting samples, the
Study is collecting DNA. We don't yet know how it will be used.
What will *that* look like to readers seventy-five years from now?

EARLY DATA ANALYSIS

By 1941, the Study had examined 211 sophomores. A debate arose as
to whether this was enough. Frederic Wells, perhaps the most established researcher employed by the Study, sent a letter to Clark Heath
imploring that he stop accepting new students so that the staff could
begin evaluating the huge amount of data that they already accumulated. But others worried that 211 cases was only a small number, after all, when the goal was to find conclusive correlations between

the psychological and anthropological. Eventually a compromise was reached, and the researchers agreed to take on one last group of students from the class of 1944. They were studied as sophomores in 1942 and completed the total Study cohort of 268. Between 1938 and 1943 the Grant Foundation had contributed $450,000 ($7 million in 2009 dollars) to the research.

But now the Study had to decide how to harvest this vast planting, and that was not proving easy. As I've said, a longitudinal study has to stay a bit ahead of itself, both in its data collecting and in its theoretical hunches. After all, it can't know in advance what will prove relevant (if it did, there would be no need for the study), and has to rely on best guesses. This can be an advantage as well as a disadvantage. But the Grant Study was a very early longitudinal study, so it didn't have the benefit of others' experience, and it began in something of a theoretical vacuum, when very little was known about adult growth and development. It had been very ambitious in collecting data, but not always very thoughtful about what it would do with the results. Many of the investigators were feeling overwhelmed, and by 1944 William Grant was beginning to express serious doubts about the administration of the Study and about whether he should continue to fund it. The trustees of the Grant Foundation put increasing pressure on the researchers to produce a summary of the Study's early results. For the next two years, the Study archives are filled with desperate efforts to discover publishable material.

Only three papers appeared before 1945 (see Appendix F), and in 1945, two monographs were hastily published to meet Grant's demand for results and a popular book that would dramatize the Study's findings.

Earnest Hooton was a Harvard professor, a brilliant physical anthropologist, and a fluent writer. He had been Study anthropologist Carl Seltzer's mentor (also my archaeologist father's). And he was firmly committed to the world of constitutional medicine. "If we wish to study the whole man, we must begin with his physique," he maintained, and he imagined that research out of the Grant Study might one day lead to "effective control of individual quality through genetics, or breeding."[16] It was he who had encouraged early researchers to expect that "[O]n the whole . . . 'normality' goes closely with a 'strong masculine component.'"[17] Hooton's summary of the Grant Study was published in 1945 as *Young Man, You Are Normal*.[18] That was a far cry from Grant's own proposed title, *The Grant Study of Social Adjustment*, which did not help to bridge the growing rift between the Study administration and its funder. Still, for the next thirty years, Hooton's was the leading book on the study.

In it he wrote, "When physique, studied from different standpoints, turns out to be so intimately related to various personality traits, it is clear that body build must also furnish clues to the social capacities of the individual."[19] Instead of adducing any experimental evidence in support of these claims, however, he simply dismissed his opposition as "crass environmentalists."[20] This is a striking term. It conveys not just simple cluelessness about the environmental considerations that are so crucial now, but real antagonism to an entire way of thinking. Yet Arlie Bock had already been wondering about the nature/nurture issue in his first visions of the Study seven years before; it's possible to see in the difference between his attitude and Hooton's a scientific paradigm nearing its tipping point.

The second monograph was *What People Are*, by Study director Clark Heath. Heath's book relied heavily on anthropometric measurements too, and also on an untested personality profiling scheme devised by William Woods, a staff psychiatrist without research training.[21]

Woods's schema (see Appendix E) scored Study members on twenty-six personality traits, many of them dichotomies such as *Vital* (warm, expressive) *Affect vs. Bland* (colorless) *Affect, Well-Integrated Personality vs. Unintegrated Personality, Verbal vs. Inarticulate, Sociable vs. Asocial* etc. An attempt was then made to correlate the resulting profiles with body build but the results were unconvincing.

Neither of these two hastily published books attracted much attention, and certainly they did nothing to halt the shift toward environmentalism in the social sciences. In the years since, somatotype categories have not proven particularly useful when matched against other independent ratings, and neither have (most of) Woods's personality traits. Even at the time, what evidence there was, was not persuasive. Worse, the early investigators were not blind to each other's ratings, so there's no way to know how much apparent early correlations between body build and personality were a function of halo effect or observer bias. In 1970 I tried to replicate the correlations reported in some of the early papers and could not.

Before we contemptuously (and prematurely) dismiss the early investigators' work as a wash, however, we have to consider the realities of statistical methodology in those days. There were no computers to absorb infinitudes of entries and magically align them along dozens of axes at the click of a key. Study members had to be listed down the left-hand side of huge ledger sheets. Scores and measurements were plotted along the top, individually, in exquisite handwriting. And that was only the beginning. To be useful, the data had to be pulled out of the ledgers, again by hand, and subjected to individual calculations, some of them quite complex. The tests that the original investigators used are out of fashion today, but analysis was laborious in a way that is unimaginable now. The Study in the 1940s did not even own a Monroe calculator (as the early electric adding machines were called). Ironically, in 1944, one of the world's first reliable computers, the Mark I, was being installed only 300 yards away. But the techniques and technologies of complex data analysis that could (and in time did) reveal unexpected correlations in the early data were not yet available to the Grant Study researchers. When the rich crop they had sown finally did come to fruition after the explosion of computer technology in the late twentieth century, it was the early investigators who had done much of the heavy lifting.

EARLY CONCLUSIONS AND SOME CORRECTIONS

The specter of World War II engaged early the Study's interest in possible military applications for its research, particularly the identification of potential officers and appropriate placements for them. John Monks was a patrician internist who joined the staff in 1946 to study the men's response to the war, and produced a well-researched if little-appreciated monograph, *College Men at War.*[22] Monks treated Woods's personality traits as if their scientific validity had been proven, with the suggestion that they be used to identify desirable traits in officer candidates. His monograph was fascinating in the stories and backgrounds it gave of the College men. It was disappointing, however, because Monks, like the original investigators, had not yet grasped the power of longitudinal study to actually test a hypothesis.

And the fresh empirical follow-up of 2010 (which came out of the questions I was trying to answer for this book as well as the Decathlon challenge) found that such promising variables as *Masculine Body Build, Well Integrated, Vital Affect,* or such foreboding ones as *Lack of Purpose and Values* and *Shy,* bore no relation at all to attained military rank. The only college trait that did predict high rank was *Political,* while low rank was predicted only by . . . *Cultural* and *Creative and Intuitive!* The scriptwriters of *M*A*S*H* might have anticipated that result, but Bock's original investigators certainly did not.

Woods's predictive schema has had more general problems than its failure to identify good officer material. Most of his traits failed to correlate significantly with any of the events in the 2010 Decathlon of Flourishing, and only one correlated significantly with more than three. That one, however, was a notable exception. *Well Integrated* (defined as "steady, dependable, thorough, sincere, and trustworthy") correlated with eight Decathlon events, and for the last twenty-five years has been a staple variable in our data analyses. It signals a bundle of traits that enable a young man "to surmount common problems which confront him such as career choice, competitive environment, and moral and religious attitudes."[23] It was assigned to 60 percent of the men, while 15 percent were called *Incompletely Integrated*. These latter men were deemed to lack perseverance, and were seen as "erratic, unreliable, sporadic, undependable, ill directed and little organized." (The remaining 25 percent of the men were unclassified for this dichotomized variable.) Half a century later, proportionately four times as many of the *Well Integrated* enjoyed good marriages than the *Incompletely Integrated*. Another finding of great interest was that as of 2012, the *Well Integrated* have lived on average seven years longer. In contrast, the two variables that the original investigators thought would be most predictive of positive outcome, *Vital Affect* and *Sociable,* were unimportant after the first ten years.

TROUBLE IN PARADISE

Writing to Bock in 1944, Grant chastised him for managing "a cooperative study with a good deal more distrust, disdain and independence of outsiders than I have ever seen productive of satisfactory results."[24] He would be willing to contribute to the Study in the future, he said, only if Bock also secured additional funds from the Harvard Department of Hygiene. He promised one more payment of $30,000 ($300,000 in 2009 dollars) to fund the project until he and the trustees of his foundation could feel confident that the Study was organized effectively. In hindsight, it appears that no one in the Study had fully understood the constraints of longitudinal studies, or the patience that they require. And while Clark Heath was a model of clinical expertise and personal warmth, administrative organization was not his forte. Ironically, there were gifted future administrators at the Grant Study, and one, at least, had an intuitive understanding of the powerful but demanding nature of this kind of research. Donald Hastings, once a Grant Study psychiatric interviewer, carried out and harvested one of the first well-designed long-term studies of neurotic outpatients when he was chairman of psychiatry at the University of Minnesota Medical School in 1958.[25] Hastings's study was one of the first prospective longitudinal studies ever published in the *American Journal of Psychiatry*, and the first that I ever clipped out of a journal as a medical student to keep in my own files. It was Hastings who showed me how prospective longitudinal study could bring order to muddled psychiatric thought, and I admired his work for years before I ever heard of the Grant Study.

Bock grumbled to Earl Bond, Grant's closest scientific advisor and future Grant Foundation trustee, "I had to tell Mr. Grant that the trustees of the foundation could not expect to run the Grant Study." After this, not surprisingly, Grant Foundation largesse dried up for almost twenty years. The embryonic Grant Foundation had no medical researchers or clinicians among its trustees, who did not yet understand either the limitations or the potential of the Grant Study's longitudinal design. An orchard like the Grant Study takes time to bear its fruit.

Even as publications began to emerge more regularly (Appendix F), the Study authors clung to the bounds of their respective disciplines, selecting the kinds of narrow topics that yield academic journal papers. They were not yet exploiting the prospective, longitudinal potential of the Study. For example, they could have tested the somatotype data more critically; certainly they had the necessary material to examine in 1946 whether this information did or didn't predict anything about military rank at discharge. But they didn't. By 1970, even after fifty publications, the Grant Study remained little known and little cited. It had also been underfunded for fifteen years, as I will describe shortly.

On the other hand, the seeds that Grant, Bock, and Heath had planted really did promise a rich harvest decades into the future. For example, as we'll see in Chapter 10, Woods's intuitively derived personality traits permitted us to predict political choice reliably over the next half-century of the men's voting lives. And forty years later, Monks's careful debriefing of the men after the war regarding the details of their combat experiences led to one of the world's few prospective studies of the antecedents of posttraumatic stress disorder.[26]

THE WAR AND ITS AFTERMATH

After Pearl Harbor, most of the Grant Study men went into the armed services directly from Harvard, and most of them performed unusually well there. But they took their first real steps into adult life in a world reeling from the stupendous social and economic sequelae of the war, which made rigorous demands on their adaptive capacities.

THE NEXT TWENTY-FIVE YEARS: 1945–1970

The end of the war saw a sea change in Western social science as well as in Western society. Hooton's ferocity notwithstanding, genetics and biological determinism were out, partly out of emotional overreaction to the Nazi misuse of eugenics in pursuit of a racist ideology. Relativism, Skinner, and cultural anthropology were in, and scientists were looking beyond constitutional endowment for the conditions of life that shape individual outcomes. In the United States, psychoanalysts were becoming chairmen of major departments of psychiatry.

Censorship issues during the war years limited communication between the Grant Study and the men to the occasional V-mail. But once the war was over, the Study began following the College cohort with annual questionnaires. (That practice continued through 1955; it's been every two years since then, with only a few exceptions.) The questionnaires are long, and designed to take advantage of the men's high verbal skills. They inquire about employment, health, habits (vacations, sports, alcohol, smoking), political views, family, and, especially, quality of marriage.[27] They have varied somewhat from decade to decade and from director to director, but inevitably they reflect changes in intellectual climate. Only in 1955, for instance, came the first inquiries about college roommates and girlfriends. A perfect prospective study, of course, would forbid retrospective data like this.

By 1948, the Grant Foundation had made good on its threat to withdraw funding. But Alan Gregg of the Rockefeller Foundation stepped in with the salary for a cultural anthropologist. After discussion with Gregg and his colleagues, Bock and the Study staff decided to rename the project the Harvard Study of Adult Development. (Since 1970, that name has covered both the College cohort of the Grant Study and the Inner City cohort of the Glueck Study.) At this point the Study was essentially supported by Harvard Health Service funds. But to avoid confusion, letters and questionnaires to the men continued to address them as "Grant Study" members, and the Study has continued to be known that way informally to this day.

Margaret Lantis was the new cultural anthropologist, trained to look for, respect, and study cultural differences. She had a doctorate in her field, and was a vastly more sophisticated interviewer, if a less cozy one, than Lewise Gregory, whose training had taken place around Virginia tea tables. In 1950, Lantis interviewed 205 Study men and their wives in their homes. In addition, in the ten years after the war, 171 men dropped by the Study to visit, and the staff was fortuitously able to glean some current information from them. Clark Heath and Gregory usually did those debriefings.

In collaboration with Charles McArthur, a social psychologist who would soon become director of the Study, Lantis administered the Thematic Apperception Test to many of the men. The TAT is a projective test designed to study affect and relationships; this was a departure from the cognitive focus of Frederic Wells. In those days the TAT was a clinical guide, not a research instrument. Twenty years later (in the 1970s), Charles Ducey, a psychologist who was skilled in projective testing but blind to other Study results, reviewed the men's TATs and their earlier Rorschachs. While the projective tests provided good portraits and corroborated many of the men's known personality quirks, they did not help the Study make accurate predictions about the men's mental health or flourishing at age forty-seven. (That's an age that keeps coming up because it was the approximate age of the men in their twenty-fifth reunion years; more on this shortly. So was seventy-two, the year they celebrated their fiftieth reunions.) In 1981, however, Dan McAdams, soon to become a distinguished researcher of Intimacy and Generativity (see Chapters 4 and 5), successfully used those same TATs to predict intimate relationships in a manner that an academic journal found worthy of publication. His construct of intimacy motivation could significantly predict success in the two fundamental criteria of the good life so often ascribed to Freud—*Arbeiten und lieben* (to be able to work and to love).[28]

In 1953, the President and Fellows of Harvard College declared that unless additional support could be garnered from private sources, Harvard would no longer continue to provide for the Study as it had over the five lean years past. And so on July 12, 1954, Drs. Bock and

Heath wrote a sad letter: "To all Grant Study Participants, We regret to inform you that financial support of the Study ceased July 1, 1954. . . . Your cooperation has been a source of great satisfaction and pleasure to us. . . ."[29]

Final farewells were averted, however. In December of that year, the desperate Study submitted a formal grant to the Tobacco Industry Research Committee for $15,880 (roughly $150,000 in 2009 dollars), suggesting that something might be learned about the "positive reasons" that people smoke. And for more than a decade tobacco became the Study's main source of support, although the documentation is sketchy.

It's worth noting that in 1954, when this *deus ex machina* materialized, the Study was already sixteen years old. It had achieved the fifteen years first envisioned for it and more, yet there was never any thought of letting it go without a struggle. Longitudinal studies were becoming better known and better understood; there was a lot of motivation to keep it going as the stakes of lifetime studies began to come into sharp focus.

Also in 1954, Dana Farnsworth, M.D., a psychiatrist committed to the mental health of students, replaced the retiring Arlie Bock as director of the Department of Hygiene, now renamed Harvard University Health Services. Frederic Wells had retired. Margaret Lantis had left in 1952 without replacement. In 1955, Clark Heath left the Study due to its financial uncertainty, and Farnsworth appointed Charles McArthur director of the Grant Study.

THE CHANGING OF THE GUARD

With the appointment of a social psychologist as director, the first great phase of the Grant Study came to an end. The early researchers had theorized that development was driven by intellectual ability, physical constitution, and personality traits that invariably resulted from them. This model was now abandoned in favor of a model of relationship-driven development. Years later, when I first became director, Arlie Bock, who had not even included psychiatrists on the full-time staff of his Department of Hygiene, was surprised that I had discovered among his "normal" Grant Study boys alcoholics, personality disorders, and manic depressives. With tongue in cheek he complained to me in the hallway, "They must have been spoiled by you psychiatrists. They didn't have problems like that when I was running the Study!"

Dana Farnsworth had attended a one-room school in West Virginia, and graduated from Harvard Medical School in 1933. He had headed the health services at Williams College and M.I.T. He was widely considered the nation's foremost expert on student mental health, and was far more sympathetic to psychiatry than Bock had ever been. As soon as he took the helm he hired five full-time and two part-time psychiatrists, and established relationships between them and each of Harvard's residential houses and graduate schools. While Farnsworth gave little direct financial help to the Study, he did provide free office space, and was personally supportive of Charles McArthur and then of me.

McArthur, a brilliant young Health Services psychologist, was completing his Ph.D. in social relations at Harvard when Farnsworth appointed him director of the virtually penniless Study. He held that position from 1955 until I, a research psychiatrist, succeeded him in 1972. When he began, he was the only member of the Study staff, and Harvard was paying him to be a clinician to the students, not to do psychological research. Between 1955 and 1967 he did manage to achieve continuity of funding by asking questions about smoking habits—"If you never smoked, why didn't you?"—a nod to another tobacco-related patron, Philip Morris. And even though he sometimes had to supplement such outside funds with his own staff psychologist's salary, he managed to carry on the work of maintaining addresses, keeping contact with the men, and sending out questionnaires. Between 1960 and 1970, however, there were only four publications, frequency of questionnaires dropped to every three years, and seventeen Grant Study men were noted to have been AWOL for more than a decade.

McArthur's research foci were smoking habits and social class differences among the Grant Study men. He was also interested in long-term validation of the Strong Vocational Test—at the time the best predictor of career choice and satisfaction available—and published several papers on his findings (see Appendix F). This was the first time that the Grant Study really capitalized on its prospective longitudinal design.[30]

In 1967 the Grant Foundation awarded McArthur a three-year grant of $100,000 ($555,000 in 2009 dollars) to bring Lewise Gregory Davies, now married, back to the Study, and to obtain secretarial help in mailing out questionnaires. Davies's return worked wonders. The seventeen nonresponders came back to the fold. In the forty-five years since then, only seven active members have withdrawn from the Study. No member has been entirely lost to follow-up, for publicly available records and alumni reports have allowed us to characterize the occupational success, marital status, and general life adjustment of those seven dropouts and the twelve who left before 1950. Only four of the nineteen dropouts are still alive; the Study has death certificates for thirteen of the other fifteen.

LATER DATA ANALYSIS CONTINUES: THE TWENTY-FIFTH REUNION YEAR

I was in my first years with the Study when the twenty-fifth reunion years of the College cohort began to roll around. These were the occasion for a major round of data analysis in which we considered the men's adaptation to post-college life and above all to life after the Second World War. The reunions were also an opportunity to compare the College men with their classmates. In 1969 I devised a questionnaire that the Reunion Committee of the Harvard Class of 1944 distributed at the time of its twenty-fifth reunion.

We already knew from freshman physicals that the Grant Study men did not differ significantly from their classmates in such attributes as height, visual acuity, eye color, hay fever, or history of rheumatic fever. Twenty-five years later, it appeared that they did not differ significantly in future occupation, either. A quarter of each class became lawyers or doctors; 15 percent became teachers, mostly at the college level; 20 percent went into business. The remaining 40 percent were distributed through such other fields as architecture, accounting, advertising, banking, insurance, government, and engineering. There were a few artists of varying persuasions. The same proportions held true of the Study men.

But by the reunion years, critical differences had appeared, and some of them are summarized in Table 3.3. The comparison isn't perfect. Alumni who have not enjoyed relatively smooth sailing in life, love, and career often decline to complete reunion questionnaires. While 92 percent of the Grant Study subjects completed this one, in the table they are being compared to a 70 percent sample of their classmates, probably self-selected for health and success. Even with this selection bias, however, the Study men seemed to have striven harder for success in college and in life than their classmates. Their mean IQ of 135 was only minimally higher than the 130 of their classmates, but a far higher percentage of them graduated with honors. They tended to take fewer sick days. A point of interest: this 1969 analysis was the first time a computer was used to analyze Grant Study data. At that time, the Harvard computer (quite a concept now!) filled

Table 3.3　Responses to a Questionnaire Distributed at the 25th Reunion of the Harvard Class of 1944

	Grant Men N = 44	Classmates N = 590
Responded to reunion questionnaire	92% Significant	70%
Graduated from college with honors	61% Significant	26%
Went on to graduate school	76%	60%
Feels job is extremely satisfying	73% Significant	54%
Occupationally less successful than father	2% Significant	18%
Takes fewer than 2 days' sick leave a year	82% Significant	57%
Believes our involvement in Vietnam should decrease (winter 1968/69)	93%	80%
Attended public high school	57%	44%
Ever divorced	14%	12%
Often attends church	27%	38%
Drinks 4 shots (6 oz.) of liquor or more daily	7%	9%
10+ visits to a psychiatrist since college	21%	17%

Very significant = p<.001; Significant = p<.01; NS = Not significant.

a whole building. Data had to be submitted on eighty-column punch cards, often carried half a mile through the snow to the computing center, and data turnaround could take several days.

Given their apparent ambition and greater striving, it wasn't surprising that the Grant Study men were more likely than their classmates to have exceeded the success of their fathers by the time they were in their late forties. However, the selection of the Grant Study men favored conventional definitions of success. Stoics outnumbered Dionysians in the Study; the achievement bias in the selection process probably worked against stable youngsters who were more interested in private contentments than in chasing brass rings, and artists whose development takes longer and is frequently less remunerative. Furthermore, capacity for intimacy had been valued less highly in the selection process than the capacity to grin and bear it. One staff member defined a "healthy" person as "someone who would never create problems for himself or anyone else," and many of the Grant men lived up to that ideal. One boasted that what he enjoyed most in life was "being beholden to no one and helping others."

Even taking this into account, the extent of the men's successes as they entered middle age was striking. Four members of the College cohort ran for the U.S. Senate. One served in a presidential cabinet; one was a governor, and another was president. There was a best-selling novelist (no, it was not Norman Mailer, even though he *was* Harvard class of '43), an assistant secretary of state, and a Fortune 500 CEO. And while the average man in the College cohort had the income and social standing of a successful businessman or physician, he displayed the political outlook, intellectual tastes, and lifestyle of a college professor. At age forty-five, the Grant Study men's average income was about $180,000 (in 2009 dollars) a year, but fewer than 5 percent of them drove sports cars or expensive sedans. Despite their economic success, they voted for Democrats more often than for Republicans, and 71 percent viewed themselves as "liberal."

I JOIN THE GRANT STUDY AND RE-INTERVIEWS BEGIN

I had arrived at the Grant Study in 1966, supported by an NIMH Research Scientist Development Award. McArthur allowed me to design the next two regular questionnaires. Perhaps the most important change I instituted at the time was to begin collecting physical exam data every five years, including chest x-rays, EKGs, and standard blood and lab work. The exams began as the men turned forty-five (1965–1967) and have continued since then, through age ninety (2010–2012). This prospective record of objective physical health distinguishes the Grant Study from the other great longitudinal studies of personality development.

My approach differed from Charles McArthur's. I wasn't a social psychologist, but an M.D., and in psychoanalytic training. My early studies of recovery from heroin addiction and acute psychosis had left me impressed by the involuntary adaptive coping mechanisms that resilient men employed. My youthful preference for blacks and whites and either/ors had given way to an appreciation of—and a serious interest in—the reality and power of incremental adaptation. Now, at the Grant Study, I could investigate these mysteries to my heart's content.

The questionnaires, concrete as they were, were an invaluable source of information. The men's idiosyncratic responses to standard questions revealed the adaptive styles and behaviors that colored all the other facets of their lives. As one man put it, "We reveal ourselves whenever we say anything." One man sent in a questionnaire two years late because he had just found it under his bed; I wasn't surprised to find evidence of passive aggression in other areas of his existence. We didn't have to depend on words or dream reports, because (as I'll show in Chapter 8) the Study's prolonged and repeated opportunities for observation allowed us to see so-called ego defenses made tangible in concrete behavior. During the years that we overlapped, Charles McArthur was an enthusiastic supporter of this interest.

In 1967 I also began obtaining two-hour interviews of the men, because no questionnaire can convey a person's flavor the way a face-to-face meeting can. This was the first systematic interviewing since Margaret Lantis's tenure in the early 1950s, and the new practice of re-interviewing about every fifteen years has continued through my directorship up to the present.

My second major research interest, maturation, came to life shortly after I joined the Grant Study. My father died when he was forty-four and I was ten. I retained an indelible memory from 1947, my thirteenth year, of his twenty-fifth reunion book, where photographs of barely post-adolescent college seniors sit next door to the portraits of mature forty-six-year-olds. When the Grant Study men returned for their twenty-fifth reunions beginning in 1967, I experienced my interviews with them as eye-openers; I myself was still a callow thirty-three, but I understood right away that I had unconsciously been waiting for this encounter for twenty-five years. After that, my interest in adult maturation competed with my interests in defense and resilience.

The single most personally rewarding facet of my involvement with the Grant Study has been the chance to interview these men over four decades. The magic of transference hasn't hurt. I am fifteen years younger than they, and when I began, I could not forget that I had been in kindergarten when they had entered the Study. I wanted to call them all "Sir." But they invariably treated me as respectfully as college sophomores would treat a Study physician. I met no condescension, and because I already knew so much about them, the interviews were often remarkably intimate. But they reinforced a belief that the questionnaires had already inculcated in me: that however they may try, people can never neutralize their personalities.

There was an intensity to many of the interviews that was both gratifying and surprising. Often talking with these men was like resuming an old friendship after a period of separation, and that made me feel a little guilty, for I had done little to earn the warmth and trust that they offered. But I soon discovered that whether the men liked me or I liked them had less to do with me than with them. The men who found loving easy made me feel warmly toward them, and left me marveling at my tact and skill as an interviewer. In contrast, the men who had spent their lives fearful of other people and gone unloved in return often left me feeling incompetent and clumsy, like a heartless investigator, vivisecting shy innocents for science.

With some of the men the interviews felt like psychiatric consultations; with some like newspaper profiles; with some like talks with an old friend. I learned to associate a man's capacity to talk frankly of his life with positive mental health. With maturity comes the capacity and the willingness to express emotion in meaningful words.

I learned quickly that the men's responses to me paralleled their ways of relating to people in general. One man, for example, evaded the first few questions I asked him, and then turned to me and said expansively, "Well, let's hear about you!" At first I thought I had just been inept, but then I saw a comment from a staff psychiatrist in 1938: "This boy is more difficult to interview than any I have encountered in the group." One of the warmest, richest personalities in the Study invited me to his home for breakfast at 7 A.M., cooked me a soft-boiled egg, and then extended the interview well past the two hours that I had requested. This in spite of the fact that he was working a sixteen-hour day, that he was moving his entire household to New York in two weeks, that his son was graduating from high school in eight hours, and that he himself had just suffered a devastating business reversal that was front-page news. Far less busy but more socially isolated men would put me off for a week and then meet me in the most neutral setting possible—two of them chose airports!

Some men came to Cambridge to be interviewed, but in most cases I went to them—to Hawaii, Canada, London, New Zealand One man—only one—seemed very reluctant to be interviewed. But once the interview began he allowed it to extend through his lunch hour, and gave me a lengthy, exciting, and startlingly frank account of his life. Other men readily agreed to see me, but retreated behind ingenious obstacles. Two took every opportunity to interpose their large families between themselves and me when I visited, while others kept their families well out of sight the whole time I was there. There were cultural differences, too. All the New Yorkers and most of the New Englanders saw me in their offices, and few offered me a meal. Virtually all the Midwesterners saw me at home and invited me to dinner. The Californians were evenly divided. Several wives were openly suspicious of the whole enterprise. One spoke so stridently into the telephone that sitting across the desk from her husband I could hear her refuse to see "that shrink" under any circumstances.

Between 1967 and 1970, I re-interviewed a random 50 percent sample of the Study classes of 1942–1944. In 1978–1979, Eva Milofsky, the social-worker daughter of Sigmund Freud's personal physician Max Schur, re-interviewed the rest. Most of the surviving men were interviewed again as they reached retirement age around 1990, almost half by me and the rest by the very gifted Maren Batalden, M.D., whose work will appear in more detail in coming chapters. When the surviving men were about 85 (2004–2006), they, and their wives, were interviewed once more by Robert Waldinger and his team. The Grant Study is probably unique in having obtained so many interviews over so many years. And since the administrative reorganization in 1970, the Study has managed to track the men from the Glueck Inner City cohort in a fashion virtually identical to our follow-up of the College cohort; both cohorts are now included as subjects in any major Study publication.

To keep reliably in touch with two cohorts over many years is a tough business, and for the last twenty years that business has been in the hands of Robin Western, MFA, our reincarnation of Lewise Gregory. Western has been a tactful and insightful interviewer and a skilled detective, finding our strays and luring them back to the Study. She has been a meticulous and skilled archivist, maintaining seventy years of collected data in manageable order. She's been a key research colleague, planning every questionnaire and wheedling the men's physical exams (with the men's written permission, of course) out of their doctors' offices and into ours. To top it all off, she's been a wonderful friend. If the Grant and Glueck Studies are now among the longest in the world, much credit goes to the perseverance and thoroughness of Robin Western.

Every time I settle down to analyze this wealth of material, yet another of the many richnesses of longitudinal studies becomes clear. No single interview, no single questionnaire, is ever adequate to reveal the complete man, but the mosaic of interviews produced by many observers over many years can be most revealing. One member of the Study, for example, was always seen as dynamic and charismatic by the female staff members, but as a neurotic fool by the males. One shy man from a very privileged background came across to a staff member from a similar milieu as charming, but a colleague from a working-class family thought he was a lifeless stick.

Some concealed truths emerged only after the Study members had been followed for a long time. One reticent man was thirty before he revealed that his mother had had a postpartum depression following his birth. This had not emerged either in his psychiatric interviews at nineteen or in Lewise Gregory's family interview. One man kept his homosexuality secret until he was seventy-five, another until he was ninety. In general, few of the men acknowledged a wife's alcoholism until they were over sixty-five. They were far more honest about their own alcoholism, extramarital affairs, and tax evasion.

In 2005, Robert Waldinger, M.D., a research psychiatrist with a particular interest in intimate relationships, succeeded me as director as the Study. From 2004 to 2006 he re-interviewed and, even better, videotaped consenting married couples and obtained their DNA. In recent years, some of the men have consented to fMRI studies, in an attempt to clarify the positive emotions involved in intimacy. I'll illustrate some of his work in Chapter 6. Some men have donated their brains to the Study. These are further generous sowings whose value will likely not be appreciated for many years.

SOME NOTES ON FUNDING

Keeping the Grant Study in continuous funds for forty years was an anxiety-provoking challenge, but one that with flexibility, resourcefulness, and some sheer good luck I was able to meet.

Since the Career Research Scientist Award that provided my salary for thirty years included no research funds, in 1971 I wrote to the Grant Foundation requesting $800 to pay the retired Lewise Gregory Davies, now in her seventies, to reenlist recalcitrant Study members once more. At that point we had perhaps seventeen nonresponders. Douglas Bond, former staff psychiatrist to the Grant Study (1942), chairman of the Grant Foundation, dean of Western Reserve Medical School, and son of the Earl Bond who had been advisor to both Arlie Bock and William Grant, came to visit me at Harvard. He listened to my ardent plea on behalf of this uniquely valuable longitudinal study, and then he wrote us a check for $50,000 ($280,000 in 2009 dollars). This astonishing munificence provided the seed money for the grant requests and infrastructure by which we obtained continuous funding for the Grant Study for the next three decades. The timing was fortuitous, because the National Institute on Alcohol Abuse and Alcoholism had just opened, and money was relatively plentiful.

Our funding from the NIAAA between 1972 and 1982 required us to pay explicit attention to alcoholism. It also supported the questions in which I had been primarily interested—involuntary coping, relationships, and adult maturation—but the study of alcoholism proved so fascinating that I have given it a chapter of its own (Chapter 9).

The first publication to emerge during my directorship, when the men were about fifty, focused explicitly on prescription drug abuse by the forty-six Study men who became physicians. I used responses from questionnaires and interviews to contrast the frequency of mood-altering drug use and abuse by these men with use and abuse by other Study participants. In this case, the Study's demographic homogeneity presented a convenient means of obtaining a matched control group for the physicians. The physicians were twice as likely to use and abuse mood-altering drugs as the controls.[31]

The publication of this paper in the *New England Journal of Medicine* brought attention to the Grant Study, and it also helped me address an important issue in longitudinal research: whether the process of being studied alters the course of participants' lives. Would the Grant Study doctors, most of whom read the *Journal,* change their self-prescribing behavior in response to my paper? Sadly for them, but happily for my trust in the validity of prospective design, we found ten years later that mood-altering drug abuse had actually increased among the physicians, but not among the controls (who were much less likely to have read the article). Despite my prediction that the doctors would modify their habits (or at least their answers), their misuse of drugs actually got worse. The controls' did not. This result also confirmed my impression that the men were being remarkably honest in their questionnaires.

In 1983, I published *The Natural History of Alcoholism,* which summarized years of research.[32] That same year, to put our next research phase on a more secure financial footing, I accepted an endowed chair in psychiatry at Dartmouth. This meant that Dartmouth would pay my salary, and scarce research funds could be reserved for purposes more interesting than keeping me housed and fed. For ten years the Study moved with its files to Hanover, although it remained the administrative responsibility of Harvard University Health Services.

In 1986 the focus of the Study turned to aging. I was still primarily interested in adaptation, but I was a true Willie Sutton of a researcher, going where the money was. As with the alcohol studies, though, once I learned a little about the subject, I found myself fascinated. And, of course, the matter of aging is absolutely intrinsic to a study of this kind—and to my interest in adult maturation—which was very soon brought vividly home to me.

My first application to the National Institute on Aging was turned down cold. The reason was simple. The proposal's age-phobic fifty-one-year-old author had offered to follow the men's aging prospectively—as progressive decline. (I'll say in my own defense that Shakespeare thought the same way in his youth.) The vigorous seventy-eight-year-old chairman of the NIA research grant committee, eminent gerontologist James Birren, made it abundantly clear that that was not how he viewed aging, and he denied my grant request in no uncertain terms. Forgiving and generative, however, he also undertook to raise my consciousness; and thoroughly chastened but now able to envision aging as a positive process, I rewrote the application. The NIA agreed to fund us, and in 2002 the resulting book, *Aging Well,* captured the inspirational reality of aging as James Birren both imagined and exemplified it. It was a study of the aging process in real time, and it put to scientific rest the negativity of mid-life popular "experts" on aging like Simone de Beauvoir and Betty Friedan. The National Institute on Aging continues to support the Study to this day, and I hope that Chapter 7 will persuade any readers who happen to doubt it that there are worse things in life than living to ninety.

In 1992, Brigham and Women's Hospital offered me secure funding, and the Grant Study came home to Boston. It remained at Brigham and Women's until 2010, when Robert Waldinger, director since 2005, moved it to the Massachusetts General Hospital.

PERSPECTIVES—THEIRS AND OURS

We at the Study have often been asked what effect the men's membership in it has had on them. When we've asked them, most have said that it had no direct effect on them, and indeed there is little to suggest that it has changed their lives in any major way. But as a group they have enjoyed their participation. In 1943 one man wrote, "The Study impressed me by its thoroughness, its interest in the little things, its ability to make the subject feel a part of it rather than a guinea pig. Now it is a gratifying reassurer and friend." A second very shy Study member, who had asked the comely but six years older Lewise Gregory to go to the theater with him (she went), wrote, "I feel my friendship with you (the Study) has more than paid back anything that I put into it." Yet another man wrote, "The very act of being a participant, receiving and pondering the follow-up questionnaires, etc. . . . has made me more self-consciously analytical about my personal development, life choices, career progress, and the like."

Staff members, of course, have our own answers about how the Study has affected us, and I will illustrate here one effect that the Study had on me. Here follows the story of my engagement with one Study member, Art Miller. It's a story of detective work, of which there is plenty in an enterprise like this. It's a story of adaptation. But more than anything, for me it was an object lesson about the dangers of judgment. Precipitous conclusions are a constant danger to incautious scientists, and the story of Art Miller is my best reminder that a long perspective is our only true protection against it.

THE STORY OF ART MILLER

In 1960, Art Miller disappeared. He came back safely from the war, and told John Monks that he had seen no heavy combat. He went to graduate school, earned a Ph.D. in Renaissance drama, and became an English professor. His published scholarly essays can still be downloaded with Google's help. But then he vanished. For twenty years he was the only "lost" member of the Study. It wasn't until 1980 that I managed to reach his elderly mother, who told me that he had quit his university job long ago and moved to Western Australia, where he was raising his family and teaching high-school drama. I called telephone information in his small town. They sent a man out to his house on a bicycle (things are different in the Outback), only to find that he had moved. The operator called up the headmaster of Miller's school in the middle of the night so as to be able to call me with the new phone number when it was daytime in Boston. When I called Miller at home and told him that I might be coming to Australia, he sounded delighted and insisted on giving me supper and a bed for the night. This evasive man seemed genuinely interested in seeing me.

When I got to Melbourne and called to schedule a meeting, it was the end of the school term, a difficult time for Art. Nevertheless, he agreed to a visit. When I arrived, his front door was open. A note revealed that he was still at the school play, but a nice dinner was laid out and a cold beer was waiting for me in the icebox. Miller and his wife got back after I had been there about an hour, and as the interview proceeded, we made a significant dent in the bottle of Scotch that I had brought as a house present. He had told me that he had moved to Australia out of disaffection with the United States. For one thing, he was afraid that drugs would endanger his children's adolescence; he had not liked "the cut" of their friends. For another, he was committedly against the war in Vietnam. There were some nonspecific reasons for his alienation as well; as he put it, "There was so much anger in the air." I found myself wondering briefly if perhaps he had been running from something.

On the phone calls that followed my visit, Miller was always courteous to me. He never returned another Study questionnaire, however, and he never revealed his whereabouts to Harvard. The last time I reached his number, his daughter told me that he had died of cancer.

I puzzled over Miller. Psychiatrists were now seeing PTSD in many of the Vietnam vets, and Miller had been in a war too. But I quickly dismissed this notion. He had reported "combat fatigue" once during the war; had he perhaps exaggerated a bit? The Woods traits attributed to him in college had been *Cultural, Ideational,* and *Creative and intuitive,* and after all, he was a drama professor. In his interview with Monks, he had said that he had not seen sustained combat, and by his own report many of the Grant men had more severe and more prolonged combat histories than he.

The years passed. I formulated various hypotheses, not all of them flattering, for Miller's elusive behavior. In 2009, I had to reread his chart for this book. There I discovered a scrap of paper entered by Clark Heath, who had traveled to Washington in 1950 to review the men's military medical records. Fifty-five years after World War II ended, I learned what had happened to Art Miller in the army. Here is his record from an Italian field hospital, beginning on June 13, 1944:

Patient saw 3 or 4 days of combat, remembers killing three Germans. The last he remembers is attacking up hill with men falling, nearby land blasts, then woke up here two days ago. He has no idea as to what happened in the intervening time. On admission, he was acutely disturbed, kept his fists clenched, and threw himself about, calling "Shells! Bombs! I'm afraid!" and no contact could be established.

There was no change on transfer here. He was restless, disturbed, over-responds to minor stimuli, crawls under covers and into fetal position at sounds of planes and there is no response obtained on questioning.

After two electro-convulsions, amnesia and agitation entirely cleared. There is some noise sensitivity, battle dreams, and fear of combat remains. He is a mildly shy, sensitive, personality type with a strong sense of duty. . . . From an emotional standpoint he will be of no use whatsoever under combat. . . . I feel that he should be reclassified to limited service, permanent.

July 3, 1945: Reclassified. Normal now.

July 5, 1945: Certificate of honorable discharge. Character excellent.

It had been easy to stand in judgment. I could point to Miller's noncompliant and passive-aggressive behavior. I could note that he had run away from his family and his country, and that his earned income was as low as any man's in the Study.

Or I could—and in 2010 I finally did—understand Art Miller's whole life as a creative example of posttraumatic growth. Many playwrights (Edward Albee and Eugene O'Neill among them) have endured traumatic childhoods or major depressions, and spun gold from this wretched straw. It's hard to imagine that Art Miller's students scorned him as a dropout or a victim; on the contrary, they must have greatly enjoyed the drama that this published scholar brought to their small-town school. But it took forty-five years for me to see that truth. And I saw it not by looking back on what Miller had or hadn't done, but by discovering the prospective entry that one very caring physician had made years before, when the trauma was still fresh and contemporary.

A SELF-PORTRAIT

Since it is I who guided the Study for more than half of its existence, it is fitting that I share some relevant aspects of my own development and the potential biases that grow out of them. I was born in New York City to academic WASP parents and educated at Phillips Exeter

Academy and Harvard College, where I majored in history and literature and was an editor of the *Lampoon*. I went on to Harvard Medical School with the internalized injunction that teaching and service were good and that business and private practice were bad. And perhaps I put a more than necessarily puritanical spin on this dichotomy, because I remember mocking my professors behind their backs for preferring research to clinical care. I have always considered the *New York Times* to be the source of truth where news is concerned, and politically I usually vote Democratic. Between 1960 and 2009 I lived mostly in Cambridge, Massachusetts. I was happily married during most of those years, despite more than one divorce. As I write this book I am recently remarried once again, living in southern California, and learning (reluctantly) to broaden my political perspectives.

As a psychiatrist, and like many of my colleagues, I pretend to belong to no "school," but it won't have taken long for my readers to realize that I am trained as a psychoanalyst and am a staunch admirer of Adolf Meyer and Erik Erikson. I began psychoanalytic training not because I thought psychoanalysis "cured" people, but because my best clinical teachers were psychoanalysts. All my life I have wanted to be a teacher, and I have always tried to teach my readers how to look for what transpires invisibly beneath their awareness—in their defenses, in their development, and in their hearts. Less obvious, but also important, is the fact that I worked for two years in a Skinnerian laboratory. There I came to believe that the experimental method far surpasses intuition as a means of uncovering truth, and that by charting behavior over time, patterns may be revealed that would interest the staunchest of psychoanalysts. Despite the strictures of my medical student conscience, I have spent more time in "selfish" research pursuits than in "noble" clinical care.

I should probably note that my interest in Alcoholics Anonymous (see Chapter 9) does not have to do with being in recovery myself. Here again I was following Willie Sutton's example. To run the Grant and Glueck Studies I needed a Harvard appointment. The only job opening at the time was a co-directorship of the Alcohol Treatment Center at Cambridge Hospital. As a condition of employment at the Treatment Center I went to AA meetings once a month for ten years. I came to scoff but stayed to admire, and as a result of my enthusiasm in 1999 I was appointed a Class A (that is, non-alcoholic) trustee of AA, and the six years that I spent in that capacity were among the best in my life.

I have always loved big questions. At ten I wrote my sixth-grade term paper on the origin of the universe. I've already told how mesmerized I was, two years after my father's death, when our family received a courtesy copy of his twenty-fifth Harvard reunion book. I marveled at how those raw college caterpillars had evolved into mature forty-seven-year-old men, and was first struck by the profound implications of adult development.

Until I was eighteen, I intended to become an astrophysicist. By nineteen, however, I had read my roommate's copy of Robert White's *Lives in Progress,* perhaps the first prospectively designed textbook on adult development, and decided to go to medical school instead.[33] In my training and in the years that followed it, I was fascinated by people's capacity to recover from apparent catastrophe and continue to develop over the course of their lives. I became a dedicated believer in the power of long-term prospective study to answer psychiatric riddles, and I read avidly about the world's great longitudinal studies. They entranced me; they were telescopes, all right, but the kind that could focus on human lives, not just inanimate stars. I was equally entranced by B. F. Skinner's idea of the *cumulative record,* a sort of behavioral cousin of the EKG that draws a "picture" of people's doings as they change over time, allowing us to visualize those changes in a single gestalt. In 1966, while still an assistant professor of psychiatry at Tufts Medical School, I joined the Grant Study and was immediately hooked.

My dream was to make the Grant Study, at the time little known, as important as the longitudinal studies that had so inspired me. One of the few memories I have of my training psychoanalysis was holding up my brand-new key to the Study files and crowing to my analyst, "I have acquired the key to Fort Knox!" But even I did not anticipate that I would still be working on the Study full-time forty-five years later.

With financial support from an NIMH Research Scientist Development Award, I worked on the Grant Study for five years before moving my appointment as Associate Professor of Psychiatry from Tufts to Harvard, and succeeding Charles McArthur as director of the Study.

At the time that I took over, Harvard was still not convinced of either the power or the importance of the Study, and was seriously considering destroying the sensitive files that so enthralled me. (They were especially sensitive just then because the year before, protesting students had broken the windows of the Health Services and occupied Harvard's administrative offices.)

In 1973, I gave a dinner as the new director. I invited various Harvard potentates and Radcliffe president Matina Horner, the sole woman. I described the Study's plight, and while I was still resisting entreaties by the Harvard authorities to reduce the case records to soulless microfiche, Horner established at Radcliffe the Henry A. Murray Center for the Study of Lives—a center dedicated to preserving longitudinal studies in their original form. One of its first and most exciting acquisitions was the records of the Harvard Study of Adult Development. Over the years (1976–2003) of the Murray Center's existence, its director, Ann Colby, was an inspiration to all who knew her, especially me.

The focus of the Study shifted again as my interests pushed it away from sociology and toward epidemiology and psychodynamics. I was especially fascinated by involuntary coping mechanisms and by the relationship between psychological stress and physical symptoms. I am increasingly convinced, especially since passing seventy myself, that psychiatry and psychology need to become more aware of people's positive emotions and experiences of spirituality. I believe too that a predilection for love and compassion are hard-wired in mammals. One early reviewer of this book said to me, "George, your view of adult development is sooooo 1970s." But you know what? The seventies weren't so bad. Those were the years of Bernice Neugarten, Robert Kegan, Jane Loevinger, Emmy Werner, and Jerome Kagan, and I do not believe that their work has so far been surpassed. And anyway, like all the Study directors, like all the men whose life voyages we accompanied, and like the Grant Study itself, I was—I am—a creature of my time. And from age thirty-three to age seventy-eight my participation in the Study has been an abiding joy.

INTO THE FUTURE

One of the points I'll be making throughout this book is that our close relationships when we are young make an enormous difference in the quality of our lives. Yet there have been few studies of intimate relationships near the end of life, let alone of how these relationships affect physical and mental health. In 2003, Robert Waldinger began to study the marriages of the Grant Study men, now in their tenth decades, and that meant inviting their wives to join the Study. Waldinger, associate professor at Harvard Medical School and a psychoanalyst, has since succeeded me as director. His fresh view of the project is now giving the Study a microscopic (as opposed to my telescopic) look at the daily lives of single and married older adults.

However, the National Institute of Mental Health has not been willing to fund such studies into the twenty-first century. Scientific fashion has once again abandoned the social for the biological. If Willie Sutton wanted grant money to investigate adult lives today, he'd have to do brain studies. Since 2005, Waldinger has obtained funding for neuroimaging studies from three sources: the Harvard Neurodiscovery Center, the Fidelity Foundation, and once again the National Institute on Aging. And for the third time since it stopped funding the Study in 1945, the W. T. Grant Foundation has come to the rescue with additional funds.

These vicissitudes of funding may be nerve-racking, but they are not an unmitigated curse. Over the last five years, they have encouraged the Study to add biological and neuroscientific data to nearly seventy-five years of information about men's lives. Most surviving Grant Study men are allowing the Study to collect their DNA, conduct sensitive tests of their intellectual capacity, and carry out neuroimaging of their brain structure and function to create an unprecedented and irreplaceable resource that someday may help us understand the links between brain and behavior in human aging. The combination of life histories and neuroscience data will allow us to investigate fundamental and pressing questions about the aging process. For example, how do genes and environmental factors—such as alcohol use or exposure to traumatic stress—interact to determine who maintains an active and vigorous mind and body into their nineties and who does not? Are there lifestyle choices and behaviors that can buffer the human brain against the ravages of old age?

As the population grows older, these questions become more urgent for families, for healthcare providers, and for public policymakers. As I've noted before, there are advantages as well as disadvantages to studying an elite sample. One of the advantages is that the unusually healthy Grant Study ninety-year-olds offer some insight into what life will look like in thirty years for the average baby boomer. The seventy-five-year-old Study may look dated to modern (or very young) investigators. But it is well to recall Santayana's warning: "Those who cannot remember the past are condemned to repeat it."

4 | HOW CHILDHOOD AND ADOLESCENCE AFFECT OLD AGE

O joy! that in our embers
Is something that doth live,
That nature yet remembers
What was so fugitive!   —WILLIAM WORDSWORTH

ONE ASPECT OF OUR PASTS that we tend to repeat unremembered are the experiences we absorb in childhood about other people and the world they embody. Novelist Joseph Conrad laid the stakes out somberly: "Woe to the man whose heart has not learned while young to hope, to love, to put its trust in life."[1] If the child really is father to the man, then it's reasonable to ask: how? This is not a simple question, obviously, but the Grant Study has been able to deconstruct it a bit, allowing us to explore the complex of childhood circumstances that shapes our sense of self and our expectations. These in turn shape the relationships that we make, and the social surround that will enrich, or not, our later years. The Study has affirmed the truth of Conrad's warning, yet it also offers some heartening news in counterbalance, and I've listed this in Chapter 2 as the fifth of the great lessons of the Grant Study: What goes right in childhood predicts the future far better than what goes wrong.

Another mitigating factor, although an elusive one, are certain "sleeper effects" of childhood, which the Study has also illuminated. These may be deep early attachments that have been lost from view through chance, tragedy, or forgetfulness, but are brought to memory's light again many decades later. Recovery of these lost loves can be profoundly healing, as we will see shortly in another chapter in the life story of Godfrey Camille. But there are less benign sleepers as well—genes for alcoholism, major depression, or Alzheimer's, for example, that are present from the outset, but do not wreak their havoc until later in life. The sleeper phenomenon points up a corollary to Lesson Five: it is not any one thing for good or ill—social advantage, abusive parents, physical weakness—that determines the way children adapt to life, but the quality of their total experience.

As we've seen already and will see again, long-term perspectives may upset established assumptions about cause and effect, especially when these have been built on the instantaneous observations of cross-sectional research. This is what makes longitudinal studies at once so powerful and so disconcerting. In cross-sectional studies, for example, two of the most powerful correlates of successful aging are income and social class.[2] But wealth and status are neither the only nor the best measures by which to judge successful aging. Furthermore, our longitudinal evidence makes clear that when mid- and late-life riches do accompany successful old age, often they are not the result of financial privilege in childhood or even of other attributes popularly considered conducive to worldly success, such as looks or an extroverted personality.[3] They are indirect fruits of a childhood experienced or recalled as warm and intimate, for it is such childhoods that give children their best shot at learning to put their trust in life. Many measures of success throughout life are predicted less reliably by early financial and social advantage than by a loved and loving childhood.

You'll remember that among the College men, even achievement of military rank correlated more highly with a warm childhood than with social class, athleticism, or intelligence. For the Glueck Inner City men, whose parents were not wealthy, admired fathers, loving mothers, and warm friendships were prominent among the predictors of high income.[4] A father on welfare, or even multiple problems in the family of origin, predicted future income and social class less robustly than a warm family.[5] The availability of warm relationships (of which parents are ideally but not necessarily a main source) was the overriding factor.

Erik Erikson was an early student of the way children acquire Conrad's learning of the heart. An artist/psychoanalyst, he spent many years working on the first major prospective study of child development, the Berkeley Growth Studies. Erikson believed that an infant's first task is to develop trust and hope; a toddler's to develop autonomy; and a five-year-old's to become comfortable with initiative.[6] I will elaborate Erikson's schema of adult development in Chapter 5; for now, however, I will concentrate on those three early necessities and what they mean for later life.

ASSESSMENT OF CHILDHOOD ENVIRONMENT

To start with, here's some statistical context. In 1970 we set out to investigate which aspects of the College men's childhoods would predict successful mastery of hope, trust, and the resulting confidence on which autonomy and initiative depend.

In assessing childhood quality, we observed several rules to minimize bias. First, ratings were based not only upon psychiatric interviews obtained when the College men were at Harvard, but also upon interviews at that time with the men's parents, especially their mothers.[7] It is a limitation of the Grant Study that the men's childhoods had to be assessed retrospectively, when they were around nineteen. But it is a compensatory strength that the retrospection was very extensive—raters had access not only to the multiple-choice questions and brief essays characteristic of retrospective investigations, but to ten hours of interviews by skilled psychiatrists as well as the parent interviews.

Second, research associates involved in any aspect of the assessments were kept ignorant of the fate of the men after adolescence.

Third, we used multiple checks to ensure accurate ratings. Each man's childhood was assessed by at least two raters, and rater reliability was excellent (that is, multiple raters were highly likely to come independently to similar conclusions).[8] When two senior child psychiatrists reviewed and rated selected cases, their ratings too concurred with, and so further supported, the validity of the assessments of the original raters.

Fourth, we had two complete and independent assessments to cross-check against each other. The men's childhoods were first assessed in the early 1940s, and were scored on a 1–3 scale, from good to bad. Then, between 1970 and 1972, thirty years after the first assessment, I organized a second independent assessment from the same material (see Appendix C). Here's how it was done. As I've said, one constant challenge of research of this kind is the need to transform intuitions and value judgments into statistically useful data. The trick by which we had assessed marital satisfaction worked its magic on the men's childhoods: we turned a set of concrete behavioral criteria into a numerical scale. This time we rated five separate questions about the men's early years, to ensure that no single one would be weighted too heavily. These questions were:

• Was the home atmosphere warm and stable?

• Was the boy's relationship with his father warm and encouraging, conducive to autonomy, and supportive of initiative and self-esteem?

• Was the boy's relationship with his mother warm and encouraging, conducive to autonomy, and supportive of initiative and self-esteem?

- Would the rater have wished to grow up in that home environment?

- Was the boy close to at least one sibling?

Based on the material collected when the men were in college, each of these five questions was answered as superior (5), average (3), or poor (1) by two blind raters, and the two scores averaged. The 5-point ratings for the five separate questions were then added together to provide a global rating of childhood environment that could range from 5 to 25. Childhoods in the top quartile were characterized as warm; those in the bottom quartile as bleak; those in the middle two quartiles as so-so. We called the men in the top tenth the Cherished, and those in the bottom tenth, the Loveless.

One potential source of bias was that the new raters were people whose parents had raised them under Spock's permissive tutelage. They were inclined to understand the much higher strictness quotient of the College men's childhoods as a sign of parental coldness, rather than simply a reflection of the typical upper-middle-class child-rearing climate of the 1920s and 1930s. I compared the new assessments with the original ones of the older raters, who had a better grasp of the mores of the time. And it turned out that I needn't have worried. The agreement between the two generations of raters was quite good.

As time went on, we were able to use this more nuanced assessment to discover, among other things, that by the time the men were into their seventh decade, the Loveless were eight times as likely as the Cherished to have experienced a major depression. We found that while even warm childhoods were often followed by heavy smoking or alcohol abuse or regular tranquilizer use, bleak childhoods were often associated with all three. And—to return to where we started, with money—we found that the fifty-nine men with the warmest childhoods made 50 percent more money than the sixty-three men with the bleakest childhoods. These were significant findings. So was the correlation between warmth of childhood and life satisfaction. But most important of all was the fact that the Cherished were four times as likely as the Loveless—very significantly more likely—to enjoy warm social supports at seventy.

Those are abstractions, however. Here are pictures of what the childhoods of the Cherished and the Loveless really looked like, and of the effects they exerted on the men's futures.

THE LIFE OF OLIVER HOLMES

Judge Oliver Holmes had one of the warmest childhoods in the Study. He received a score of 23 out of a possible 25 on our Childhood Rating Scale. The important people in his life had made his childhood a pleasure, and their affectionate and empowering warmth lived on in his own parenting and even his children's. I have told much of Holmes's story in *Aging Well,* but I will reprise selected portions here because it is such a vivid illustration of the power of love.

Holmes's childhood placed him firmly amongst the Cherished. His parents were well-off, and they spent their money on music lessons and private schooling for their children, not on luxuries for themselves. At fifty, Holmes wrote, "My parents gave me wonderful opportunities." His relatives were nurses, music teachers, YMCA directors. No one in his family was known to be mentally ill. The Holmeses were Quakers, but Mrs. Holmes never stood over her children and made them say prayers; she felt, as she explained to Lewise Gregory in 1940, that "religion must be taught by example rather than by words." Besides, she went on, "Oliver was always cooperative and reasonable and almost never had to be punished. He has a delightful sense of humor."

Gregory saw Judge Holmes's mother as "a rather serious person with a great deal of kindness and gentleness," and she noted an example.

> During the interview Holmes's younger brothers were in and out of the living room, playing football on top of us, aiming BB guns at our heads, but it didn't take Oliver's mother long to have the situation well in hand. I would call her a wise person, an intelligent mother, and calm in emotional makeup.

Oliver got into frequent fights with other boys when he was in grammar school. One time he was knocked unconscious with a rock, but—ever the optimist—he reassured his parents that the other fellow had gotten the worst of it! In college, even while he was a member of the Pacifist Society, he still enjoyed aggressive games, and belonged to the judo club and to a debating team. It is easier to grow up to be a Quaker if when you are young your parents can accept your assertiveness—and have a sense of humor. It's easier to grow up to be a confident and competent individual, too.

Holmes's father had put himself through medical school. He became a top orthopedic surgeon, and Mrs. Holmes said about him that "his fellow physicians don't see how he can be so genuinely concerned about his patients. . . . He has so much sympathy for people who are ill and in trouble." Holmes described his father as "generous and never encroaching on a person's individuality."

The Holmeses were a close-knit family, and Holmes considered his parents among his best friends. After Oliver completed law school, his father bought him a house in Cambridge, just fifteen blocks away from his own. Oliver's wife's parents lived even nearer; for just as greatly as Holmes admired his father did Cecily admire hers. At age sixty-five, Judge Holmes described his brother's family as "so nice that one would have to be a stone not to enjoy it." Yes, his brother lived in Cambridge too.

I can hear my readers muttering, "I knew it. This guy is a patrician judge. His father bought him an expensive house. No wonder he's a picture of contentment in his old age! What did he ever have to worry about?" This reaction is exactly why social science needs statistics as well as evocative case histories. We found that contentment in the late seventies was not even suggestively associated with parental social class or even the man's own income. What it *was* significantly associated with was warmth of childhood environment, and it was very significantly associated with a man's closeness to his father. Holmes was very fortunate, there's no doubt about that. But the best part of his good fortune was not financial.

When I re-interviewed the Holmeses at seventy-eight, Judge Holmes was still working several days a week in an effort to reform the Massachusetts judicial system. He and his wife had recently sold their Cambridge house and were living in a retirement complex. Their living room was warm and welcoming, with an upright piano and a working fireplace, and the walls were covered with Cecily Holmes's very good watercolors and photographs of the children and grand-children.

Judge Holmes in old age was a tall, slightly balding man. At first he reminded me of a distinguished but very uptight professor of neurology I had once known, but as he talked the depth and complexity of his humanity came through. Holmes always focused on the positive side of people—not as someone who sees the world through rose-colored glasses, but in the way that a clear-eyed but loving parent appreciates his child. I was struck by how comfortable I felt with the Holmeses. Our interview lasted three hours, but they betrayed no impatience. They laughed frequently, and it wasn't nervous or social laughter, but the kind that comes straight from the belly.

This was a responsible and venerated judge, still working at seventy-eight. He knew how to stand up for himself. But he knew how to play, too, and he knew how to depend on others, as his interactions with his wife made very clear. Unlike many of the Study men who could barely list one intimate friend, Holmes listed six with whom he shared "joys and sorrows."

Seven years later, at eighty-five, Judge Holmes still experienced his health as "good" and his energy as "very good," although he could no longer walk for two miles or climb stairs without resting. He had to get up three times to urinate during the interview, but his description of his medical problems displayed what his college interviewers had recognized as "whimsical humor" sixty-five years before. It was his prostate, Holmes explained dryly. "My doctor admires its size." When Holmes was eighty-nine, and I was summarizing for this book the hundreds upon hundreds of marriage ratings we had collected over seven decades (see Chapter 6), his marriage scored among the four best in the Study. For the last ten years Judge Holmes in counterpoint to Cecily's painting has been writing beautiful love poetry to his wife.

Holmes's is as good an example as we have in the Study of a warm childhood. How do we know that it predicts success, though? And how can we know how much it was the security of warmth and how much the security of money that allowed Holmes to become what he did? Furthermore, what contribution do genes and hormones (oxytocin, for example) make to mental health and the capacity for empathy and love? We don't know exactly (yet) what the interplay among these factors is. Certainly good fortune in any aspect of the "nature" lottery may exaggerate the effects of a warm childhood on later outcome, and I will use Study statistics to untangle some of these threads as we go on. In the meantime, suffice it to say that the powerful effects of a caring family can sometimes be discerned best in their absence, as the following life story shows.

THE LIFE OF SAM LOVELACE

When he first entered the Grant Study in 1940, Samuel Lovelace was scared. "Sam's anxiety is far in excess of the average Grant Study man," the examining physician wrote. Even at rest, his pulse was 107. The Study staff described him as an immature lad who "tires rather easily," and they were disturbed by his diffidence and his "inability to make friends."

Six years into the Study, a staff member summed Lovelace up as "one of the few men to whom I would assign the adjective *selfish*. . . . It was as if he were looking out of a very small gun barrel." That was a warning; "selfishness" in the Grant Study men appears to have been the result not of too much love in childhood, but of too little. More on this shortly.

No one looking at Sam Lovelace's whole life could possibly regard him as selfish. Even when he was in college, there were some Grant Study staff who were touched by him. They called him a "nice boy . . . intelligent, warm, open chap . . . with potential ability that has not yet found an avenue through which he could direct it." But his childhood environment score was 5 out of 25, a score that put him in the bottom 2 percent of the Study for childhood warmth. Even Godfrey Camille's childhood got a 6.

Sam Lovelace entered life as the result of an accidental pregnancy in a family that was solidly middle-class, even if it didn't possess quite the wealth of the Holmeses. But remember that while warm childhood correlated very significantly with a high Decathlon score, social class was statistically unrelated. And class was the least of Sam's problems. By his mother's own testimony, he never got much attention, even of the most fundamental and presumably nonnegotiable kinds. Asked by Lewise Gregory what she would do over again in bringing him up, she replied, "I would try to provide better care and feeding during infancy . . . and more companionship." She also said of her two sons that she had "always expected them to be adults." When asked at thirty what he wanted his children to have that his childhood had lacked, Lovelace replied, "A richer environment in terms of stimuli."

Sam grew up feeling that he "didn't know either parent very well," and recollected "little demonstration of affection from either of them." They, in turn, dismissed him as "too dependent." But the most independent and most stoical men in the Grant Study were the men who came from the most loving homes; they had learned that they could put their trust in life, which gave them courage to go out and face it. I use the term *stoical* to indicate people who use the adaptive involuntary defense of suppression, which I'll discuss in Chapter 8. As a child, Sam spent most of his time with his dog, and he "grew apart" from his only brother. His mother believed that "people like Sam more than he likes them."

Sam got all A's in high school and was editor of the school newspaper. To the Grant Study staff he appeared "graceful and coordinated." But his parents focused on their belief that he was "not good at sports, and he hates those that he does." For many years Sam Lovelace was a stranger to games; only late in life did he learn how to play.

In the early interviews of his college years, Sam described his mother as "very moody, unpredictable, and given to worry. . . . I don't feel too close to her." He had little respect for his father. At forty-seven, he remembered his mother as "pretty tense and high-strung," and his father as a "remote, anxious, tired" man and a bigot. He visited his parents often as an adult, but not so much out of love as out of his fear that they would soon die. Often they ended up in arguments over politics, which reduced his conservative parents to tears and left the more liberal Sam feeling once more on the other side of an unbridgeable gulf. He took little vacation time—"play" still wasn't easy for him—and spent most of it in dutiful visits to relatives.

When I met Samuel Lovelace at fifty, he was a distinguished-looking man, neatly dressed in suit and bow tie. His hair was gray; as in college, he looked older than his years. During the interview he smoked incessantly and gazed out the window, making no eye contact. Since I never got a smile from him or even a direct gaze, it was hard to feel connected with him. The reasons were not hard to understand; he didn't feel connected either.

He was finding love as hard to come by in adult life as he had in childhood. At nineteen, he had said of himself, "I don't find it very easy to make friends," and at thirty that it was "difficult to meet new people." At fifty, nothing had changed. He described himself as "sort of shy," and told me that he didn't socialize much. At work (he was an architect) he felt bullied and manipulated by his boss. I asked him who his oldest friend was, but he told me instead about a man whom he greatly envied. In the twelve months preceding our interview, he and his wife had invited no one over to their house.

Lovelace's lack of hope and trust in other people made him extremely vulnerable to loneliness, which was one of the enduring issues of his adult life. His marriage to a chronically alcoholic woman was desperately unhappy. His mother had nagged him into going to church when he was young, but he had quit as soon as he escaped to college—another opportunity for community lost. His discomfort in social situations kept him from joining organizations, and so did his apparent dislike of games (which, as his later life demonstrated, probably reflected social discomfort far more than the lack of skill his parents so insistently attributed to him).

Lovelace approved of hippies on the grounds that "I'm for anything that will shake up the adult world"; the only positive feeling that he harbored about the status quo was the hope that it would change. At eighteen, such a philosophy is healthy, but at fifty, as I noted in a paper on the twenty-fifth reunion poll of the class of 1944, it is correlated with social isolation and with seeking psychotherapy.[9] Indeed, other people were so lacking in his life that Lovelace's chief source of comfort at that time was the psychiatrist whom he had visited for fifteen years.

The unloved often have a special capacity to identify and empathize with the pain and suffering in the world, and so it was with the self-doubting, pessimistic, hippie-loving Lovelace, who had opposed Joe McCarthy, supported Adlai Stevenson, and believed American military involvement was no solution to the problem of Vietnam. While many conservative men with happy childhoods wished to slow down the process of racial integration in the country, Lovelace wished to hasten it. Still, an essential passivity and sense of isolation kept him from acting. "Although I take a liberal stance at cocktail parties, I can't follow it out onto the streets," he said. Sometimes he didn't even vote.

In Freud's psychoanalytic worldview, *orality* characterizes the earliest stage of psychosexual development. It is the period during which, allegedly, the mouth is the infant's primary erogenous zone and the focus of all libidinal interests and conflicts. But orality is more usefully seen as a metaphor for the longing of hearts that have not learned to fill themselves with hope and love, as Erikson recognized when he wisely reframed its developmental correlate as lack of basic trust.[10] It was perfectly true that Lovelace was a man of many oral habits. He used stimulants to start the day, inhaled several packs of cigarettes to see him through it, and drank eight ounces of bourbon to soothe its end. Then he took three sleeping pills to go to sleep. It was also true that as a child he had bitten his fingernails, and he still occasionally put his thumb into his mouth as he talked with me. But none of this was the point. The main legacy that Sam Lovelace's parents had left him was not literal hunger, but a profound lack of trust in them or anyone else, including himself. What was spoiling Lovelace's life was fear.

At thirty-nine he wrote, "I feel lonely, rootless, and sort of disoriented." He said of his painful marriage: "No matter how hollow a marriage it may be, it still gives one a home and a place in society. . . . Despite even some hatred, it's easier to suffer with Janie than to suffer without her." He confessed that he was "afraid of giving up marriage and standing on my own," and one reason for his fear was that his wife was much richer than he was. Clearly he wanted something better, but he was afraid to try to find it. Initiative and autonomy are very difficult when the world is such a perilous place, but Lovelace berated himself for his inadequacies.

I interviewed Lovelace again when he was fifty-three, after the death of his alcoholic wife. I wanted permission to publish his story, and I thought it was important to make that request in person. With the burden of his agonizing marriage relieved, his life was one important degree less dangerous, and he was no longer the cadaverous, worried man I had met three years before. He—almost—had a glow in his face. He was now president of the Louisville Home for the Aged, and had sublimated his fears of aging and death (as marked in him as in any man in the Study) in pride at giving the older residents of his city a better quality of life. His social life had improved, if only slightly. He gave me permission to publish his bleak history, which was certainly an altruistic act.

Even having seen the changes in Lovelace over his first three years as a widower, however, I was not prepared for the man I interviewed just before his eightieth birthday. He was much less depressed. He looked younger than he had twenty-five years before, and when I said something about that he told me that I was not alone in thinking so. Except for the antidepressant Zoloft, he had given up the tranquilizers, sleeping pills, and stimulants of his youth. He had had girlfriends after his wife died. Like many of the Loveless men, it was hard for him to let love in, and for many years he had eluded any serious attachment. But at sixty-three he finally succumbed and remarried. Compared to the other men's, his second marriage scored in the "unhappy" category. Nevertheless, at seventy-nine he told me that his wife was the best thing that ever happened to him. As we'll see in Chapters 5 and 7, brain maturation leads to an increased capacity for intimacy, especially when that has been retarded by emotional privation. Even this ambiguous marriage was an improvement over anything he had ever had before, and he could appreciate and rejoice in that.

In his chronic withdrawal and lack of energy, Lovelace had had little room for personal interests. But now, warmed perhaps by his new marriage, the man who had been a "failure" at sports throughout his life was (his wife assured me) a fabulous dancer. Lives change, and things can get better. But the people who don't learn to love early pay a high price in what Conrad calls woe. And, as I'll show in a brief statistical foray, Lovelace remained to the end one of the most loveless men in the Study.

IMPORTANT LINKS BETWEEN CHILDHOOD AND OLD AGE

The men at fifty.   When the College cohort was about fifty, I explored in a series of analyses the relationship between childhood environment and adult outcome. The results confirmed the qualitative message of Holmes's and Lovelace's stories—that for good or ill, the effects of childhood last a long time. Half the men from the best childhoods, but only an eighth from the worst, had made what we then considered optimum adult adjustments. Only a seventh of the Cherished, but fully half of the Loveless, had at one time been diagnosed as mentally ill, under which rubric we included serious depression, abuse of drugs or alcohol, and a need for extended psychiatric care (more than 100 visits) or hospitalization. The Loveless were five times as likely as the Cherished to be unusually anxious. They took more prescription drugs of all kinds, and they were twice as likely to seek medical attention for minor physical complaints. They spent five times as much time in psychiatric offices.

Some readers may object to my classifying a hundred visits to a psychotherapist as a statistical manifestation of mental illness. There are lots of reasons that healthy people go to psychiatrists, you might point out. What about young psychotherapists who seek analysis as part of their training or professional development, for instance? Here I'll avoid a he-said she-said argument, and let the Grant Study's telescope provide a statistical resolution. Of the twenty-three men with the lowest Decathlon score, nine had seen a psychiatrist one hundred times and only seven never. Of the thirty men with Decathlon scores over 6, only three had ever seen a psychiatrist at all and none more than ninety-nine times—a very significant difference. Of the five psychiatrists in the Study, all had seen a psychiatrist over one hundred times. Even if for the sake of argument we accept that as irrelevant, four had also used harmful amounts of psychoactive drugs and three had experienced a major depression. This is not to say that having once been considered less than mentally robust by the Study a man was doomed to carry that label forever; Godfrey Camille did not. Nor does it mean that people can't turn their lives around; Godfrey Camille did. But it does indicate that people who seek a great deal of psychiatric care have something in their lives that needs turning around. People whose lives feel just fine tend to eschew psychiatrists.

In individual case records, we frequently observed the following unhappy sequence. A poor childhood led (first) to an impaired capacity for intimacy, and (second) to an above-average use of mood-altering drugs. Warm childhoods clearly gave the more fortunate men a sense of comfort with and acceptance of their emotional lives that lasted into old age, while bleak early years left their survivors with (third) an enduring apprehension about trusting and facing the world by themselves. The fourth and cruelest aspect of bleak childhood was its correlation with friendlessness at the end of life. The Cherished were likely to be rich in friendships and other social supports at seventy—five times more likely than the Loveless who trusted neither the universe nor their emotions, and remained essentially friendless for much of their lives. The Loveless Adam Newman, for example, who insisted that he never knew what the word "friend" meant, had managed to find a wife who fully satisfied his limited emotional desires. But she was his only friend, and he was all too uncomfortably aware that her death would leave him essentially alone.

The Cherished were eight times more likely to report good relationships with their siblings, and more likely to achieve the wider social radius that Erikson called *mature* or *generative*. I will admit that I feel some uncertainty about whether childhood environment is solely responsible for this; I suspect that there are genes for loving natures, undiscovered though they yet be. Here again, time will tell.

Let me turn now to the corollary of this chapter's lesson—that it is the quality of their total experience that determine the way children adapt to life, not any single piece of good or ill fortune. The Grant Study turned up no single childhood factor that predicted well-being (or its opposite) at fifty. It was the number, or perhaps the constellation, of positive and negative childhood factors that predicted mental health risks, not any particular ones or any pattern among them. But the total mattered.

At about age fifty, the men were asked to respond to 182 true-or-false statements in the Lazare Personality Scale.[11] Eight of those statements very significantly distinguished the men who thirty years later would receive the lowest Decathlon scores from the men who received the highest. That is, they distinguished the lonely, the unhappy, and the physically disabled from the happy, successful, and physically well. Remarkably, the same eight midlife questions also very significantly distinguished the men who had been classified as Loveless and Cherished *thirty years prior to testing.*

- My behavior with the opposite sex has led to situations that make me anxious.

- I have often thought that sexually, people are animals.

- I usually feel that my needs come first.

- Others have felt that I have been afraid of sex.

- I easily become wrapped up in my own interests and forget the existence of others.

- I put up a wall or shell around me when the situation requires it.

- I keep people at a distance more than I really want to.

- I have sometimes thought that the depth of my feelings might become destructive.

What does this have to do with childhood? These eight revealing questions address a fundamental discomfort with the emotional aspects of life, and a resulting self-doubt, pessimism, and fearfulness. Men with warm childhoods subscribed to very few of these statements, but (very significantly) the less fortunate men often subscribed to four or more. And the more at ease the men were with their feelings, the more successful they were at the rest of their lives.

The College men who did not achieve successful or gratifying careers revealed a life-long inability to deal with anger, which I have documented in *Adaptation to Life* and also in a statistical examination of the work lives of the Inner City men.[12] Anger is a delicate balancing act. No one can avoid anger forever and hope to prosper in the real world, but discomfort with aggression is a developmental challenge not only for women and overeducated men; it was for the Inner City men, too.

So how does a child learn to trust what he feels and other people's responses to his feelings? When you're just getting the hang of grief, rage, and joy, it makes all the difference in the world to have parents who can tolerate and "hold" your feelings rather than treating them as misbehavior.[13] Maybe Mrs. Holmes's willingness to let her two sons act up in the living room was a coincidence, but I doubt it.

And what if you don't have parents like that? How do you learn such things if you can't engage comfortably with other people? How do you tackle the world confidently, or risk finding someone to love, or even find room among your fears to relax and pay attention to things other than yourself? There was no man in the Study who had a bleaker childhood than Sam Lovelace. All the circumstances in his constellation conduced to a heart that didn't know how to be satisfied. His Childhood Environment score was a rock-bottom 5, and he alone answered "True of me" to all eight of the questions. His Decathlon score was 0, and at age seventy there were only two men in the Study with fewer social supports than he. It's not enough to be loved; you have to be able to let love in. As Sam's mother had told us when he was eighteen, "People like Sam more than he likes them." But liking—*trusting* is probably more the issue—has to be learned, and Sam didn't learn it at home.

HEREDITY VERSUS ENVIRONMENT

In 1977 I wrote, "If isolated trauma did not affect adult life, chronically distorted childhoods did affect adult outcome. Although the mental health of relatives was not related to subsequent psychopathology in the men, the mental health of their parents was. The Worst Outcomes were twice as likely as the Best Outcomes to have a mentally ill parent. This effect would seem to be mediated environmentally."[14]

In retrospect, though, I can see that environmentalism in the post-war social sciences was just as extreme as the pre-war hereditarianism had been. This is a good example of the way scientists change science and then are changed by science in their turn. Note too that both the turn to environmentalism and the recent reaction away from it in favor of the neurosciences are encompassed in the lifespan of the Grant Study. Longitudinal studies are inescapably subject to such developments; this is one of their infuriating complications and one of their invaluable advantages. So as the years passed, I had to reconsider some of my nature vs. nurture convictions. Once again, I turned to prospective study to provide accurate data, and to reveal and redress the cultural biases of previous investigators (including myself).

For instance, data that I gathered in 2001, when the men were turning eighty, indicated that the mental health of relatives (not only parents) was an important fork in the road leading from a warm childhood to adult mental health. The presence of mental illness in the family was a wild card with the potential to trump even the most auspicious nurturance. That pointed to hereditary, rather than environmental, influence.

Table 4.1 summarizes some of that analysis. It shows mental health plotted against three conditions: bleak childhood environment, familial mental illness, and personality rating at age twenty-one. I've described already how we assessed childhood environment; the personality assessment at age twenty-one was part of the original investigation of the men during their college years. The scores for the second condition, familial mental illness, were established this way. In a process that will be familiar by now, we reduced four concrete indicators of genetic vulnerability—alcoholism, depression, poor familial longevity, and, surprisingly, early death of the maternal grandfather—to 3- or 4-point scales. (It can be surprisingly difficult to assess familial longevity accurately, but we were able to do it. The men's parents knew the dates of death of *their* parents, and the Study lasted the necessary six decades until the last parent of a Study man died in 2001. Another gift of lifetime studies. Familial longevity was estimated by summing the ages of the oldest parent or grandparent on the maternal and paternal sides.)

Because these four disparate variables correlated significantly with each other, they were summed to yield a heredity score of 0 to 12, which, as Table 4.1 illustrates, did have some predictive power. These scores did not reflect any kind of genetic testing or even precise diagnoses. They were pragmatic and crude estimates of the presence of three broad syndromes in a family—alcoholism, depression, and short life. Longevity of the maternal grandfather was included individually because its absence had a very significant—and also a fascinating and difficult to explain—association with both neuroticism and depression; much more on this in Chapter 10. It's worth noting that a high heredity score was very significantly correlated with a bleak childhood, and my work on alcoholism later demonstrated that heredity accounts for almost 100 percent of familial transmission of alcoholism (see Chapter 9). Holmes was blessed with a heredity score of 0, while

Table 4.1 Important Associations of Heredity and Childhood Environment with Adjustment to Life

	Bleak Childhood Environment	Familial Mental Illness	Neuroticism / Extroversion Age 21
. Mental Health			
Major depressive disorder	Significant	Very Significant	Significant
Psychiatric visits	NS	Significant	Very Significant
Lifetime mental illness	Significant	Very Significant	Very Significant
Adjustment, age 21	Very Significant	NS	NS
Maturity of defenses	Significant	Significant	NS
Adjustment, age 30–47	Very Significant	Significant	Significant
Decathlon	Very Significant	Significant	Very Significant
Adjustment, age 65–80	Very Significant	NS	Significant
. Social Health			
Relationships, age 47	Very Significant	Very Significant	Significant
Social supports, age 50–70	Very Significant	NS	NS
Eriksonian maturation	Significant	Significant	Significant
Relationship with kids, age 50–70	Significant	NS	NS
. Heredity More Important Than Childhood			
Alcoholism	NS	Very Significant	NS
Vascular risk variables*	NS	Significant	NS
Smoking	NS	Very Significant	NS

Very Significant = p<.001; Significant = p<.01; NS = Not Significant.
* This was a sum of five variables leading to vascular and heart disease (high diastolic blood pressure, type II diabetes, obesity, heavy smoking, and alcohol abuse). The vascular risk variables are discussed in chapter 7.

both Camille and Lovelace had significant genetic loading for mental illness.

The first two columns of the table illustrate the correlations of childhood environment and heredity with various maturational and situational circumstances in later years. A warm childhood predicted later social and love relationships better than hereditary factors did. But heredity was the better predictor of later health-related developments like alcohol abuse, smoking, and, most dangerously, the vascular risk variables that I'll discuss in Chapter 7. Why heredity was less associated with adjustment in old age is puzzling. One factor may have been selective attrition by early death among the alcoholics and depressives.

The third column of the table reflects another curious chapter in the history of Woods's equivocal personality trait schema—an arresting tale of a scientific ghost and the inestimable value of long perspective. In the mid-eighties, Paul Costa and Robert McCrae, longtime senior investigators at the National Institutes of Health and the Baltimore Longitudinal Study of Aging, established an inventory of personality traits that became very well known under several names: the NEO, the Big Five, and the Five Factor model.[15] That first name is an acronym for the first three of its five traits—*Neuroticism, Extraversion, Openness, Agreeableness,* and *Conscientiousness;* the second and third are self-explanatory. Among the statistically inclined, the Big Five has emerged as a robust model for understanding personality, but clinicians, including me, have found it less useful. One of my reasons is that McCrae and Costa have used it to argue that personality does not change over time.[16] This is a point of view clearly not in accord with my Grant Study experience, and one that has been sharply contested in other quarters as well.[17]

In 1998 Stephen Soldz, a psychologist expert in statistics and in the Big Five, thought to look back to Woods's earlier twenty-six-trait classification and the way it was applied to the College men in the 1940s. As I've said, Woods's methodology was not all it might have been, and the usefulness of his schema has been slight, with the one startling exception that I described in the last chapter. Nonetheless, out of the material from that time, Soldz was able to extrapolate five traits that correlated, with high agreement among seven independent raters, with the Big Five. In his hands, these five traits were highly predictive of most of the outcomes in Table 4.1, a result that forced me to reconsider my distrust of the Big Five's predictive power. There was also a high correlation between Soldz's traits and the College men's results on the Big Five proper, which was administered to the College men at the age of sixty-seven.[18]

It was fascinating to watch a measure from the very earliest days of the Study—one that for a long time looked almost laughably useless and that redeemed itself only by the skin of its teeth with the surprising success of *Well Integrated*—receive a whole new lease on life in the hands of a truly sophisticated statistician. Woods tried something. It mostly didn't work, and the material that he amassed wasn't much use in his own context. But if he hadn't tried, his material would not have been available once a larger context appeared. His reach exceeded his grasp, but that overreach is an important attribute of longsighted and imaginative science.

In another "However," though, recent twin studies at multiple universities, especially those by David Lykken at the University of Minnesota, have shown that there is a major genetic component to Big Five test scores—perhaps up to 50 percent.[19] This implies that many personality characteristics that people tend to attribute to family and societal influences (including ones as unlikely as spirituality) are at least in part genetic. It also means that the "Bleak Childhood Environment" in Table 4.1 reflects contributions from hereditary factors as well as environmental ones. This is a major complication for the analysis of our data, and another issue that we will have to wait for time to sort out.

The third column in Table 4.1 refers to the NEO-related scores. It is more speculative than the other two, but it suggests that a high score on the Big Five trait of *Extraversion* (that is, thriving on challenging environments, social interactions, and keeping busy) and a low score on the trait of *Neuroticism* (that is, anxiety, hostility, depression, and self-consciousness) predicts a high Decathlon score. Indeed, the association (for the statisticians, rho $= .45$) was as strong as the benchmark correlation between people's height and their weight.

Some of the traits that the men had been rated for in college— *Humanistic, Vital Affect, Friendly*—had been significantly associated at that time with a soundness rating of A. Others—*Inhibited, Ideational, Self-conscious and Introspective, Lack of purpose and values*—had been significantly associated at the time with a college adjustment rating of C, or unsound.[20] In midlife, however, none of these adolescent traits were associated positively or negatively either with psychopathology or with good adjustment. In retrospect, the traits associated with poor college adjustment seem to have been part of normal adolescence (which adults do sometimes view as pathological!). Surprisingly, the one trait that proved to be strongly associated with healthy midlife adjustment was a trait rather uncharacteristic of adolescents, called *Practical, Organized*. It included the ability to delay gratification, and proved to be strongly associated with healthy midlife adjustment, but not after age seventy-five. On the other hand, we found a "sleeper" trait—one that didn't appear to be important in my first foray when the men were forty-seven, but became notably so afterward. This was our old friend *Well Integrated,* which I defined in Chapter 2, and which, while not so important in midlife, significantly predicted a physically active and cognitively intact old age. The reason for this appears to be that *Well Integrated* was the one Woods variable that (like a warm childhood) independently predicted the absence of vascular risk factors like smoking, obesity, and elevated blood pressure. *Self-starting* was another trait not important early in midlife but very important at the end of life.

What does it mean that a variable is predictive at one time of life but not at another? It could mean that what once looked significant really wasn't, or vice versa. But it could also mean that some things really are significant at certain times, but not at others. This makes prediction a much more complicated issue, and it also points up a crucial aspect of lifetime studies. They never let us forget that something that is true at one time of life may not necessarily be true at another. The longer our view, the better a chance we have to figure out why some correlations endure and others don't. Sometimes the missing link is in our data; sometimes it's in our science; sometimes it's in our intuition. But even during the (admittedly uncomfortable) periods when we know a shift has taken place, or a predictor has stopped predicting, but we don't know why, at least we know that it's happened. It is only by studying lives over time that we become aware of changes like this, which may be markers of important developments, such as physical maturation in the brain. The vicissitudes of associations and correlations that we can follow only in longitudinal studies remind us to keep our eyes and our minds open and to guard against premature closure.

MOTHERING VERSUS FATHERING

Here's an example of how these vicissitudes look. When I first began to analyze the effects of childhood upon adulthood, it looked like the total childhood environment was more important than the maternal relationship per se; as a psychiatrist, I found this surprising. After eighty, however, the men's childhood relationships with their mothers look more significant—another "sleeper" effect. What its meaning might be we don't yet know.

The Study found some facets of adulthood in which a good relationship with one parent or another exerted the more important influence. As the men approached old age, their boyhood relationships with their mothers were associated with their effectiveness at work, but their relationships with their fathers were not. A man's maximum late-life income was significantly associated with a warm relationship with his mother, as was his continuing to work until seventy. His military rank and his inclusion in *Who's Who* were also marginally significantly associated with a warm relationship with his mother.

A warm childhood relationship with his mother was significantly associated with a man's IQ in college, and, more important, with his mental competence at eighty. A poor relationship with his mother was very significantly, and very surprisingly, associated with dementia. For example, of the 115 men without a warm maternal relationship who survived until eighty, 39 (33 percent) were suffering from dementia by age ninety. Of the surviving men with a warm maternal relationship, only 5 (13 percent) have become demented—a Significant difference. In the Grant Study, dementia has not been significantly associated with vascular risk factors. One senior colleague of mine insists that this finding must be wrong, on the grounds that no one has noted it before. He forgets that seventy-year prospective studies are as rare as hen's teeth. Only time—or replication—will resolve the matter.

None of these issues were even suggestively associated with the quality of the man's relationship with his father. However, warm relationships with fathers (but not with mothers) seemed to enhance the men's capacity to play. Men with warm paternal relationships enjoyed their vacations significantly more than the others, employed humor more as a coping mechanism, and achieved a very significantly better adjustment to, and contentment with, life after retirement. Counterintuitively, it was not the men with poor mothering but the ones with poor fathering who were significantly more likely to have poor marriages over their lifetimes. All five of the men who reported that marriage without sex would suit them had poor paternal relationships, but these men were evenly distributed as to the adequacy or inadequacy of their mothering.

Men with good father relationships also manifested less anxiety—

a significantly lower standing pulse rate in college, for example, and fewer physical and mental symptoms under stress in young adulthood. Men with poor father relationships were much more likely to call themselves pessimists and to report having trouble letting others get close. And good father relationships very significantly predicted subjective life satisfaction at seventy-five, a variable not even suggestively associated with the maternal relationship.

Nonetheless, a mother who could enjoy her son's initiative and autonomy was a tremendous boon to his future. Judge Holmes was by no means the only successful man whose mother admired his assertiveness. Other mothers of successful sons boasted: "John is fearless to the point of being reckless"; "William could fight any kid on the block. . . . he was perfectly fearless"; and "Bob is a tyrant in a way that I adore." Not all the men were so lucky; certainly Frances DeMille (in Chapter 8) was not. Yet as the men moved into adulthood, their mothers, as if by magic, were remembered as progressively weaker, and their fathers as progressively more dominant, childhood figures.

RECOVERY OF LOST LOVES

No one whom we have loved is ever totally lost. That is the blessing and the curse of memory. As Tolstoy wrote, "Only people who are capable of loving strongly can also suffer great sorrow; but this same necessity of loving serves to counteract their grief and heals them."[21] There's a distinction to be made here between privation and *deprivation*. Privation—never having loved or been loved—leads to psychopathology, not to grief. Deprivation—the loss of those whom we have loved or been loved by—leads to tears, but not to sickness. Grief hurts, but it doesn't kill.

The psychodynamic work of mourning has less to do with saying good-bye and letting go than with bringing down from the attic the old pictures that re-evoke forgotten moments of intimacy. I believe, though it has not yet been proven, that the primate brain is constructed to retain love, not to relinquish it. It is much easier to conceptualize the searing pain of loss than the subtle and almost imperceptible ways that we absorb the people we love into ourselves, and perhaps this is one reason that some psychotherapists emphasize the pain of grief over its work of remembrance. But dynamic psychotherapy must attend as carefully to the recovery of lost loves as it does to the experience of pain, or, worse, the retrieval of old resentments.

The recent years of the Grant Study have shown that our lives when we are old are the sum of all of our loves. It is important, therefore, that we not let any of these go to waste. One task of the last half of life is to recover the memories of the loves of the first half. This is an important way that the past affects the present. The rediscovery of a love once lost, or of the power to forgive, may be a source of great healing. To participate in the recovery of others' lost loves is yet another joy of conducting a longitudinal study.

I will revert once more to pictures to convey this joy. Godfrey Camille, whom we met in Chapter 2, wrote when he was eighteen, "I am sorry that my family was so terribly strict with me from the ages of 5 to 12." He was charitable about this, but realistic: "They can't be blamed for it because they were brought up in 1890 New York. . . . Dad and I have never gotten on very well."

Over the years, as Camille uncovered his past through psychotherapy, he tried to help me understand what that uncovering meant to his life.

Is it the re-finding of the love that matters? Or is it the chance to re-celebrate and to re-celebrate again and again the bond that holds? An image comes to mind—the empty wine bottle used as a candlestick. The wine may have been the initial warmth in life, but once it is gone only the cold

and empty glass bottle remains—until in memory we re-
light the candle of conviviality, and as it drips, it transforms
the symbol of the warmth that was spent into a differently
shaped, differently colored object—alive with new warmth.
Early loves may be "lost" because they were taken for
granted and never reinforced by review. Recollection and
retelling have a way of making them increasingly visual and
real. It is the visual perception that's particularly effective
when it comes to learning.

Long before he wrote those words, however, Godfrey Camille
had engaged himself in the process of turning the cold, empty bottles
of his life into vessels of new light, hope, and strength. At sixty-five he
reminded me (as if I could forget) that "empathy was not my father's
strongest suit," and he offered an old, but newly remembered, example. Having climbed a cherry tree to look at the blossoms, he lost his
balance and fell twelve feet to the ground. He got no comfort, only a
spanking for disobeying a parental injunction not to climb that tree.
But now he was remembering something else. A few days later his father had picked him up in his arms, and . . .

Once beneath the shimmering canopy of the cherry tree, my
father pulled a small branch towards me and let me hold it in
my fist while we traded glances and smiles. Before he drove
off to work, he used his pocketknife to cut a sprig, which he
put in a tumbler of water, placing this on the little table
where I ate my meals. . . . I knew I was both understood and
forgiven. . . . This was my first mystical experience.

Camille had always had an abstract interest in genealogy, but late
in his life he translated that into a new source of actual love. He discovered that his father had a whole network of French cousins, and

through correspondence and travel he made relationships with them. Out of this relational archeology he created the warm extended family his painful childhood had denied him, mining memories of time past and investing them in real people in time present. This is one form of resilience. In his renewed religious involvement and his new relatives, he was finding fresh sources of strength, and fresh connections with his father. Creative writing and extensive psychotherapy promoted his discoveries of these lost loves; he is not alone in making good use of those activities to that end.

Despite such happy transformations as Camille's, however, at the end of the day Conrad is more right than wrong. When we contrasted the lives of the men whose childhoods were bleakest—the Loveless—with the lives of those whose childhoods were sunniest—the Cherished—poor childhood was very significantly associated with poor later adjustment. Still, while being unloved as a child is painful, it isn't the bottom line. The bottom line is what being unloved does to a child's capacity to be loved, and to be loving, later. Some children manage to develop this capacity in spite of everything. But the ones who do not learn to love, or cannot let themselves be loved, do indeed live lives of woe.

SURPRISING WAYS THAT CHILDHOOD DID NOT AFFECT OLD AGE

Prospective studies of normal development are only now attaining the many decades they need to accompany adolescents into maturity, let alone old age. But in the process they are demolishing many cherished assumptions. We all "know," for instance, that childhood affects the well-being of adults. But recent scientific reviews reveal that many popular explanations of how this happens—for example, through the loss of a parent—are rather less conclusive than we have long thought.[22]

It's not hard to explain the way people turn out—however they turn out—after the fact. I can evidence a crazy aunt, a rejecting mother, a clubfoot, a bad neighborhood, or any other circumstance I like to explain a poor outcome. And I can prove anything at all with single case studies. That is why I accompany the biographies here with their statistical context, and why I harp on how important it is to keep looking for statistical validation, even though we may not always like what it shows us.

Now, therefore, having used Study data to illustrate prospectively assessed childhood contingencies that were statistically important, I want to discuss some commonly held beliefs that are not supported by the data on the College (and sometimes the Inner City) men. It is very important for developmental psychologists to keep careful track of the interface between theory and fact, and to strive to bring the two into ever closer agreement. That is our best hope of keeping the children of the future from the devastating effects of destructive childhoods. It is true, and it is a hopeful truth, that some of the men with the worst childhoods enjoyed the period of end-of-life. Still, sixty years is a very long time for a Sam Lovelace to have to wait for happiness and even then his happiness was only relative. Furthermore, whereas a warm childhood, like a rich father, tends to inoculate a man against future pain, a bleak childhood is like poverty; it cannot cushion the difficulties of life. Yes, difficulties may sometimes lead to post-traumatic growth, and some men's lives did improve over time. But there is always a high cost in pain and lost opportunities, and for many men with bleak childhoods the outlook remained bleak until they died, sometimes young and sometimes by their own hands. We must get better at intervening when childhood conditions bode ill for the future, and that means finding ways to recognize them, not by sentiment but by data-based understandings of development and its antecedents.

The Grant Study, which approaches development from the front rather than from the back, has called many popular theories into question. The early Study staff did examine the effects of strict vs. liberal toilet training, and Earnest Hooton concluded in *Young Man, You Are Normal* that toilet training seemed to be entirely without significance for future behavior.[23] Hooton's book was published in 1945; sixty-five years of further study has done nothing to prove him wrong. Freud's theories about the deleterious effects of strict toilet training were based on retrospection, in the form of the memories of his usually middle-aged patients; yet again, prospective studies rule!

More recently, other childhood conditions that have been thought at times to have powerful implications for the future have failed to correlate significantly with the events of the Decathlon. For example, the pre-Spock Grant Study men were disciplined strictly by their parents. Eighty-six percent of them were breast-fed, but more than 50 percent were out of diapers by eighteen months. Thumb-sucking was inhibited with aluminum mitts and bitter aloes, even sometimes by tying an infant's arm to his body so as to keep hands away from mouth. None of this mattered. Physical health in childhood didn't hold up. Neither did the distance in age between the subject and the next child. Birth order was insignificant too, except that being the eldest child was correlated with occupational success. Even the death of a parent was relatively unimportant predictively by the time the men were fifty; by the time they were eighty, men who had lost parents when young were as mentally and physically healthy as men whose parents had lovingly watched them graduate from high school.

Even that old standby, the cold, rejecting mother, failed to predict late-life emotional illness or poor aging. It was good not to have mentally ill relatives, since genes so often trump environment. But basically it wasn't the absence of a loving mother that made all the difference, but the presence of one, or of an adored father, or of an otherwise warm childhood surround. This is what I mean when I say that our lives are shaped more by what goes right than by what goes wrong.

Another common assumption, based on retrospective "evidence," is that alcoholism is the result of an unhappy childhood. Certainly both alcoholics and clinicians may point blaming fingers in that direction after the fact. But prospective evidence suggests that this may be a reversal in which the results of alcoholism are being misconstrued as its cause. Alcoholism is passed on genetically, not environmentally, which means that an alcoholic parent appears in many alcoholics' backgrounds.[24] Alcoholism in a parent certainly ups the likelihood that childhood will be unhappy, but in that case the unhappy family is the cart; the horse constitutes the genes conducive to alcoholism. And as we shall see in Chapter 9, alcoholics' memories of their childhoods may be unwittingly altered either by physiologic changes due to alcoholism or defensively, to reduce guilt.

Another example. Ten years ago, I believed that the Loveless died sooner because they were careless of their own well-being. I wrote in *Aging Well*, "The Loveless, believing that they belonged to no one, failed to remember the old song's advice to 'button up your overcoat when the wind blows free.'" In this I was wrong; today's data reveal that the Loveless and the Cherished differ only slightly in the way they take care of themselves, at least with regard to the common vascular health risk factors of smoking, blood pressure, weight, and diabetes.

CONCLUSION

First, contrary to what might be called the developmental failure model of psychopathology, in the Harvard Study of Adult Development it was the men's successes, not their failures, that predicted subsequent mental health. What they did with a loving or bleak childhood had as much to do with future success as the childhood itself. Of the twenty-six personality traits that Woods assessed when the Grant Study men were in college, it was the one called *Practical, Organized* that best predicted objective mental health at ages thirty through fifty. The Terman Study also found that prudence, forethought, willpower, and perseverance in junior high school were the best predictors of vocational success at age fifty.[25] It's hard not to think that these are precisely the traits people need to find ways around failures, and make the most of successes when they come along.

Second, it was not a disturbed relationship with one parent as much as a globally disturbed childhood, like those endured by Camille and Lovelace, that affected adult adjustment. Men from bleak childhoods were more likely than the others to be pessimistic and self-doubting; perhaps it was this that made many of them unable to take love in even when it was offered, or fearful about offering love themselves. These dynamics are illustrated to some degree by Sam Lovelace, and even more grimly by Peter Penn in Chapter 6 and Bill Loman in Chapter 8. The Gluecks' and other prospective longitudinal studies have shown that the children of multiproblem households are likely to be severely impaired in their later capacity to work.[26] For the upper-middle-class Grant Study men, however, inability to love seemed to be a more sensitive reflector of bad nurturing than work problems. Their childhoods were not so badly disrupted as those of the less fortunate Glueck men, and even relatively poor nurture did not in any way impair the capacity of many of these College men to excel in their jobs. (Of course, these particular men were all preselected as good workers; if they hadn't been, they wouldn't have been at Harvard.) But it did impair their capacity to love and be loved.

Men from poor childhoods were less able to deal consciously with strong emotions, either pleasurable or distressing ones; perhaps for this reason—that they couldn't grapple with strong feelings directly—they were more likely to turn to drugs to soothe themselves.

So, while an isolated trauma or a bad relationship need not in itself condition adult psychopathology, the Harvard Study of Adult Development makes clear that global disruptions of childhood have strong predictive power, none of it good. Children who fail to learn basic love and trust at home are handicapped later in mastering the assertiveness, initiative, and autonomy that are the foundation of successful adulthood. Prevention will be best served when seriously troubled families can be accurately identified and made the foci of special concern. Mental illnesses and alcoholism create conditions that can ravage entire families, and can destroy children's futures for decades to come.

5 | MATURATION

It is well for the world that in most of us, by the age of thirty, the character has set like plaster, and will never soften again.

—WILLIAM JAMES

I FIRST INTERVIEWED the Grant Study men in 1967, when I was thirty-three and they were in their late forties. I've always respected William James, and for a while those early encounters seemed to confirm his contention that character is set by thirty. But watching the men change in real time quickly persuaded me that about this, at least, he was mistaken. Lesson Four of the Grant Study is that people really do grow. Having learned it, I now argue with friends not about whether personalities change in adulthood, but about how best to measure the changes, and about what a person should be at the end of life that he was not at the beginning. That is, I ponder models of maturation, and hope that the Grant Study will, in the fullness of time, contribute to the construction of a better one than any we yet possess.

Developmental psychology textbooks still suggest, by omission if not by active assertion, that maturation stops at twenty—thirty at the latest. One major work maintained as recently as 2010, "It appears that individuals change very little in personality (either self-reported or rated by spouses) over periods of up to thirty years and over the age range 20 to 90."[1]

It's true that an oak tree's leaves don't vary much as the tree ages—their shape and character are inherent fixed traits. But that doesn't mean that the tree itself remains the same. On the contrary, the grandeur and complexity of a well-growing oak do develop with age, enhanced increasingly by time and circumstances until the tree is damaged or dies. The color of the wine in a bottle of Château Margaux doesn't change much as the years pass, either—but its character sure does, and so does its price.

Over the last seventy-five years, the Study staff has been influenced by at least six models that try to encompass developmental elaborations such as these. The first was the *lieben und arbeiten* model famously attributed to Freud by Erikson, which equated maturation with a deepening capacity for love and for work.[2] The Menninger Clinic's brilliant research psychologist Lester Luborsky turned Freud's epigram into the model of development used in the clinic's thirty plus-year Psychotherapy Research Project.[3] His scale, slightly tweaked, was eventually enshrined in DSM IV as Axis V of the American Psychiatric Association's definition of mental health.[4] When I was doing those first interviews, I used it to assess maturity in the Grant Study men, and reflections of it remain in the Decathlon and the Adult Adjustment scales (Appendix D).[5]

There's a problem with *arbeiten und lieben* as a comprehensive model of adult development, however. Maturation means development forward over time, but as the decades passed, the men's standing in those two realms was as changeable as the weather. Only five (9 percent) of the fifty-five men who scored in the top third in Adult Adjustment (which included assessments of success in love and in work) at forty-seven remained in the top third for all of the subsequent ratings (at fifty to sixty-five years, at sixty-five to eighty years, and for the Decathlon). Sixteen (29 percent) actually scored in the bottom third during one of those next three periods. Periods of success at work and at love came and went unsystematically and accord-

ing to circumstance, and there were no universally discernible processes of "deepening" with age and experience. This model did not
tell us enough about *how* people "grow up."

At thirty-four I was only just becoming aware of how complex a
concept maturity is. Furthermore, it means different things at different
times and to different people. As my acquaintance with the Grant men
(and with myself) advanced, Erik Erikson's model of psychosocial development seemed increasingly more to the point:

> Human personality, in principle, develops according to steps
> predetermined in the growing person's readiness to be
> driven toward, to be aware of, and to interact with a widen-
> ing social radius.[6]

We worked with this model a great deal for my first ten years
with the Study, and it is the one that most deeply informs this chapter.[7] Maturation along Eriksonian lines implies a growing capacity to
tolerate difference and a growing sense of responsibility for others; it
is the evolution of teenage self-centeredness into the disinterested
empathy of a grandparent. Unlike success at love and work, Erikson's
developmental achievements are relatively independent of surround-
ings. They appear and endure in predictable ways. Barring the kinds of
organic damage that diminish functional capacity, they do not get
lost; people are not seen to revert to earlier stages of maturation. This
makes Erikson's model, and the ones that follow, a more convincing
conception of adult growth processes.[8]

My third model of maturity focuses on the development of social
and emotional intelligence. This is the model that best addresses the
way our involuntary coping styles develop over time, becoming, if we
are fortunate, ever more empathic and less narcissistic—that is, ever
more conducive to rewarding relationships. It makes room not only

for the expanding effects of experience and psychological growth, but also for certain relevant aspects of brain development. The development of adaptational capacity is a major interest of mine; it was one of the interests that brought me to the Grant Study, and it dominated my research and writing through my sixtieth year. I will discuss adaptational capacity in detail in Chapter 8.

A fourth model of maturity, honored in Hindu cultures, holds that the task of the grandfather is to retire to the forest to tend to his spiritual life, turning secular preoccupations over to his son. Adam Newman's story illustrates how a man's emotional life can evolve away from personal passions and toward trust, love, and compassion. The attractiveness of this model has been enhanced by recent advances in brain imaging and neuroscience, which locate the emotions traditionally linked with spirituality—awe, hope, compassion, love, trust, gratitude, joy, forgiveness—in their neuroanatomic and evolutionary contexts as biological reality, not abstract sentiment.[9] This work was becoming known as I was approaching seventy and beginning to contemplate my own retirement. I have found it increasingly compelling since.

My fifth model is the one espoused by developmental neuroanatomists such as Paul Yakovlev and Francine Benes.[10] It focuses not on socioemotional growth per se, but on the emotional changes that accompany changes in the brain, the development of which does not come to a halt at sixteen. Neuroanatomists have demonstrated that there is increasing myelinization in the brain between twenty and sixty years of age. (Myelin is the substance that insulates neurons, increasing the efficiency of their electrical functioning.) The net effect of this development is that cognition and the passions work increasingly in concert. That is old news to car rental agencies. They know very well that forty-year-old drivers are more likely, and more able, to think before they act than eighteen-year-olds. Developmental science has now caught up with them enough to know that this is thanks to continued integration of the impulsive limbic system and the reflective frontal lobes, which advances in brain imaging have newly made accessible to study. Robert Waldinger, the current director of the Grant Study, is pursuing this model of development.

The sixth and most recent model of maturity envisions the development of wisdom as the goal and pinnacle of adult development. Sociologist Monika Ardelt at the University of Florida has used the lifetime studies of the College men to study empirically the age-dependent nature of that mysterious and fateful quality.[11] More about this shortly.

The life history of Charles Boatwright, which appears later in this chapter, encompasses all six of these models. And so it should be, for all of them apply to all of us, if not all equally usefully at all times. Maturity means something different to us in youth than it does later, whether we are researchers or just individuals living our lives. Certainly different aspects of maturity came to salience for me as I watched the Grant Study men aging. But I can't separate that from the changing awareness that came with my own increasing years and, I hope, wisdom. Or from the fact that science is growing and maturing too; even now it is capable of feats that were unthinkable ten years ago, and its lifespan is much longer than ours. In this chapter, however, I will concentrate on Erik Erikson's conception of adult development, and some Grant Study–related revisions to it.

ERIKSON'S MODEL: THEME AND VARIATIONS

In *As You Like It,* Shakespeare portrayed everything after middle age as decay. Freud, like many psychologists, ignored adult development entirely. It took Erikson, who as a young man wanted to be an artist, to envision all of life as forward motion and growth. He imagined adult development as a staircase, ascending from the adolescent task of Identity formation to the young adult's movement away from dependence on parents to Intimacy with peers, and then from there to the preoccupations of older adults—Generativity (looking after others) and eventually Integrity (maintaining equanimity in the face of death).

Carol Gilligan, a scholar of women's development, has suggested to me a provocative alternative image of this process.[12] Gilligan thinks not of a staircase, but of the expanding ripples produced when a stone is dropped into a pond. Each older ripple encompasses, yet never obliterates, the circles emanating from the younger ones. Her image is as vivid as Erikson's, but it engages a less goal-directed—a less stereotypically masculine—concept of how people grow, and a compelling evocation of Erikson's model of adult development as an ever-widening social circle and moral compass.

I have made two modification of my own to Erikson's model, distinguishing from his stages two others, which I have called *Career Consolidation* and *Guardianship* (or, in previous publications, *Keeper of the Meaning*).

Erikson established his model, now very well known, in *Childhood and Society*.[13] But brilliant though it was, Erikson's "stages," like most other outlines of adulthood, were armchair intuitions. It is only in the twenty-first century that adult lives and lifetimes, observed prospectively from their beginnings to their ends, have become available for viewing as developing wholes.[14] The *in vivo* empirical study of adult development dates back just forty years, to when Jack Block, Glen Elder, Robert White, and Charles McArthur and I began (in four separate efforts) to look at prospectively studied lives of adults in midlife.[15]

I must make clear too that in adult development, *stage* is a metaphor. It is a popular one, thanks to its wide use by Erikson and others. But it is not descriptively accurate. Clearly defined developmental stages can be seen in embryology, in endocrinology, and perhaps in the process of cognitive development that Piaget and his students observed in children. But adult development as we saw it in Study subjects, and as I will describe it here, is a much less tidy process.

Let me therefore begin my exposition of Erikson's model by noting that *developmental task* is a more useful concept than *stage,* and that—in a process that I'm sure is familiar by now—I assessed mastery of the so-called stages of adult development by tracking the specific accomplishments of psychosocial maturation that I believe reflect and underlie them.[16] I was trying to concretize in an enduring and statistically useful way life situations that are usually described as abstractions, intuitions, and value judgments.

Identity.   Erikson calls the first developmental stage of adulthood *Identity vs. Role Diffusion.* For working and research purposes, I modify Erikson's terminology to *Identity vs. Identity Diffusion,* and define it as follows: to achieve Identity is to separate from social, economic, and ideological dependence upon one's parents. The specific tasks I used to define the achievement of Identity were: to live independently of family of origin, and to be self-supporting.

Identity is not egocentricity. Nor is it a simple matter of running away from home, acquiring a driver's license, or even getting married, as the adolescent trope would have it. There's a world of difference between the instrumental act of running away and the developmental achievement of learning to distinguish one's own values from surrounding ones, and remaining true to them even when life is at its most contradictory and confusing. Identity does not imply rejection of one's past; on the contrary, it derives very much from identification with and internalization of important childhood figures and surrounds, as well as from independent experience in adult life. But it does involve choices about where and how one places one's loyalties.

Some of the Study men never achieved separation from their families of origin or the other institutions that formed them. In our research, we considered this a failure to achieve Identity. In middle life, these individuals remained emotionally dependent on childhood supports, and never moved far enough out into the world to embark upon such voluntary new loyalties as an occupation, an intimate friendship, or a love partner. Although they did not usually come to psychiatric attention, many of them, as they grew older, viewed themselves as incomplete. Some possessed full insight into how little they shared the usual adult preoccupations of guiding the young and trying to keep the world spinning smoothly on its axis. Francis DeMille in Chapter 8 struggled for a long time with identity issues. But the window of opportunity on Identity stays open for a long time, and its ripple keeps expanding into old age. Separation and individuation are lifelong processes.

Intimacy. Erikson calls his second adult stage *Intimacy vs. Isolation.* I defined the specific task of Intimacy as the capacity to live with another person in an emotionally attached, interdependent, and committed relationship for ten years or more. According to that criterion, a man could achieve Intimacy at any point in his life, and indeed, there was a lot of variation in when this task was accomplished. One of the issues I'll address as we go on is what the variability of Eriksonian achievement means in adult development. We'll also consider at some length in the next chapter the difference between Intimacy as a developmental task and intimacy as a relational aptitude. For now, however, I'll note that you can't establish an attached, committed, and interdependent relationship until you've moved out of your parents' world and ensconced yourself in the world of your peers. Mastery of Intimacy depends on first mastering Identity.

A few words of explanation about my criterion. True intimacy is notoriously difficult to measure, as we are all shockingly reminded when a friend's apparently solid marriage goes on the rocks. Ten years of committed, interdependent, laughing marriage is a reasonable approximation of Intimacy for most people living here and now, but marriage itself is not the point. There are intimate friendships; there are nonintimate marriages; there are love relationships that cannot be consummated in marriage. In stable homosexual relationships, or in highly interdependent institutions like convents, where rules for communal living take the place of dyadic bonding, criteria for Intimacy other than marriage may be needed. When issues of that kind arose, I took them into account. I also took into account the fact that the Grant Study came out of an intolerant era, which made the achievement of Intimacy difficult for some. Only two of the Grant Study men achieved stable homosexual relationships, which were counted as achievements of Intimacy; the other five who acknowledged a homosexual preference did not make lasting intimate commitments. For several single women in the Terman sample, Intimacy was achieved with a very close lifelong woman friend, who may or may not have been a sexual partner.

The arbitrary determination of ten years was a piece of pragmatic reductionism designed to facilitate the assignment of a numerical score. On the one hand, nothing lasts forever; on the other, ten years is long enough to distinguish a "real" relationship from a clearly illusory one.

Career Consolidation. I observed repeatedly that some men who successfully established identities within (and then beyond) their families of origin nevertheless failed to accomplish the same thing in the world of work. Accordingly, I took the step of distinguishing career maturation from the other aspects of identity formation to create a separate stage that I now call *Career Consolidation vs. Role Diffusion.* As I have explained in detail elsewhere, Erikson conflated the tasks of mastery of identity formation and mastery of career identification.[17]

I defined the specific tasks of Career Consolidation as *commitment, compensation, contentment,* and *competence.* These four words distinguish a career from a job. By this I do not mean that a lawyer has a career and a janitor has a job. Prestige has nothing to do with the presence of those four characteristics. You can be a competent and well-compensated physician like Carlton Tarryton (below), but if you are so contemptuous about the value of medicine that you turn to Christian Science for your own care, you do not have a career. Men and women in hunter-gatherer societies can bring commitment and enjoyment to the tasks at hand, and lawyers in our society can work without either.

Career consolidation engages a paradox. Selflessness, in the conventional sense of magnanimity or altruism, depends on a sturdy and reliable sense of self. Commitment to a career (and establishing a robust sense of career identity) is an essential developmental task, and it is in large degree a selfish one. That is why society tolerates the ineffable egocentricity of graduate students, young housewives, and business trainees. Only when developmental "self-ishness" has been achieved are we reliably capable of giving the self away as professors, mothers-of-the-bride, and managers. Career Consolidation has roots in Intimacy as well as Identity. Like Intimacy, it combines the necessary self-absorption of young adulthood with the commitment to others necessary to merit a paycheck.

The process that leads today's medical students through internship, residency, and fellowship to professional autonomy is not so very different from the medieval guild structure that took a youth from apprentice to journeyman to master weaver. Young people have always been taught their craft by older practitioners, and then encouraged to make their own individual contributions. That sense of unique competence was considered the culmination of professional formation; it is the culmination of Career Consolidation as well. As one of the Terman women, a writer, put it, "Being home and being married wasn't sufficient. . . . I wanted to secure my sense of competence, to be good at something, acquire a measurable skill, something that I could say 'I have learned this, I am good at this, I can do this, I know this.'"

In tracking the achievement of Career Consolidation I had to take into account that societies constrain the way men and women achieve this sense of competence. The Terman women, for example, were born around 1910; they were in middle school before their mothers could vote. There were limited occupational possibilities open to these very gifted women. I therefore deemed a woman to have mastered Career Consolidation if she was committed, competent, and content at her work, even if she was not always compensated for it in money. Similarly, in the twenty-first century there are men who have consolidated careers as househusbands. The story of Charles Boatwright, coming up in a moment, offers yet another take on Career Consolidation.

Generativity.   Erikson's third stage is *Generativity vs. Stagnation,* which I define as the wish and the capacity to foster and guide the next generations (not only one's own adolescents) to independence. The specific task by which I defined Generativity was the assumption of sustained responsibility for the growth and well-being of others still young enough to need care but old enough to make their own decisions. Generativity, of course, may also include community building and other forms of leadership, but not, to my mind, such pursuits as raising children, painting pictures, and growing crops. These are valuable and creative tasks, but they do not demand the ego skills required to care for "adolescents" of any age. Consider the incredible sensitivity required of famed Scribner's editor Maxwell Perkins, trying to protect the self-centered and self-destructive F. Scott Fitzgerald from his folly while at the same time nurturing his literary genius. As my school-principal aunt once explained: "You can always tell who is headmistress by who moves the most furniture."

Guardianship. In this second modification of Erikson's schema, I organized distinctive aspects of Generativity or Integrity (as he saw it) into a separate stage that I have discussed in earlier works as *Keeper of the Meaning vs. Rigidity,* but am now calling *Guardianship vs. Hoarding* (see Chapter 11). Andrew Carnegie, a Guardian, built libraries with his life's savings; the Pharaohs built pyramids.

Generative people care for others in a direct, forward-oriented relationship—mentor to mentee, teacher to student. They are caregivers. Guardians are care*takers.* They take responsibility for the cultural values and riches from which we all benefit, offering their concern beyond specific individuals to their culture as a whole; they engage a social radius that extends beyond their immediate personal surround. They are curators, looking to the past to preserve it for the future. We tracked this developmental achievement by such curatorial activities, which will be exemplified in many of the life stories that follow.

In his writings, Erikson sometimes fails to distinguish between the *care* that characterizes Generativity and the *wisdom* that characterizes Guardianship (and which he has ascribed, I believe incorrectly, to Integrity). Generativity has to do with the people one chooses to take care of; Guardianship entails a dispassionate and less personal world view. It is possible to imagine care without wisdom, but not wisdom without care—and indeed, in adult development, the capacity to care does precede wisdom. Wisdom requires not only concern, but also the appreciation of irony and ambiguity, and enough perspective and dispassion not to take sides. These are ego skills that come relatively late in life. They are the fruit of long experience and they sometimes conflict with the more generative forms of caring, which may imply sticking up for one person against another. Guardianship is the disinterested even-handedness of the judge who protects the processes of the law in the interests of all of us, as opposed to the advocacy of the generative lawyer, who uses those processes in the service of the client he protects. This is the difference in attitude between the successfully generative Reagan's relentless demonization of the "Evil Empire" and the guardian Lincoln's Second Inaugural plea for malice toward none and charity for all.

Wisdom is often defined in abstract terms: discernment, judgment, discretion, prudence. But it is a developmental achievement, just like the capacity to thrive away from one's mother, or to live harmoniously with a spouse, or to be a forbearing parent to one's entitled teenagers. The task of the Guardian is to honor the vast competing realities of past, present, and future, and to find, as the judge does in *The Merchant of Venice,* the true wisdom that is a fusion of caring and justice.

One fifty-five-year-old College man described in a letter some aspects of his development into Guardianship. He said that he was feeling a sense of broadening. "I have finally come through to a realization of what is of critical importance for our future—that we finally come to live in harmony with nature and our natural environment, not in victory over it. . . . The earlier period was one of comparative innocence and youthful exuberance—a celebration more of my physical powers, of unfettered freedom. Those powers I now celebrate are more of an intellectual variety, somewhat tinged by experience of the world; in a way my lately-acquired knowledge . . . is not an unalloyed blessing, but a burden in some respects."

Integrity.   Erikson called his final life stage *Integrity vs. Despair.* Integrity is the capacity to come to terms constructively with our pasts and our futures in the face of inevitable death. It is a demanding achievement that requires the embrace of contradiction: How do we maintain hope when the inevitability of our end is staring us in the face? The very old have less control and fewer choices than they had when they were younger, but in confronting this reality they may become great masters of Niebuhr's beloved prayer: "God grant me the serenity to accept the things I cannot change; courage to change the things I can; and the wisdom to know the difference."

Integrity differs from the other Eriksonian stages in that (while Integrity issues are perhaps most common in the very old) it is not associated exclusively with any one chronological time of life. It is a developmental response to the anticipation of death, which illness or ill fortune may bring to the fore at any age.

Integrity is a life task I have not yet experienced, and instead of trying to describe it I will recuse myself and let the Study members show what it feels like to them. One Study member asserted, "I think it is enormously important to the next generation that we be happy into old age—happy and confident—not necessarily that we are right but that it is wonderful to persist in our search for meaning and rectitude. Ultimately, that is our most valuable legacy—the conviction that life is and has been worthwhile right up to the limit." Another man dying of prostate cancer explained to me that whenever he had a sudden pain, he could never know if it was "simply old age or another metastasis. . . . But I am a fatalist; when it comes it comes. . . . Each of us was born of this earth, nurtured by it; and each of us will return to the earth."

Perhaps it was a Terman woman, very near death, who embraced the concept of Integrity most succinctly in an explanation of what life was like for her since becoming bedridden: "My accomplishments since then have been to stay alive and alert and to be thankful for all the blessings that have been mine." A recent questionnaire had inquired about her aims for the future. Instead of checking whether it was important for her to "die peacefully," she wrote beside that box, "Who has a choice? Death comes when it comes." Instead of checking whether it was important for her "to make a contribution to society" she wrote beside the box, "In a minor way I have already done so." Instead of fruitlessly complaining that she was of no further use, she—empathically to those around her—reflected the inner peace that came of having paid her dues. That's not chutzpah, that's wisdom.

AN ERIKSONIAN LIFE

My first efforts in 1971 to unravel adult development focused on Erikson's stages of Intimacy and Generativity.[18] At that time, my favorite illustration of adult development was a fifty-year-old Study member, George Bancroft. Asked how he had grown in stature between college and middle life, Bancroft replied, "From age twenty to thirty I learned how to get along with my wife; from age thirty to forty I learned how to be a success at my job; and since I have been forty, I have worried less about myself and more about the children." That summed up succinctly everything I (then) thought there was to adult development. It was the map of world as I knew it at thirty-seven: grown-ups mastered Generativity, and then they died.

But as the decades passed I was increasingly aware that this was not an accurate life map. It took me a while to see what was going on; I was a man in my early fifties studying men in their late sixties, and the fact that they were still growing was more than I (or the fifty-year-old Erikson before me) could readily appreciate. We all needed more time. This was the gift the Grant Study gave over and over again. The ensuing years afforded me both a wider theoretical compass and an expanded experience of my own development, as well as many more chances to observe the men. One man in particular. As I sailed off toward *terra incognita,* my favorite guide to adult development continued to be Professor Bancroft.

At fifty, as you've seen, Bancroft was a generative Mr. Chips. He looked after his own children, and he looked after the history students at his small college. But not so long after that he became dean of the college, and, I noticed, he suddenly had all the students to attend to, to say nothing of the care of his entire young faculty. I could also see that that was a different order of responsibility; it was no longer Bancroft's job to manage individuals' day-to-day development, but instead to establish the kind of atmosphere in which everyone could thrive. His social radius had expanded greatly, and he was no longer functioning as a parent, but as an elder.

For a while Bancroft was still too busy "worrying about the children" to write books. At research universities, writing books is a top priority, because career development and ultimately tenure depend on it. But at small colleges and high schools, book-writing tends to be a retirement activity, like genealogy and town histories. And so it was for Bancroft. At seventy he retired. His focus, which had shifted from teaching history to tending his school now shifted again—this time to writing history. His attention moved even further outward as he began to fill the very role for which, in my opinion, evolution permits grandparents to survive. Once older adults can no longer procreate children, their task has been to preserve the culture, to become what anthropologists call "firestick elders." That is what Bancroft did. Between seventy and ninety he wrote five books that for generations to come will bring America's past to life for her new citizens. He had turned his attention to recollection and preservation, and at last I was able to recognize this, in him and in others. Henry Ford, at age seventy, founded the Greenfield Village Museum to preserve the beauty of a style of life that his creation of the Model T assembly line had helped to destroy. Charles Lindbergh devoted his life after seventy to the preservation of Stone Age cultures that his charting of intercontinental air routes had begun to obliterate. They too became guardians of what seemed to them most meaningful in life.

Bancroft always did have a special way of making development real and visible to me. In 2010, when he was eighty-eight, I asked him that same question of almost forty years before, about how he had grown in stature. (I was fishing for clues about how his transformation into Guardianship had come about.) We were talking on the phone, and my question caught him by surprise. But his answer surprised me even more. "You learn a little more about yourself and you learn how to be alone . . . in part so that you can face death without fear. As the saying goes, 'When you grow old, you get to know women and doctors.' All my male friends have died. . . . You let your wife learn about you. . . . I have to go for a driving test, to see if the world will be safer if I give up my license." And there you have it—an extraordinary description of Erikson's final task of Integrity.

For Bancroft, as for every other American adolescent, the acquisition of that driver's license had once been a critical first venture into adulthood, part of the mastery of Identity and the world outside his parents' house. Seventy years later, the developmental task of Integrity was about being able to give the precious license up, and, if necessary, to adopt with equanimity Job's mantra: "The Lord giveth, the Lord taketh away. Blessed be the name of the Lord."

THE LIFE OF CHARLES BOATWRIGHT

I didn't think about wisdom as a model of maturity until I was over sixty-five. I didn't think much about Charles Boatwright, either. When I was doing my work on good and bad outcomes in the 1970s, his marriage was a shambles, and even in his mid-fifties he seemed to me occupationally feckless. At that point in my life I didn't see much to exemplify optimum adult development at that point in his, and I let him slide gently off my radar.

In 2009, however, when I was seventy-five, Monika Ardelt pointed out to me that Charles Boatwright had scored higher than any man in the Grant Study on her measure of wisdom.[19] That got my attention, all right, and I looked at him closely for the first time in years. Boatwright's Decathlon score was 7 out of 10; only 3 percent of Grant Study men scored higher. In fact, late in life he was scoring high on every measure of maturity I had ever devised. Here was yet one more demonstration that my black-and-white, good-or-bad predictions at age forty-seven were unreliable. What on earth had happened?

I thought back to those early years. I'd found Boatwright's file tedious going. His lack of career commitment was one of the reasons I'd scored him as a potential "bad outcome" in my old black-and-white days. It was easy to marshal other evidence of failure, too: a divorce, an estrangement from his daughter, a drifting son. Yet he wrote constantly of his good fortune in leading such a wonderful life, and exclamation points punctuated his enthusiasm.

Here's how he answered a Study questionnaire when he was forty-nine. How had he grown in stature and matured between the ages of twenty and fifty? Between twenty and thirty, Boatwright said, "I learned humility and how to work hard and to dedicate myself to others. I learned to love." Between thirty and forty, "I went to graduate school and matured rapidly in business and in the community. I became an important cog in the community. I was a flaming do-gooder. I learned further to take responsibility." Between forty and fifty, "I feel a marked change has come over me. I have learned to be more kind, and have more empathy. I have learned to be tolerant. I have a much better understanding of life, its meaning and purposes. I've left the church, but in many ways I feel more Christian. I now understand . . . the old, the meek, the hard worker, and most of all children." Reading this in 1974, I was inclined to roll my eyes. I thought I recognized the kind of premature selflessness that cloaks a lack of clear identity—a series of early efforts to deny his own needs and project them onto others—and Boatwright's failures at Intimacy and Career Consolidation, like Camille's, seemed to confirm that. However, I was dead wrong.

Dr. Maren Batalden, the astute internist who did some of the post-retirement interviews for us, had a similar reaction to Boatwright many years later. She visited him when he was seventy-nine. When she asked him about his mood, he gushed, "Optimistic, optimistic. Pollyanna, Pollyanna." Even Boatwright did not seem quite able to believe his own words.

But Batalden changed her mind, which should have alerted me even before Monika Ardelt did. She recounted that while transcribing the tapes of her interview with the seventy-nine-year-old Boatwright, she had felt "critical of his contradictions and his 'methinks the lady doth protest too much' assertions of his good luck." But as she began to write her report, she suddenly found that she had to contradict herself. "In fact, my experience with Boatwright was delightful . . . interesting and interested, gracious, charming, and engaged. I was impressed with his voracious hunger for learning, which obviously keeps him vital. He is, I think, remarkably effective in actually getting what he wants. After fifteen years of gradually progressive discontent in the staid corporate world, he had leaped into debt and boldly returned to the rhythm of life he felt himself most suited to. Without really consolidating a career, he seems to be in a state of unequivocal Generativity."

Still, it wasn't until ten years later that Ardelt's comment made me look again at Boatwright's record. And when I did, I recognized that it wasn't he who had suddenly matured. It was I. I had finally learned that hope and optimism are not emotions to be dismissed lightly; perhaps this was a reflection of some spiritual growth of my own. And while it's easy to scoff at Pollyanna stereotypes, the wisdom of Pollyanna herself—the real young heroine of the book named for her—is nothing to laugh at.[20]

My years of studying adaptive styles had shown me that projection sometimes evolves into altruism. But when? When does a miserable Albanian teenager turn into Mother Teresa? Would Nelson Mandela have served himself better, locked away on Robben Island, by bemoaning his helplessness and planning revenge? Or was he wiser to do just as he did—tell his captors funny jokes while maintaining the invincible hope that they'd "walk hand in hand some day"? It can take a lifetime even to formulate questions like that, and for me, where Charles Boatwright was concerned, it did.

His life illustrates five of the Study's six models for maturation: increasing capacities for working and loving, widening of the social radius, development of mature defenses, attention to the spiritual as well as the material, and growing wisdom. The sixth model, brain maturation, will have to wait for a lifetime study of neuroimaging—a study that probably won't be feasible until close to the end of the twenty-first century.

Boatwright came from a distinguished New England academic family. Both his parents had taught at the college level before assuming more conventional 1920s careers. His father became a stockbroker, and his mother was a homemaker who was active in unpaid social service work. Throughout his life, Boatwright enjoyed a warm and loving relationship with his father, mother, and younger sister, and with a close-knit extended family. By 1940 the family was wealthy and owned three houses, but Boatwright inherited a commitment to social welfare from his mother, and he boasted that his father had "started out from absolutely nothing in the way of a job and worked his way up to the top." The hardworking father was not remote. He made sure that he spent time with his family.

From the very beginning of the Study, Boatwright appeared to be well adjusted. His childhood received high marks from independent raters blind to his future, even though his father would manifest very serious mental illness by the middle of Boatwright's adolescence. When Boatwright was twenty, his mother described him as "very affectionate, sensitive but with a great deal of courage and determination. Gets on remarkably well with both young and old. . . . Has never been a problem in any sense of the word. Good self-control. . . . Always made friends easily but was always able to amuse himself." When he was a child his mother would spank him and put him in a large closet for punishment. When she returned, she would find that he was "entirely content and had usually found something to play with." Sophisticates may scorn optimists—consider the scorn Voltaire heaped upon Dr. Pangloss—but the Grant Study suggests that Martin Seligman's research is right on target. Optimism is far more often a blessing than a curse.[21]

After college, Boatwright was rejected for military service because of poor eyesight. He found employment in the Brooklyn Navy Yard, equipping ships with radar and repairing radar equipment. In the Study questionnaires he described his work with great satisfaction. Then he moved to Vermont as "assistant manager," but really caretaker, of a tree farm owned by his father.

At nineteen Boatwright said of his father, "We do everything together, practically," and of his family, "We all get along beautifully together." This statement was typical of his tendency to see the glass as always half full. From Boatwright's fifteenth year through about his thirty-fifth, his father was racked with manic-depressive illness, and became a very difficult and sometimes cruelly critical man.

Boatwright's marriage was a similar story. He married at twenty-two, and stayed married for thirty years. And he always rated his marriage as happy until his early fifties, when I was deep in my process of classifying the men as good or bad outcomes. That was the year he wrote to the Study that his wife had been very unhappy living in Vermont, that she was in love with an old friend, and that she had decided to leave. "It's all so confining. I feel wasted and unused," he wrote. "I don't have really much love or feeling for her, and that is terribly hard on her too." A year later he was divorced, and I was dismissing him once again as a Pollyanna and a master of denial.

Thirty years later, I have changed my tune. Now I can see that Boatwright remained committed to his wife as long as he could. Even in the throes of divorce he relied on empathy rather than blame: "She is a wonderful person, but so negative with everything." We learned much later that over those last five years she had been increasingly incapacitated by alcoholism, but Boatwright did not reveal this fact until he was over seventy. (In retrospect, his temporary difficulties with his children may have been a result of their mother's alcoholism.) An observer might complain that he was in denial, but he was not blind to the situation. It would be more precise to say that, true to his character, he was taking responsibility for his wife's failings. Resentments, however justified, are rarely a source of happiness, and Boatwright was a natural adherent to the principle that forgiveness is better than revenge.

Similarly, Boatwright stayed close to his father, remaining appreciative and protective of him throughout his illness. On his deathbed, Boatwright's father said to him, "I don't understand why you were always so nice to me." Batalden asked him about that too, and Boatwright replied, "*He* was nice to *me*. He took an interest in me, and everything I did. He meant to be a good man. He really did." For Charles Boatwright, lemons were mostly the *sine qua non* of lemonade. Gratitude, and the mature adaptive coping style of sublimation, came naturally to him. Some Grant Study men went to look after Germany and Japan after the end of World War II; they had to be sent home because their anger broke through their ersatz altruism. But there was nothing ersatz about Boatwright.

While in Vermont, Boatwright had been active in building up a lumber cooperative, a farmer's cooperative, an egg cooperative, and a central high school. He made extra money as journalist, milk deliverer, carpenter/painter, accountant for a filling station, and artificial inseminator of cattle. At that time, the early 1950s, Clark Heath noted that Boatwright had "persistent difficulty establishing a career," but he thought him to be, nevertheless, one of the "most stable and successful men" in the Study. Heath was a very wise man and nearing his own retirement, and this was a paradox that it took the Study (or at least me) a lifetime to resolve.

Now I understand that community-building is a career of its own—one of the really great ones. But when I began with the Grant Study in my thirties, I was too deep into the "selfish" phase of my own career consolidation to understand what Charles Boatwright was about. I could see that he worked hard throughout his life, but as he moved from one job to another, it was hard to tell where his commitment or his competence lay. Clearly he didn't see his work as a career, yet—I realized once I stopped dismissing his optimism—he found meaning and success in whatever he undertook. It took me a long time to understand that the career that Boatwright consolidated was looking after others more needy than himself. Even in college his chief extracurricular activity had been Phillips Brooks House, Harvard's social service organization. His career was not all about him. Pollyanna's wasn't about her, either.

In his fifties, abandoning caution, Boatwright left the corporate world of his post-Vermont forties, and borrowed the money to follow a dream. He bought a boatyard. At the age of fifty-six, he married the widow (and mother of three sons) of his business partner in that venture, who had died suddenly and tragically the year before. According to both spouses, this marriage has been happy for the last thirty-five years. In the 1980 biennial questionnaire, he wrote characteristically of his second marriage, "Her children needed me very badly. So in January 1978 we were married. It has been perfect for me. No one has ever healed me with such love. And in return I have come to love her completely. We have been enormously happy."

Characteristically too, Boatwright devoted himself to his stepsons. "Being a stepfather has an enormous number of problems, but I seemed to have coped well. They all call me Dad. We are a very loving and close family. I'm an enormously lucky fellow." True, Boatwright had told us that he was happy with his first wife the first time around. But this time his second wife, who has had her own private interview, confirms how happy she is with him. When Boatwright was sixty-one, an interviewer asked him what pleased him most about his wife. "He said, 'She loves me,'" the interviewer wrote. "He beamed, and his face lit up."

When he was seventy-nine, Boatwright told Batalden that he and his wife gave more to charity than they should. Most of their giving, he explained, goes into land conservation—that is, preserving the past. Nevertheless, he was in touch with the future. He worked as the town auditor in Stowe, Vermont, where he had a vacation house. This was a volunteer position requiring a month's labor every winter. He also, on a largely volunteer basis, managed the town offices, transitioning them all to the computer.

At eighty-three, Boatwright was still working twenty-eight hours a week: "I'm pushing nonprofits to be all they can be. . . . I'm the guy who says we have to try." The hope of men like Mandela and Boatwright springs eternal. At eighty-five he believed that his most cre-ative activity was "to inspire people to see all sides of a problem"—a hallmark of wisdom.

By the time he was eighty-nine there was no doubt that Boat-wright was an old man. He still exercised two hours a day, but cross-country skis and tennis racquets had been given away; now he con-tented himself with slow walking. He admitted that he felt tired and was plagued with minor ailments: two bad knees, two bad shoulders, shingles, cataracts, and ankle edema. But he still took no medicines and called his health "excellent." When asked what he did now that he hadn't done a decade ago, he growled, "A hell of a lot less!" He had reduced his volunteer work to three hours a week, spending his time instead with his grandchildren, and visiting shut-in and dying friends. The task of Integrity is not to set the world on fire, but to come to terms with reality and maintain one's sense of life in the face of death. At age ninety Charles Boatwright is still very much alive.

Psychologist Laura Carstensen and her Stanford colleagues have documented that in late life emotions often take the place of think-ing.[22] As Boatwright explained it to Batalden, "With age, you acquire more understanding. The things you felt so passionate about when you are young, you learn to let go of. You realize that all those things you thought you were going to be, you ain't. As I have often said, at this stage in life it's not what you've accomplished in a day, but how the day felt."

A questionnaire when the men were about seventy-six asked what they were proudest of, and what they wanted to be remembered for. Boatwright's response: "I don't give a damn if I'm remembered for anything. I've enjoyed my life and had a hell of a good time. I'm more proud of those times I've helped others." He added at eighty-three, "I know I'm a Pollyanna, but it's better than being a pessimistic grouch." Maybe Pollyanna and Aristotle have something in common on the subject of the good life. And maybe the fans of "selfish" genes are selling Mother Nature short.

Boatwright knew how to love and to work. He was capable of a long and happy marriage, and could look tenderly after the well-being of children (and others) in need of his care. He was truly gifted in the mature adaptational device of sublimation, and he seemed to be ever more absorbed, as he grew older, in explicitly spiritual rather than worldly fulfillment. When asked what he had learned from his children, Boatwright replied without hesitation, "Oh my gosh, an infinite amount. Much, much more than they've learned from me, I'm sure. . . . They keep me up to date; they keep me young. I'm infinitely grateful to them for keeping me on the positive side of life." So there's that question again. Are compassion and gratitude Pollyannaish? Or are they the beginning of wisdom? A paradox, a paradox, a most ingenious paradox! But Professor Monika Ardelt, who has been studying wisdom for years, identified him as the wisest man in the Study, and she should know. Certainly her identification, which called him belatedly to my more respectful attention, had a hand in raising my own wisdom quotient, too.

Boatwright's story brings up another aspect of adult development—both mine and the men's. When I was thirty-three, most of the Study men looked depressed to me. In fact they were not clinically depressed, but at age forty-seven they had become comfortable about acknowledging their depressed emotions. I had not. Yet. Age has long been observed to be a factor in emotional experience, even outside of the context of healthy development. In manic-depressive psychosis, for instance, mania often dominates the picture in the years between twenty and thirty, while between forty and fifty the depressive component is more prominent. Delinquents and addicts tend to be more able to admit previously denied depression into consciousness after the age of forty, feeling it instead of acting it out or projecting their emotional pain. The same pattern occurred with the maturation of normal men in the Grant Study.

Their capacity to tolerate more conscious depression in midlife than earlier in their lives (or than I could as a still-young man) meant that they were also less incapacitated by the frustrations of life, their own and others'. What had happened to permit such changes? When I was forty, I attributed them to maturing defense mechanisms. Later I learned (from Washington University developmentalist Jane Loevinger, who collaborated with the Study for a time) that the ability to pull differentiated emotions up into awareness is another marker of ego development.[23] Advances in brain science now suggest that the biological capacity to bring emotional valence into consciousness matures as the brain's tracts become more efficiently insulated (better myelinated) with increasing age.[24] In fact, what we understand as maturation depends, in part, on this fact of brain physiology, which allows better integration of the "emotional" subcortical brain with the "planful" frontal cortex. Is there any reason to dismiss any of these considerations? I don't think so. I grow older and wiser, and science does, too.

Over the last thirty years there has also developed an increasing awareness that people become less—not more—depressed between fifty and eighty.[25] Selective attrition probably accounts for some of the declining prevalence of depression, but it is also due in part to what Laura Carstensen has called *socioemotional selectivity:* that is, the tendency of the old to remember the pleasant in favor of the unpleasant.[26] The lives of the Grant Study men support that theory. If Adam Newman had remained twenty years old all his life, he would have been a basket case at the end of it.

WHEN PEOPLE DON'T GROW

What happens, however, when people *don't* accomplish Erikson's life tasks? What happens to adults who, as they grow older, fail at work and at love, or who are trapped in unempathic, isolating coping styles?

Peter Penn was one of those. He was married for almost forty-five years; he was a tenured professor of English; he published a book in his field. And yet he never really entered the world of adulthood; in most ways he never left home. He didn't meet the criteria for depression, but there is no evidence that he ever felt joy. This is his story. It's a sad one. Developmental failure is always sad.

Penn was a frightened little boy. Until he was seven, he couldn't fall asleep without his mother in his room. He was very inhibited; the closest to profanity that he ever came, by his mother's report, was a slip of paper in his high-school knickerbockers that said, "Gosh dang it." His mother actually confronted him about this, and he explained that he had been furious at his teacher, and had written the words to dissipate his rage.

His childhood was very bleak. His mother was a nervous worrier, and his father a very distant man. Nevertheless Peter was president both of his church group and of his seventh-grade class. That was perhaps his finest hour. Once puberty began, his emotional growth ceased. I have no good explanation for this. One colleague has suggested undiscovered abuse, and another has wondered if perhaps Penn was a closeted homosexual. But there's no evidence for either of these suppositions; they are facile speculations and they tell us less about Penn than about how easy—and tempting—it is to manufacture theories after the fact.

There was nothing the matter with his intelligence. Fascinated by

religion and by history, he majored in American history and literature—an elite major. His father had attended business college, and his
mother was a high school graduate, but neither cared much for reading. Still, Penn planned to become an English professor.

In college, he seemed to lack a sense of identity and even a life
narrative. He could not describe his relationships with his parents;
when asked to describe himself, he could only tell stories, and they
mostly didn't involve him. Interviewers wrote, "He was pleasant,
cheerful and extremely boring," and "He was passive in the way a
sponge is passive." He remained that way as an adult—passive and very
dependent. The zest for life that makes our adolescent children want
to leave even loving parental homes was not in Peter Penn's reper-
toire.

Lewise Gregory described the sophomore Penn as "ponderous
and lumbering." His college life was barren. The literary magazine
The Advocate didn't offer him a position on the staff. He did no dating
because, he explained, he was "too busy" and had "no car or money."
He didn't like dances, and had few friends. He did not take part in
sports. Like Sam Lovelace he had a high resting pulse rate. His only
activity was the Glee Club.

Penn won writing prizes and graduated *magna cum laude;* even so,
Harvard's Graduate School of Arts and Sciences turned him down.
He was one of only six Grant men who finished the war as a private,
and he returned from World War II with a good conduct medal.

After the war, he returned to his hometown to pursue a Ph.D. in
English. At thirty he confessed, "It is so easy to live at home and eat
Mother's cooking." Like many second-year graduate students, he became a teaching assistant in a freshman basic writing course. The
scholarly essays that he submitted to academic journals were not accepted. He wasn't mentally ill. He never saw a psychiatrist; he took no

mood-altering drugs. He just remained essentially what he had been in seventh grade—a good and not very imaginative seventh-grader.

He was unadventurous in other ways, too. At twenty-nine he explained, "Some honest doubts, unpropitious circumstances, and perhaps a bachelor's wariness have prevented me from facing up to the question of marriage. . . . My relations with women are more satisfying socially than sexually. . . . I can rationalize my single state by pleading the comfort and security of life at home." At thirty-two he had still never had a serious girlfriend. His mother was the one with whom he talked about personal problems. At age thirty-five, his thesis still incomplete, he had to leave his university for a community college, and he stopped answering questionnaires until his twenty-fifth reunion year.

When Penn began returning questionnaires again at age forty-seven, the Study learned that he had finally received his Ph.D. at thirty-seven, and had married the same year, still a virgin. His wife was five years older than he. Penn acknowledged that they fought all the time: "Her attacks on me are so savage they reduce me to tears. . . . I think I love my wife." Ouch. Despite all his years of marriage, the Study never gave Penn credit for completing the task of Intimacy. Although he and his wife remained stably together, there was no evidence that Penn enjoyed his marriage any more than he did his students or anything else, and he spoke about his wife as if he had finally married the mother he had lived with for so long. (His mother died when he was forty-three.)

The acquisition of his doctoral degree did not add to Penn's excitement about his career. When he took early retirement thirty years later, he recognized to his dismay that he had spent his whole teaching life in almost exactly the same entry-level job—teaching remedial writing. He was unhappy at work, and never in his life did he express any pleasure in teaching. What Penn liked best about his teaching job was its "tenure and security." He preferred small classes because they were less work.

During his fifties, Penn took a lot of sick days. He was hospitalized three times, but no problems were ever discovered. He published his thesis as a book. It sold a few copies and was then remaindered; Penn bought fifty copies himself. It received no attention and has long been out of print.

Penn retired early, at sixty-three. His most frequent daydream was that he might yet do something of importance. But what? He had no interests or hobbies, only hope. He still remembered his pre-teen days as the happiest of his life. His sister lived in an adjoining state, but they hadn't seen each other for two years, and he hadn't seen his best friend in three. When I visited Penn for his retirement interview at sixty-five, he rarely made eye contact. He lectured to me impersonally about this and that, but pedantically and without charm. His social skills had apparently not improved since he was a sophomore.

In setting up the interview, Penn had told me that he needed to ask "Mrs. Penn" if I could come to visit. "We live between frustration and despair," Agatha Penn quipped as she (reluctantly, I thought) opened the door to me, and she pointed to two prints dedicated to those painful topics hanging on their living room wall. She didn't laugh, however, and she avoided me for the rest of the time I was there, unlike many of the wives of my experience who—out of friendliness or curiosity or both—made their presence known from time to time with coffee or cookies.

Penn told me that he had retired out of "disillusionment." After years of teaching remedial English in an inner-city commuter college, "I got a little fed up."

"I have a very good singing voice," Penn told me, "but since I married, I have kept my voice on ice." His wife had been jealous of his accompanist. It was also she, it seemed, who had forbade him to return many of the Study questionnaires.

When Professor Peter Penn died at eighty-one of cancer, he had been happier for the first half of his life, living with his mother, than he had ever managed to be in the second half. He always worked hard, and like an obedient Boy Scout or soldier he never did anything to endanger his good-conduct awards. He did not abuse alcohol or cigarettes. He stayed married for forty-four years and spent his life teaching disadvantaged college kids. But his first year of junior high school was the high point of his life, and the closest that he ever came to Identity. What happened? Penn's college adjustment was A. There was never a hint of depression, either in his heredity or in his life. His ancestral longevity was in the top 15 percent of the College sample. Yet his score on the Decathlon was 1. The only clue—and the clue itself is a mystery, as I'll discuss in Chapter 10—is that his maternal grandfather died very young. Despite superior ambition, excellent verbal skills, a fine singing voice, a deep wish to be a scholar, and the possession of 8,000 books, he could find no purpose or meaning in his life, professionally or otherwise. He just never grew up. A tragedy. Reviewers have called me heartless and unempathic for publishing such an unhappy tale. But I offer it not out of lack of compassion, but to show convincingly how tragic developmental failures really are.

For a long time that was the end of Peter Penn's story, as far as I knew it. But in April 2012, as I was readying this manuscript for the publisher, I discovered by chance that Penn had written a series of poems to his wife, and that she had underwritten their publication many years after his death. I raced home with the book, wondering (hopefully, this time) if continued follow-up would once again prove me wrong. Alas, it did not. These poems were loving, but they might have been written by a sixteen-year-old with a stiff upper lip. After ten years of marriage he had said, "Her attacks on me are so savage they reduce me to tears. . . . I think I love my wife." And by the book's testimony, little had changed. For forty-four years Penn wrote ritual poems to his wife on birthdays, anniversaries, holidays, and Valentine's Day. They say nothing about the relationship between them, only about his thoughts and wishes, and his unquenchable hope that if he just said the word "love" often enough it might come true. I was moved by his hope and by evidence of this unexpected mastery of sublimation as a coping style. But hope is not always enough. Nelson Mandela and Charles Boatwright used indomitable hope to get out of prison. Peter Penn used it to bear an endless incarceration.

THE LIFE OF BILL DIMAGGIO

Consider the contrasting life of one of the men of the Inner City cohort, Bill DiMaggio. He came from an "underclass" family; as a child he had shared a bed with his brother in an apartment without central heating. His father, a laborer, had been disabled since DiMaggio was a teenager, and his mother died when he was sixteen. With a Wechsler Bellevue IQ of 82 and a Stanford reading IQ of 71, Bill completed ten grades of school with difficulty.

Nevertheless, at age fifty, Bill DiMaggio was a charming, responsible, and committed man. If he was short and noticeably overweight around the midsection, he still retained plenty of youthful vigor. His face was expressive; there was a twinkle in his eyes; and he had a good sense of humor and a healthy appetite for conversation. He maintained eye contact easily and answered questions directly and frankly. He told his interviewer that by consenting to be a continuing part of the Study, he felt that he was contributing something to other people; he felt that this "little thing" was important.

For the first fifteen years of his adult life, DiMaggio worked as a laborer for the Massachusetts Department of Public Works. When a position opened for a carpenter, he got it by seniority. He didn't have any carpentry skills, but he learned them on the job. "I like working with my hands," he said. Now he took special pride in his role in maintaining some of Boston's historic and antique municipal buildings.

Asked how he handled problems with people at work, DiMaggio replied, "I'm the shop steward in the union, so they lay off me. I'll stand up to them if I feel I'm right." If he thought that a job was dangerous, for example, he would not allow his men to work on it. Under union rules, management had to listen to him, and he had learned to speak with authority.

His bosses had been trying to get along better with him over the last year, too, DiMaggio explained. He was one of the few really experienced men on the job, and they were depending on him more. Management also depended on his experience to help teach other men; DiMaggio had a lot of responsibility. (The Study has found that after the age of 40, IQ as measured by a school-oriented test like the Wechsler Bellevue does not count for much, and DiMaggio was an excellent illustration of this.) But, he continued, "It's only a job. I'm more concerned about my wife and kids. Once I leave work I forget about it."

He was in fact very interested in family matters. He described taking his sons on fishing and other trips as they were growing up. "We spent a lot of time together." His own father—chronically un employed—had never taken him fishing or gone out with him much.

Asked what his greatest problem had been with his children's growing up, DiMaggio replied with a smile, "Do you have about six months?" But he added seriously, "Being worried about drugs." He knew that his kids smoked marijuana; and he said he and his wife could accept that. But they would not let them smoke in their house. They accepted the fact that their youngest son had moved in with his girlfriend, and they made no moral judgments about it. They chalked it up to his being "very immature," and felt that he would become more mature as time went on. One vital ingredient of Generativity is hope, but hope is only possible if one's mind can encompass the concept of development. Bill DiMaggio's greatest ambition in the next ten years was to see his children living independently. The capacity to accept a generative balance between care and letting go requires a lot of maturity of one's own.

DiMaggio belonged to the Sons of Italy. He was quite active in that organization, and he regularly helped out with Bingo night, which meant running games for all the women who came on Wednesdays to play. He and another member, a friend, regularly cooked for the club's Saturday morning lunches. DiMaggio enjoyed that; he liked people being happy with his food. Through the Sons of Italy, he also did volunteer community work. On the Fourth of July he helped host a big party for the kids in the neighborhood, with games and refreshments, and he was active in various other club activities for children throughout the year.

DiMaggio and his wife had signed up to work for a candidate who was challenging the old-guard city boss for mayor. And he was active in the "Council of Organizations," an umbrella organization for the charitable clubs in Boston's North End. This socially and intellectually limited schoolboy had matured into a leader of leaders, a wise man, and a Guardian.

You don't need a Harvard degree, or even a high IQ, to grow up into a *mensch,* to do your bit for the next generation, to work in community building. Nor do career and family commitments have to conflict. Sadly, DiMaggio died young from a heart attack.

CATERPILLARS AND BUTTERFLIES

In 1980, when I was forty-five, I wrote with presumptuous authority that maturity is not a moral imperative.[27] At that time, my motto for

the Grant Study was that life is a journey, not a footrace, and that butterflies are neither better nor healthier than caterpillars. The men had
to develop for three more decades, and I had to gather a lot more data
and experience of life before I could contradict in my late seventies
what I believed twenty years before. But now I do. Peter Penn and the
other men like him endured lives of desperation, whether quiet or
not. The Terman women who failed to master Generativity were only
one third as likely to be orgasmic as those who had. There is a time for
being a caterpillar, but it is brief. Psychosocial maturity may not be a
moral imperative, but its lack is very painful.

And it is a matter of life or death. In 2011, only four of the 31
Grant Study men who failed to mature past Erikson's stage of Intimacy are still alive. Fifty of the 128 men who reached Generativity
are—a very significant difference. In fact, the Generative died eight
years later on average than those who never mastered Career Consolidation. The men who achieved Generativity were three times as likely
to be enjoying their lives at eighty-five than men whose lives were
still centered on themselves. Adult development is not just a descrip-
tive convention; it is part of healthy growth.

Furthermore, much of the joy in adult life grows out of these developmental accomplishments. At least for these men, to fail at Intimacy implied other painful lacks. In the Inner City cohort, two-thirds
of the never-married were in the bottom fifth in overall social relations, 57 percent were in the bottom fifth in income, and 71 percent
were classified by the Study raters as mentally ill.

Some men took their developmental steps late. Occasionally this
turned their lives around sufficiently that they lived long and died
content. Certainly late is better than never, and many of the advantages that the Cherished enjoy in young adulthood can become available to the Loveless grown old—if they can learn in the meantime
how to find love. Still, these men spent lonely and sometimes wretched

years in the waiting. And although there's no statute of limitations on development, some specific opportunities—to have children, for instance—do not wait forever.

DEVELOPMENT THWARTED: THE LIFE OF ALGERNON YOUNG

Maturation is not an inevitable byproduct of aging; it can be derailed. Drought can blight ripening wheat; corkage can destroy a fine Bordeaux; shin splints can turn a Derby prospect into a shambling nag. An organic insult to the brain can destroy or reverse the normal maturational process and leave an individual an insecure youth forever. Major depression, alcoholism, and Alzheimer's disease are the most common culprits. But they aren't the only ones. A certain amount of good luck is involved in growing old without accident, disease, or social catastrophe. The more we know about ideal development, the more easily we can recognize the forces that hinder or disrupt it, and possibly counteract them.

The life of Algernon Young is a cautionary tale, a sharp reminder of the vulnerability of developmental processes and of the pain of an undeveloped life. Young was a gifted man, well loved, from a family that the original Study staff had ranked as upper class. (Judge Holmes's family had been ranked as upper-middle.) Young started out with everything going for him. He was not an alcoholic. Yet his outcome was poor, and much of his life was misery. He is one of the Study's best rebuttals to the argument that Oliver Holmes's money ensured his success.

Young's mother grew up socially privileged, and her father was a sixth-generation Harvard alum. His father was a Harvard grad, and through Algernon's childhood he was the headmaster of a Denver prep school.

Algernon's mother depicted a cheerful and competent childhood.

Her son was "a grown man when he was two years old," she said—a haunting description, given how things turned out. "The children adore each other," she told Miss Gregory, "and have always gotten on well." Algernon attended his father's school, where he did well academically and was one of the leaders of the student body.

His intellectual gifts were as great as any man's in the Grant Study, and like Peter Penn he received an A on psychological soundness from his college raters. Miss Gregory thought he had "good social ability," and so did the Study psychiatrist, who judged him "well-adjusted socially." His college summary noted that "he has no conflict with his parents. . . . His early life was happy and, for the most part, he was a member of the group." Only Clark Heath saw him as immature; he did not say how.

But when Young was twenty-nine, the Study anthropologist noticed that he was still "closely tied to his mother, unwilling to make new associations." From thirty to thirty-seven he dated a woman who was a heavy drinker; he proposed to her, but she turned him down. At forty-two he finally got married. His new wife was still attached to a high-school sweetheart of whom her parents had disapproved. Three years after her wedding to Algernon her mother died, and she ran off with her old love. At forty-nine Young was single again, and living just a few blocks from his parents. His life revolved around his pets, who kept him "too busy" for other relationships. "Catering to six cats can be a big affair," he explained. He found engagement with other people demanding and frightening; when he had to move outside his tiny circle of work and family (and sometimes within it, too) he deployed the coping style of an obsessional: isolation of affect, reaction formation, and displacement. His social life consisted of waving to fellow commuters in their cars.

Young's work life was no more satisfying than his personal life. At twenty-nine he wrote, "The brevity of my answers leads me to believe that my life must have been pretty substandard. It still may be. . . . My fondness for my current work has grown less. I seem to be stuck in a rut."

When he was in college, Young had wanted to be an automotive engineer. But he never rose above a low-paying position in a Denver heating and plumbing firm. In thirty-five years there, his greatest responsibility—installing furnaces—never changed. He took pleasure in his job only because he could build things, which had been one of the pleasures of his childhood. He told one interviewer in detail and with real enthusiasm about furnaces, but he communicated no pride in or commitment to his work role, and no sense of connection to the people he worked with.

At forty-six, when classmates far less privileged and intellectually gifted than he were firmly established in the upper middle class, Young wrote, "I feel deeply inadequate. . . . I have always felt that I could never sell myself." At fifty, he remained unpromoted and discontented. He was still working twelve hours a day, and often on Saturdays. His work was "the same old rat race," and his earned income was the lowest in the Study. At sixty, after thirty-five years at it, he said that what he liked best about his job was "the good benefits and the lack of stress."

For a very long time Young did not involve himself in any community service—or indeed, in any community. His philosophical rationale for this at fifty was, "I know I can't be my brother's keeper," but it appeared to me that in fact he felt too perpetually out of control in his own life to feel safe extending his attention beyond it. This of course reduced even further any chance of his establishing supportive relationships. He was asked on a questionnaire what contributions he had made that would benefit others. "If there is one, I can't imagine what it would be," he responded. In mid-adulthood, he was not advancing in his job, and he was too overwhelmed to emerge from self-absorption toward Generativity.

I met Algernon Young when he was forty-seven, two years after his wife had left him. I was struck by the contrast between his Brooks Brothers tweed jacket with the stylish leather patches at the elbows and his cheap shoes and workingman's hands. Like Heath, I felt that he was immature. He was one of only two Grant Study men who felt continuously overwhelmed by bills, and he told me that he ate alone in diners. He acted like a deferential prep-school boy being interviewed by a teacher, while in fact he was fourteen years my senior.

What stunted the development of Algernon Young after such a promising adolescence? The answer seems to have been a one-two punch of tragic fate. When Young was eleven, his mother was psychiatrically hospitalized for crippling anxiety. Her symptoms turned out to be due to hyperthyroidism, and after thyroid surgery they never recurred. But her temporary loss of control was enough of a shock to Algernon that he quit believing in God. And then, when he was halfway through college, the other shoe dropped.

This time it was his father who was hospitalized, and the circumstances were terrible. After twenty-five years of loyal service to his school, he was fired, and a whiff of scandal attended the firing, concerning a possible mismanagement of funds. The facts were that a major depression had compromised his ability to handle the complex responsibilities of his office.

Like his mother, Algernon's father recovered from his depression and obtained a prestigious job at another school. But the son never recovered. For the rest of his life he felt (as Willy said about the death of his father in *Death of a Salesman*) "kind of temporary about myself." Algernon Young seemed to have lost all faith in everything, including himself. As gifted in math as he was, the moment his father began having trouble handling the school finances, he failed algebra. Then he dropped out of Harvard. He took a job in a factory to support his family, and never returned to college. All of his positive college rat-

ings were made before the Study accepted that he was never coming back.

In our forty years of follow-up we saw no evidence of major depressive disorder in Young himself. Nor did he ever seriously abuse
alcohol. He just kept away from people and confined himself to the
rational predictability of furnaces. His parents' sequential abandon-
ments under such frightening circumstances seem to have left him
trapped in a fearful need; he couldn't trust them any longer, yet he
wasn't ready to let them go enough to form new relationships where
trust was possible. He tried in his first marriage, only to be abandoned
again. At age forty-nine, Young could still say, like a ten-year-old boy,
"My main interest is in things mechanical."

It is not only changes in ourselves that drive us to essay new roles;
our interactions with others transform us too. But maturational transformations take place from within. They are the fruit of internalization and identification, not of instruction or even socialization. They
happen only when we can metabolize, as it were, those influential
others, taking them into ourselves in a profoundly intimate and struc-
tural way. There's a world of difference between knowing something
in one's head and knowing it in one's gut. When internalization does
not take place, one of our main avenues of growth is foiled. As a child
Algernon Young was given plenty of love, but after the twin catastro-
phes of his youth—which, however innocent and inadvertent, must
have felt to him as twin failures of the two most important people
in his life—he seems to have developed an emotional malabsorption
syndrome—he could no longer take anything in. His identificatory
capacity gave out in his mid-teens, and so did his growing. Young did
not have Asperger's syndrome, but the jury is still out on the other
possible causes of his failure in maturation. It appears that he stopped
being able to make use of the love that came his way; contrast his life
course with that of Godfrey Camille, who was given so little love at

the beginning that even the usually nurturing Grant Study staff dismissed him as a "regular psychoneurotic." Yet Camille left no stone unturned until he found the love he needed, and then he absorbed it greedily.

When Young was fifty, one of his very few friends died. He told the Study that he had lost a piece of himself, and added, "Do not ask for whom the bell tolls; it tolls for thee." He was addressing a question that more usually preoccupies those who survive into old age: How, over the long years of the adult lifespan, are losses replaced? How do we face the losses that we can never replace? But Young wasn't old. He was only fifty, and he had stopped growing at eighteen.

In his later years Young's life improved somewhat, and by the time he died he had technically met my criteria for Intimacy and partial Career Consolidation. His renascence began at fifty-one, when he gave up the agnosticism he had maintained since the early days of his mother's troubles and rejoined his parents' church. Probably through the church, Young met and married a widow with two children. When his mother died three years later he coped well, despite his prior dependence on her. His second marriage endured until he died at sixty-six, but it was the partners' mutual involvement with their church that gave it substance. Study men with richer professional lives and more grandchildren often grew away from the church. But men who were needful of connection late in life sometimes sought it in religious affiliation.[28] Looking back, it appeared at first as though Young was returning to the point of crisis, allowing his mother and father to become sources of identification once more, and resuming, slowly, his own truncated development. Certainly the death of parents may release fresh growth in adult life. But even that hope would not be fully realized for the timid Algernon Young.

While Camille, newly returned to his church, gave communion to housebound parishioners and befriended the entire congregation,

Young headed straight for the account books and became the (unpaid) church treasurer. At fifty-eight, as a born-again evangelical Baptist, he wrote the Study, "My God has supplied my needs material and otherwise." At sixty-four he told the Study, "My labor of love to the Lord is being treasurer for the First Baptist Church," which, he explained, was his greatest interest in life. He was still primarily interested in "mechanical" things. By the time he died at sixty-six, he could finally write, "I am a success with my family." Camille at the end of his life fell in the top third of the Decathlon, but Young, like Lovelace, received a 0. Admittedly, Camille had had fourteen more years in which to develop.

THE LAST YEARS

At age seventy-five, the College men were asked to define what they considered wisdom to be. Here are some of their definitions.

"Empathy through which one must synthesize both care and justice."

"Tolerance and a capacity to appreciate paradox and irony even as one learns to manage uncertainty."

"A seamless integration of affect and cognition."

"Self-awareness combined with an absence of self-absorption."

"The capacity to 'hear' what others say."

We also asked these seventy-five-year-old parents of the Sixties Generation a provocative question: "Taboos on obscenity, nudity, premarital sex, homosexuality, and pornography seem to be dead or dying. Do you believe this is good or bad?" We often received black and white answers. But one Episcopalian minister wrote: "NEITHER. What human beings need are limits to their behavior and freedom to realize their true selves—we really need a societal consensus on limits balanced with freedoms. I think these limits and freedoms and the balance between them change with the culture." He had put aside absolute convictions about faith, morality, and authority in favor of a new appreciation of their relativity and mutability.

But remember, he was seventy-five when he wrote that reply. In his younger years, according to his daughters, he had been more judgmental. As we mature, we learn from experience how important a dimension Time is, and how profoundly it determines the shape of our reality. As we understand the relativities and complexities of life more deeply, the immature need to believe becomes a mature capacity to trust, and religious ideology makes room for spiritual empathy.

Students of late-life development, like Jane Loevinger at Washington University in St. Louis and Paul Baltes at the Max Planck Institute in Berlin (who was a pioneer in wisdom research), share views very similar to that minister's.[29] Loevinger understands the most mature phase of adult development (which she calls *Stage 6* or *Integrated*) to be characterized by tolerance of ambiguity, reconciliation of inner conflict, and the ability to cherish another's individuality while respecting interdependence.[30] For Baltes, it was "an awareness that all judgments are a function of, and are relative to, a given cultural and personal value system.[31]

Sociologist and colleague Monika Ardelt, who brought Charles Boatwright to my attention as the wisest of the Grant Study men, has spent her professional life trying to operationalize Baltes's views.[32] Ardelt examined personality inventories that the men had taken in 1972 and 2000 (see Chapter 3), and derived from the results three components that she considers intrinsic to wisdom and essential to it.[33] These were: the cognitive capacity to grasp the deeper significance (both positive and negative) of transient phenomena; the reflective capacity to consider issues from multiple perspectives; and the affective capacity to care deeply about the well-being of others.[34] Thus measured, wisdom correlated with maturity of defenses and absence of mental illness at midlife, with close friendships, and with late-life adjustment.

CONCLUSION

Ultimately, to age successfully is to transcend decay. The task of Integrity is to retain human dignity despite the ravages of mortality. This is not a task particular to old age; it is the developmental challenge of all those who face imminent death. At fifty-two, Grant Study member Dr. Eric Carey knew already that he was doomed to an early death from complications of polio. From his wheelchair he articulated the challenge that faced him: "The frustration of seeing what needs to be done and how to do it but being unable to carry it out because of physical limitations . . . has been one of the daily pervading problems of my life in the last four years." But three years later, he had answered his own challenge: "I have coped . . . by limiting my activities (occupational and social) to the essential ones and the ones that are within the scope of my abilities." To be able to honor life's essentials and simultaneously bow to its realities is Integrity in a nutshell.

At fifty-seven, Dr. Carey told the Study that the last five years had been the happiest of his life: "I came to a new sense of fruition and peace with self, wife and children." He spoke of peace, and his actions portrayed it. He understood that, whenever possible, legacies (both concrete and metaphoric) should be bestowed before death.

At sixty-two, he talked about the risky anesthesia he had recently required for an operation. "Every group gives percentages for people who will die: one out of three will get cancer, one out of five will get heart disease, but in reality one out of one will die. Everybody is mortal." He was dead a year later from pulmonary insufficiency.

As the Study member I quoted above pointed out, it's the old who can teach us that life is worthwhile "to its very end." That is a lesson that it took me, young and arrogant maven of adult development that I once was, thirty years to learn.

6 | MARRIAGE

If you have someone who loves you, you've got it made.

—CHARLES BOATWRIGHT, Study Interview

MENTAL HEALTH AND THE CAPACITY to love are linked, but the linkages are elusive. We can't weigh love on a scale, or examine it with special lenses. Poets can encompass it up to a point, but for most of us, psychologists and psychiatrists included, it's something of a mystery. The importance of intimate, warm, mutual attachment (not just sex, and not even the biological/instinctual drive often called *Eros*) is the third lesson of the Grant Study. But no aspect of human behavior is assessed more subjectively, or measured less easily, than intimacy.

Fortunately this doesn't stop us from enjoying love anyway—not only in its passionate aspects, but also in the enduring warmth and comfort of close relationship. Here, for instance, is Charles Boatwright at age eighty-five on the pleasures of his second marriage: "Really just being together. Share each other's lives and our children's lives. Snuggle on cold nights." Jim Hart, who at eighty-one participated with his wife, Julia, in director Robert Waldinger's Study of Marital Intimacy, told the Study that she was the essence of his life, and called their relationship "a lovely, lovely partnership." What are his hopes for his marriage? "I want it to stay like it is," Jim says. "Period. It can't get better." Julia's view of it? They're best friends, and there's a physical relationship, if not quite what it was when they were young. But the main thing is: "I adore him. More than I ever did. We laugh a lot. We laugh at ourselves. . . . You can't take yourself too seriously. . . . I don't know how we got here, but it's wonderful."

This kind of pleasure in the company of another person is quite different from Eriksonian Intimacy as we dealt with it in the last chapter, and throughout this chapter I'll be making a distinction between the two. Eriksonian Intimacy, like puberty, is a developmental task. It comes later to some people than others, but most of us get there. As fledglings fly from the parental nest, we all must leave our parents' homes and establish ourselves emotionally in the peer world, sharing space, money, decisions, plans, and other issues of mutual interdependence. The Harvard Study of Adult Development defined Eriksonian Intimacy operationally as ten years of living in an interdependent and committed relationship. But that commitment can take very different forms.

Eriksonian Intimacy is an intimacy of physical and practical proximity. The emotional intimacy of deep relatedness is different. Some couples have a shared emotional economy. There's a constant circulation between them; they're under each other's skin, and happy to have it so. This isn't the blurred boundary of codependency; it's a mutuality based on a clear sense of self and other. But it is a knack, like the huge hand-span of a great pianist. Not everyone has it or wants it, and it is not necessary for a fulfilling life.

This chapter will address four questions. What does emotional intimacy look like? What can we learn from marriages that have endured for fifty years or more? What can we learn from marriages that don't endure? And what do intimacy and mental health have to do with each other?

Investigation of the Grant Study marriages was an eye-opener for me. Although I stand by my rash assertion to *The Atlantic* that relationship (that is, the capacity for loving attachment) is what matters most in life, I can't quite say the same for this pronouncement from my first book: "In the Grant Study, there was probably no single longitudinal variable that predicted mental health as clearly as a man's capacity to remain happily married over time."[1] In 1977, I firmly believed that divorce boded statistically ill for future development and future happiness. But it has since become clear that this was yet another premature conclusion.

Before I proceed, here in Table 6.1 is a summary of the Grant Study men's marital histories. We'll be referring to it as we go on.

As you can see, these numbers include all marriages either through 2010, or until one of the partners died. One hundred and seventy-three of the men's first marriages remained intact, including fifty-one happy ones, seventy-three average ones (which we called "so-so"), and forty-nine unhappy ones. Seventy-four men, including twenty-three of the sixty-two who once divorced, contracted very happy remarriages that have stayed that way through 2010, or through the death of

Table 6.1   Life Course of Study Marriages (1940 through 2010, or until the death of one partner)

MARITAL CATEGORY	N=237*	PERCENT
Still married; very happy intact first marriage**	51	21%
Divorced but happily remarried	23	10%
Still in a so-so intact first marriage	73	30%
Still in a poor intact first marriage	49	20%
Divorced; now single or unhappily remarried	39	16%
Never married	7	3%
Total	242	100%

*26 of the 268 original Study members were excluded from Table 6.1. Four of them died in the war, three never married, and 19 withdrew from the Study.

**One man who lived happily all his life with another man in a close interdependent relationship was classified as "Still married; very happy."

a partner. The thirty-seven remaining divorced men either did not remarry or remarried unhappily.

The mean length of marriage for the surviving couples still in their first and only marriage was over sixty years. That's a nice statistic, if not a surprising one. But the mean length of remarriage for the twenty-three divorced but happily remarried men was almost thirty-five years, a finding that required me, as I will recount in a moment, to rethink *de novo* my assumptions about divorce, mental health, and the capacity for intimacy.

RATING MARRIAGES

How did we know who was happily married and who wasn't? The short answer is: the men told us, repeatedly. Cynics may wonder how many of them were Pollyannas like Boatwright, or possibly even out-right liars. But we had decades of prospective follow-up in our files; there was plenty of objective information against which to test their subjective reports. I don't contend that men in the Study never lied to us or to themselves about their marriages, but it is hard to maintain deception for fifty years and more.

Where did our information come from? Qualitative responses to questionnaires and interviews were a major source. The men's wives were sent questionnaires, too. Most couples were interviewed together at about age thirty. After that there were no planned couple interviews (although there were some adventitious ones) until Study director Robert Waldinger began interviewing the surviving men and their wives together and separately on videotape. The recordings—another forward-looking investment—are being saved for future study.[2] In scoring responses, we always used multiple independent raters, each blind to all other information about the men. That way we avoided the halo effects that can plague longitudinal studies, where raters' assessments are contaminated by what they know about successes or failures in the past.

A second source was multiple-choice questions. To quantify elusive intuitions about marital satisfaction, we asked husbands to complete a simple multiple-choice scale on five questionnaires between their thirty-fifth and seventieth years. Wives did the same three times, between the ages of forty-five and sixty-five. The four questions whose answers correlated most highly with marital satisfaction were:[3]

1. Solutions to disagreements generally come: 1=easily, 2=moderately hard, 3=always difficult, 4=we go on without a solution

2. How stable do you think your marriage is? 1=quite stable, 2=some minor weaknesses, 3=moderate weaknesses, 4=major weaknesses, 5=not stable

3. Sexual adjustment is, on the whole: 1=very satisfying, 2=satisfying, 3=at times not as good as wished, 4=rather poor

4. Separation or divorce has been considered: 1=never, 2=only casually, 3=seriously

The scores on these four questions were summed into a global assessment of marital adjustment: the lower the score, the happier the marriage.

The men's own subjective ratings were a third source of data; three times between the ages of seventy and ninety they rated their marriages on a scale of 1 (very unhappy) to 6 (very happy).

At about sixty, the husbands and wives filled out a chart in which they retrospectively rated their marriages in five-year chunks as:

1=Very enjoyable, 2=Not one of the very best periods, 3=Rocky, 4=Divorce considered.

Our final way of assessing the Study marriages was to look for comments the men made about them outside of specifically marriage-related contexts. This allowed us to test yet again the reliability of their multiple-choice statements. Here are some spontaneous comments made by men who scored their marriages poorly:

"She has an inferiority complex."

"I am more affectionate than her."

"She likes her beer."

"It's easier to suffer with her than without her."

"We sleep in separate rooms."

"When she throws the plates, I catch them. I never throw them back. When she hits me, I never hit her back." (This man added, however, "Although I've slapped her to bring her to her senses"—a good example of a fifty-plus-year marriage that was scored as chronically unhappy.)

The following quotes were characteristic of happy marriages:

"My wife is the kindest and most considerate person I have ever known."

"Our marriage is completely challenging, completely exciting."

"Tennis doubles with my wife is my greatest enjoyment."

"I am very proud of her."

"I love and admire her; she is my best friend."

"Our marriage is GREAT. My wife has been the best thing
that ever happened to me."

A dichotomy between good and bad marriages may be hard to
establish when feelings are looked at moment to moment, but it is real
enough over time.

There were two areas in which our data were weak. The men
were loath to return questionnaires that inquired too specifically into
their sex lives. Too much curiosity on our part was firmly ignored, and
we had to settle for global generalizations. More on this in a bit. Also,
the worse the marriage, the less information was forthcoming about it
from either husband or wife, which limits my capacity to illustrate a
very troubled marriage in a life study. Agatha Penn, for instance, declined to return any of the questionnaires sent to her, or otherwise
participate in Study contacts. She also discouraged her husband from
responding to inquiries sent to him.

A THIRTY-YEAR CHALLENGE

In 1977 I handed in the manuscript of *Adaptation to Life*. My editor at
Little, Brown, Lewellen Howland, took issue with my contention that
divorce was a serious indicator of poor mental health, and suggested
gently, "George, it is not that divorce is bad; it is that loving people for
a long time is good." I liked his sentiment, but I didn't believe him,
despite the fact that I myself was in a happy second marriage at age
forty. (We're all the exceptions to our own rules.) The numbers I'd
been working with for the previous ten years didn't look promising at
all. By 1967, seventeen men had divorced. By 1973, fourteen of them
had been remarried for longer than a year. Of those fourteen second
marriages, eight had already ended in divorce again—you'll hear

about two of those in a little while—and four more showed weaknesses that kept them securely out of the good marriage category. In other words, of the fourteen remarriages, only two looked to be anything like happy, and they were still too new to be trusted. Louie's a romantic, I thought. All I have to do is wait for thirty years and I'll be able to show him his error.

Fifteen years later I was still right and Howland was still wrong. When I checked marital history against the best and worst Adult Adjustment Outcome determinations that I had established for that 1977 book, all of the fifty-five Best Outcomes had gotten married relatively early and stayed married for most of their adult lives. (And by the time those men were eighty-five, we learned later, only one marriage had ended in divorce.) In contrast, among the seventy-eight Worst Outcomes, five had never married, and by seventy-five years of age, thirty-five (45 percent) of the marriages had ended in divorce. Proportionately three times as many of the Best Adjusted men enjoyed lifelong happy marriages as the Worst Adjusted.

But as the first decade of the twenty-first century wound to a close, the men were well into their eighties and the Study was still going strong. And so were a bunch of second marriages. I could no longer get away with my flippant dismissal of Louie's rebuke. I was also intrigued by a growing sense that as the men got older they talked about their marriages differently. So in 2010, after many years of concentrating mostly on aging, I took another look at marriage. This time I was armed with a great deal of information about alcohol use among the men and their wives (which I'll detail in Chapter 9). And it turned out that Lewellen Howland was a very wise man.

Once again, the long picture was quite different from the shorter-term one. Not about everything. At eighty-five, twenty-six of the twenty-eight men with consistently happy first marriages reported that their marriages remained happy. Marriages that had been poor to start with tended to remain that way, whether they endured without divorce for fifty years or ended, still unhappy, in death. Of the thirty surviving men who had had unhappy marriages between twenty and eighty years of age, only five reported happy marriages after eighty. Four of these were new marriages, undertaken after the first wife had died. None of this was very startling (with the exception of the fifth husband's mysterious report that he and his wife were "still in love, mutually dependent and the best of friends"). But it was *very* startling, to me at least, that twenty-three of the twenty-seven surviving divorced and remarried men reported that their current marriages were happy—and had been for an average length of thirty-three years!

What magic had occurred in those final years to shed such a different light on the early statistics? None. It was just that a new calculation had cleared away a lot of obscuring underbrush. On second thought, though, maybe it was magic after all—the magic of lifetime study.

That calculation showed that the single most important factor in the Grant Study divorces was alcoholism; thirty-four of the divorces—*57 percent*—had occurred when at least one spouse was abusing alcohol. I didn't uncover this fact until I included the wives' alcohol use in my marital calculations, and that information had been a very long time in the gathering. The men were much less forthcoming about their wives' drinking habits than their own. Statistics trickled in slowly, and as they did I had been sequestering them on yellow-lined paper in a folder all their own. One problem with the huge volume of longitudinal data is that it can drown even a computer if you dump everything in at once. So I didn't enter the wives' drinking data until they reached critical mass. The minute I did, however, alcohol leaped out at me as a new and extremely important consideration.

When I made my categorical 1977 statement on divorce, the data had suggested that marriages fail due to immaturity of coping mechanisms, poor relationship skills, and evidence of mental illness. Divorce itself, it appeared, was evidence of poor mental health that would surely make itself felt again in future attempts at marriage.[4] (Here as always I'm using our standard rough gauges of mental health problems—immature coping style, abuse of alcohol, heavy recourse to psychotherapy or other psychiatric care, use of antidepressants and tranquilizers, and so on.) There were very powerful associations between all of these and divorce. But as always, association and causation are not the same thing. In 2010, when I controlled for alcoholism for the first time, the three alleged 1977 causes of bad marriage ceased to be significant. In fact, it looked very much as though alcoholism in a marriage often caused not only the divorce, but also the failed relationships, the poor coping style, and the evidence of shaky mental health. Yet alcoholism does not even appear in the index of two landmark 500-plus-page books on marriage, Lewis Terman's 1938 *Psychological Factors in Marital Happiness* and John Gottman's 1994 *What Predicts Divorce?*[5] These works, separated by more than fifty years, are important exemplars of how influential researchers think about marriage, one from the time of the Study's inception, and the other from the days of its maturity. Alcoholism is still, arguably, the most ignored causal factor in modern social science, and it took the Grant Study sixty-eight years to notice that it was the most important cause of failed marriage.[6]

So it was back to the drawing board for me. In 1977 I had viewed divorce as a failure in the capacity for intimacy. But by 2010 I knew (and will demonstrate in Chapter 9) that alcoholism is not "a good man's failing," but a chronic and relapsing disease; not a matter of personality, but very often of genes.

Once it is clear that divorce sometimes (indeed, frequently) reflects factors other than the partners' emotional immaturity, it is easier to understand that is not necessarily a predictor of future troubled relationships, and that previously divorced men are capable of long and happy marriages after all.

These findings raise some interesting issues, which will undoubtedly be tested once the necessary lifetime studies are available. From my point of view, however, the take-home lesson is the great power of the Grant Study's unique approach to diagnosing alcoholism over time by means of behavioral longitudinal data. You can't measure obesity by asking someone how much he ate last week; people with weight problems are often on diets. You have to get them on a scale. The Study didn't try to make quantitative estimates of how much or how often the men drank. But as time passed, repeated questions about what problems the men had with their drinking made abuse of alcohol ever more clear.

The wives' issues with alcohol, as I've said, were trickier; a gentleman doesn't reveal his wife's weight, and apparently he doesn't say much about her drinking, either. And we had no direct behavioral information from the wives themselves. So it took a long time for this dramatic correlation to come into focus. But once we did, it was very clear that alcohol is the 800-pound gorilla in any study of marital failure. I will mention it in passing here when it comes up, but Chapter 9 is devoted to an extensive study of alcohol abuse and its hugely destructive effects on individual and interpersonal development.

These two related findings (the implication of alcohol and the exculpation of divorce) point up yet again the crucial importance of long-term follow-up in any research where the passage of time is at issue. As far back as 1951 researchers had begun to acknowledge that their studies of marriage were limited by the cross-sectional nature of the available data, and had sent out a call for longitudinal investigations. But as we've seen repeatedly, prospective material seldom comes easily, and it never comes fast. Field and Wieshaus pointed out that "[u]ntil recently marriages that endured 20 years or longer have been considered long-term. . . . It is very likely that a marriage of 40 years differs considerably from the same marriage at 20 years."[7] That comment encapsulates perfectly the truncated expectations of so much psychological and sociological research, and why, when a long perspective on marriage does become available, it is too valuable to neglect. The Grant Study's perspective is long enough to show us that at *sixty* years a marriage can turn into a partnership different from what had been twenty years before, for better or for worse; Leo Tolstoy's story is a notable example of a great marriage that turned sour after many decades and thirteen children.[8]

THREE MARRIAGES

Now let me try to make these intangibles tangible. I'll use case histories, commentary from independent observers, and long-term follow up to illustrate a very happy and lasting marriage, a very happy and lasting remarriage, and a long marriage that endured in the absence of either conflict or tenderness.

Fredrick and Catherine Chipp: Even the mosquitoes. Of the hundreds of responses we got when the Grant Study sent questionnaires out to the men's children in 1986, there was only one family in which all the children checked off that their parents' marriage was "[b]etter than my friends.'" That marriage was Fredrick and Catherine Chipp's. One daughter even scribbled in, "Much better."

By the time Fredrick Chipp was eighty, he (and his wife) had been giving their marriage rave reviews for six decades. The first time he met Catherine, the sixteen-year-old Chipp went home and told his mother, "I met the girl I'm going to marry." It took him a few years to get her to see it that way, but from then on, he says, "I've lived happily ever after." Not that there weren't changes along the way. When he was seventy-five, Chipp described in an interview how their relationship has evolved. "She has become more confident, and I have learned to adjust to that."

On their fiftieth wedding anniversary, he dug out the diary of his teenage years, in which he had described his future wife as "simply swell." The two of them started out sailing together. The day after our interview, sixty years later, they were planning to take off on vacation—two weeks of sailing. "I do the skilled and the nautical parts. I take charge—it's just instinct," Chipp said. Catherine enjoyed the aesthetics. Every year for decades they went canoeing together in Nova Scotia, too. Chipp told me solemnly, "That is important time." I've had readers complain that marriage isn't all vacation, and that stories like this don't tell us much about what the Chipps' life together was like. My answer to that is: to the Chipps, all of life was a vacation. They enjoyed every bit of it, including the mosquitoes, as long as they were battling them together. You could feel it when you were with them. (And, as Eben Frost's story below will remind you, in some marriages even the vacations aren't idyllic.)

Yet they didn't live in each other's pockets. Once Chipp retired (he was a successful schoolteacher and administrator), I asked him what it was like being home so much. He said that he and Catherine led different lives and had different passions. They shared what they shared, but "I do not impinge on her work." They had supper and breakfast together, but they ate lunch separately. When I asked how they collaborated, he pointed to their lush gardens. Catherine did the planting and harvesting, Fredrick said, and he did the heavy labor. They walked together, three miles a day. They read to each other; the Chipps carried out even the (usually solo) act of reading in relationship. This year they had gone camping in Florida for two weeks; all three children and their families, including eight grandchildren, had come along too.

When I interviewed the Chipps for the first time they were forty-seven, and I was struck by how attractive Mrs. Chipp was. "The lines on her face and her facial expressions were all happy," I wrote in my notes. It made her unusually pretty. More than that, it was clear that humor, one of the most adaptive coping tools of all, was a great Chipp favorite.

They teased each other constantly. At seventy-five Chipp called his wife into the room to help him retrieve a name. "My mind's a blank," he told her. "So what's new?" said she. It reminded me of an encounter with another Grant Study wife in a warm marriage. I asked the routine Study question about whether divorce had ever been considered. "Divorce, never," she replied without missing a beat. "Murder, frequently."

What about serious disagreements, though? I asked Mrs. Chipp how she resolved them. "There's nothing like losing your temper once in a while," she said. "It clears the air." But she said it with great good humor. Her husband had told me separately that he believed in bringing conflict out into the open. They got angry occasionally, yes. But covert hostilities had no place in their relationship, while passive aggression is the coping technique most associated with bad marriages. The conflict-avoidant Penns, in contrast, just endured their mutual despair.

When Chipp was eighty-three, he and Catherine participated in Robert Waldinger's study of marital intimacy. This recent development of the Grant Study is an example of how twenty-first-century social science is making attachment visible. Waldinger's team has done extensive neuropsychological testing on both members of consenting couples; they have also videotaped couples in discussion of a conflictual topic. Afterward, they telephone the partners individually every night for a week, to study their responses to the interaction (the "daily diaries" noted in Chapter 3). Waldinger is also compiling fMRIs on these couples, using the imaging study to visualize brain responses to positive and negative emotional stimuli.

Waldinger taped the Chipps as they discussed a pressing conflict situation concerning two very disabled relatives and what the Chipps should do about their care. Fredrick Chipp is a taciturn person; his wife is verbal and outgoing. But they didn't waste time or effort struggling over whose point of view would prevail; every comment either one of them made was aimed at the problem of their relatives, and moved them closer to a solution. They were a couple dancing beautifully together. The follow-up questions about the conflict discussion included a query about "trying to get my partner to understand." "We don't try; we do!" Mrs. Chipps interjected. For them, even conflict was filled with laughter.

After this intensive restudy, both Mr. and Mrs. Chipp at age eighty-five were rated as "securely attached"—attachment theory's healthiest classification, which implies among other things the kind of trust and comfort in a relationship that enable it to withstand the stresses of conflict and separation.[9] They also did well in assessments of how they behaved to each other, scoring low on such variables as "derogation of partner" and high on "stated satisfaction," "care giving," and "loving behavior."

When that interviewer asked Chipp about his relationship with his wife, he replied, "I always felt this is somebody who has a lot of depth. Who has a soul within her that is something I'm able to plumb, frequently. When I read a book or when she reads a book, for a second you may be deep in one yourself, but you want to share it with that other person."

Mrs. Chipp put it this way: "I can't imagine not having him there.

And this doesn't mean that we do everything together. . . . I would drive him crazy and he would drive me crazy. But we always know the other one is there and that we have this relationship and that makes us both very content. . . . We're very different. We're not different in what we like in life. We're very, very close in what we like in life."

Rereading my own interview notes, I found a comment to myself: "He was perhaps the happiest man in the Study." I recalled the end of my interview with him five years before, when I had asked Chipp how he and his wife depended on each other. He looked off into space, choked up, and fought back tears. "Gosh," he blurted out, "just by being there. If she goes first, it would be pretty traumatic."

She did, and it was. Losing a spouse to death is often the most painful blow of old age.

John and Nancy Adams: Try, try again. Here's a happy marriage that took longer to happen. It's the story of the proper Bostonian I'm calling John Adams, Esquire. In the first half of his life he was married three times, but those marriages lasted only nine years *in toto.* He met his fourth wife at the age of forty-five and has spent the forty-two years since then living with her very happily indeed. In contrast, I'll offer a few notes on the life of Carlton Tarryton, M.D., whom we met briefly in Chapter 5, and who also had three divorces and four wives.

Tarryton and Adams resembled each other a lot in their first twenty-five years. They came from unhappy and isolated childhoods, with parents who spent a lot of time away. (Blind raters reported that they would not have wished to endure the childhood of either of these men.) They were bad losers. They manifestly lacked self-discipline; Adams often played hooky, and they were heavy drinkers and irresponsible students in college, earning grades in the bottom tenth of the Grant Study spectrum. They were both among the small

minority of men to whom Study investigators ascribed the traits
"Poorly Integrated" and "Lack Purpose and Values."

There were two big differences between them, though. John Adams had a sibling with whom he was close, and in his teens there had been a stepfather whom he deeply loved. At twenty he said about his stepfather, "I really liked him more than any person I have ever known." Tarryton was an only child and never had a parental figure whom he admired.

By the time they were thirty, their lives were beginning to diverge. Tarryton, who had gone to medical school, was often drinking a fifth of whiskey a day. He was extremely careful to remain sober when he was working, but his engagement in his career was limited. He maintained that he believed in Christian Science and that he had become a doctor only for the paycheck; in actuality, however, he made less money than any other practicing physician in the Study. He became an active alcoholic, and destroyed his brief and allegedly happy fourth marriage when he fell off the wagon one time too many at age fifty. In despair, he gave up and took his own life.

Adams went on drinking heavily too. But he was always able to maintain enough control over his alcohol use that it didn't lead to problems in his academic life or elsewhere. He was on the Law Review at Stanford. By forty he had won an award for public service, and made partner in an excellent firm in Boston. It wasn't all good news. He married and divorced three women. But he had some insight into his marital problems, which were due, he thought, to his own "emotional immaturity coupled with the habit of gravitating to persons whose emotions were less stable than my own."

After his third divorce he began drinking less and exercising more. And he met Nancy, who became his fourth wife. A highly competent and emotionally supportive person, she could stand on her own and even offer Adams some of the caretaking he needed instead of wanting him to keep her together all the time. (Some Study members with unhappy childhoods achieved gratifying lives by devoting themselves to those more needy than they, but this otherwise useful altruistic coping strategy served less well in marriage than in the rest of life. Marriage seems to work best when both parties can gratefully allow the other to bring them breakfast in bed.)

In his fifties, after seven years of marriage to Nancy, Adams was feeling so good that he confessed to "incurable optimism." He stopped answering questionnaires for a while, but when he resumed at the age of sixty, he reported his marriage as "perfectly happy" and suggested that it was "maturity that makes this marriage work." He explained that in his previous marriages he had been "attracted to weakness, and that all my previous wives had major flaws of a psychiatric character." And thirty-two years after he and Nancy married, Adams wrote that it was not his career but his marriage that had "turned my life into an absolute joy."

Adams retired young and began writing semi-autobiographical short stories, transmuting much of the pain of his early life into art. I was intrigued to note that the interviewer who met him when he was thirty and still very unsettled had perceived this fledgling lawyer as "sensitive, bright—like a writer of fiction."

When Adams was eighty he was interviewed by Maren Batalden, who mistrusted his optimism; she granted that the marriage was "genuinely happy," but gave his wife credit for this. But at eighty-eight he was as cheerful as ever. It was clear to me, too, that there was some exaggeration going on; his optimistic assertion that there had been no decline in his physical activities couldn't possibly be true.

But so what? There were three times as many optimists among the happily married as among the unhappily married—a very significant difference. Samuel Johnson seems to have had a point when he quipped that a second marriage is the triumph of hope over experience. Besides, I reflected, if at eighty-five you're still playing tennis, your only medicine is Viagra, and the happiest seven-year period in your life is "now," neither your marriage nor your physical health can be all that bad, even if your glasses are a bit rose-colored.

It came as a surprise to me that bleak childhoods were not always associated with bleak marriages. As we saw in Chapter 4, warm childhood predicted both future trust in the universe and future friendship patterns. And it was very significantly associated with life flourishing as measured by the Decathlon. Still, Camille finished in the top quartile in that test. With the exception of a man's closeness to his father, childhood environment did not predict stable marriage, and even where a warm paternal relationship was lacking, good marriages could be made—eventually. Indeed, marriage seemed to be a means for making good on a poor childhood. After almost fifty years of following disadvantaged youths, psychologist Emmy Werner noted that "the most salient turning points . . . for most of these troubled individuals were meeting a caring friend and marrying an accepting spouse.[10] Childhood trauma does not always blight our lives, and the good things that happen to us keep on giving. Restorative marriages and maturing defenses are the soil out of which new resilience and post-traumatic growth emerge.

So it was for both the College and Inner City men. The Chipps and Oliver Holmes had very happy childhoods. But for the men who didn't, good marriage seemed to have been part persistence and part the luck of the draw—Gilbert meeting Sullivan. My mother used to say that two people who are healthy can easily have a good marriage, but a great marriage takes a lucky match between two neurotics. John Adams made one such reparative marriage; Adam Newman made another. Others from the Study have been illustrated extensively elsewhere.[11]

Eben and Patricia Frost: The limits of containment. Fifty-four of the Grant Study men celebrated golden anniversaries in "so-so" marriages—marriages that were neither particularly conflicted nor particularly loving. Their relations with their children and siblings were as warm as those of the happily married; so were their general relationships and social supports. Their childhoods were no less caring, their adaptive styles no less good, and they remained sexually active no less long. So what was different? These men reported that they were "very content" very significantly less often than the happily married men in all aspects of their lives, and this general tendency might have influenced the way they perceived and reported on their marriages. They had been significantly less close to their fathers. Their objective health on average was not quite as good as that of the men in happier marriages, and they were somewhat less likely to have close and confiding relationships of any kind, in or out of marriage. But most critically, as Eben Frost's story will make clear, not everyone brings to marriage the desire (or is it the capacity?) for intimacy.

Frost grew up in a family that was scratching an annual income of $1,000 1939 dollars out of a Vermont hillside. He walked three miles each way to a two-room schoolhouse in the winter, and in the summer he worked on his parents' dairy farm. He knew that he was lonely and that he needed a job where he could talk to people. By the time he was ten, in silent rebellion against endless chores, he had secretly decided to leave the farm forever. He would go to college and then to Harvard Law School. And that is exactly what he did.

There was little outward warmth in the Frost family. The Frosts cared for each other, but they never exactly said so. The father was described as "unruffled" and "a great man of character and old-world morality." The mother "never let things prey on her mind." She told the Study little about Eben except that he was nice to live with.

"When we had a job around the house to be done with neatness, we called in Eben." But feelings were not discussed. Eben told the Study psychiatrist, "The family bonds have been almost entirely lacking insofar as the tie of intimacy is concerned."

Eben was valedictorian of his high school class; he "ran the school" during his time there. But he felt that his greatest strength was his "ability to make friends." At college he preferred team sports, and what he enjoyed most about the law was always contact with his clients. When he first entered the Grant Study, he was observed to be "well-poised, extremely friendly, very active, forceful, and energetic."

Even at eighteen, Frost knew that he paid a price for the self-sufficiency that got the jobs done around the house and then took him to Harvard. He knew the limits of his talent for friendship, too. He admired the kind of marriage in which couples become completely absorbed in each other, but he knew it wasn't for him. "Although I hate nobody," he said, "I'm sometimes afraid I can love nobody." At twenty-five, Frost said of himself, "I'm a very easygoing character," and twenty years later, "I'm the most reasonable of men." That sweeping globality of expression—a very easygoing character, the most reasonable of men—remained typical of Frost over years of interviews and questionnaires, and seemed to serve as something of a Distant Early Warning Line. It signaled pride in the extent of his extreme containment, but also an undertone of threat: thus far and no further. And he did appreciate that his sweet reasonableness could be infuriating to others. "I'm so self-sufficient, or something, that I actually don't have any rough spots. This does not mean that things are perfect, it's just that nothing really bothers me—and I'm well aware that this is not necessarily an ideal constitution—I've been this way all my life."

When he was in college, one observer called Eben "supernormal." Time did not produce much change in that regard. When I met him at his twenty-fifth reunion I thought he was supernormal too, except in matters of intimacy. Frost exemplified and embodied the feeling tone of the "so-so" marriages, which were faithful, harmonious, and enduring, but lacking in warmth. The Study, as I've said, made many and repeated queries about the men's marital situations, and this included engagement and wedding plans. Frost's answers were utterly unrevealing until at twenty-four he wrote with a laconic nonchalance worthy of Calvin Coolidge, "Married on furlough." Nothing more, even though it turned out that his marriage had been in the works for many months.

As Study queries continued, Frost went on replying in his characteristic compliant but circumspect way. Asked about marital problems, he wrote, "Only the stupid or the liars will say 'none,' but that is my answer." The years of follow-up suggest that he was neither dim-witted nor dishonest, but he was limited. At age thirty he wrote that his wife was the person in his life whom he most admired; at forty-seven he regarded his marriage as stable. His wife's personality, he said, produced no problems for him. But he knew that his permanent reserve was a problem for her. This concerned him, but—as he saw it—there was nothing to be done about it; it was just the way he was.

When I met him, Frost presented himself as a charming, extroverted, happy, outgoing man who was very interested in people. He was calm, incisive, and unambiguous, and he had a contagious ease about him. He was a good-looking man with a fine sense of humor, and at forty-seven he looked far more like an old-money museum trustee than an impecunious farm boy who had somehow grown up to practice law. He described all his activities in terms of other people. It was only when I tried to push him into discussing close relationships that the No Trespassing signs appeared. He balked like a horse, becoming less cooperative and more abrupt until I returned to less charged material.

When he was eighteen, the Study internist noted that he was

in "extraordinarily good health." This remained true. At forty-seven, Frost had never spent a day in the hospital or missed work due to illness. For over thirty years he described his overall health as "excellent." He wasn't ignoring things, either; he went for regular physicals that turned up nothing worse than hemorrhoids. So much for the supposed dangers to one's health of keeping feelings locked away; Eben Frost never in his life found, or wanted, anyone to confide in.

When he was forty-seven, I ranked Frost among the top third in the Study in adjustment; later, in the Decathlon, he scored in the top tenth. According to him, his life was as smooth as silk. "My practice doesn't have pressure; I chose a career without tensions." And he added with his characteristic lordly sang-froid, "I am the most unpressured person that exists."

Once in a midlife assessment, Frost complained that he was unsatisfied in his law career, and was "doomed to frustration." He expressed a wish to have been a creative artist. But as always, he knew how to comfort himself in his frustration, and he knew how to take care of himself, too. He designed his own house. He became president of his suburban PTA. And by the time six years had passed, he had altered his professional life to give him the personal, caring contact that he craved. He was now responsible for training young associates and could say to me with the enthusiasm of the generative, "My job turns me on."

But for all his repeated assertions that he was "as easy to get along with as anyone who ever lived," things weren't quite so silky at home. He reported that his wife "engaged in a certain amount of nagging." When they went on vacation, she spent her time on the beach while he played golf. He understood very well that his avoidance of mutual dependency was a loss to both of them, and that the greatest stress in his marriage was "her frustration over my unwillingness to let her get inside of me." He himself didn't crave that kind of connection, though. He knew that she sometimes felt shut out, but his view was that she let her problems "grow and grow unnecessarily." He didn't take kindly to weakness. He said of himself that "[m]y greatest weakness is not leaning on anybody." Insight was always one of his strong suits, but it didn't change him. In contrast, here's Fredrick Chipp at the same age, responding to a question about what he did when he was upset about something. "I go to her," he said of his wife. "I talk with her. It's very natural and easy and helpful. I certainly don't keep it to myself. She's my confidant."

One of Frost's daughters wrote that in her parents' years together, she had never seen them fight; equally revealing, she had never seen them affectionate. He had always been more loving with his grand-children than with his own children, she said. His other daughter summed up the situation: "Frosts never say 'I love you' to their chil-dren." But people who can't hug their peers can hug puppies. And grandkids.

I interviewed Frost after he retired. He told me that he made sure to be out of the house all day and never "home for lunch." When I asked how he and his wife depended on each other, he explained that his wife was "self-sufficient. She has different interests." So had he, both recreational and relational. Besides occasionally going on trips, "we don't get involved in what the other is doing." He liked art, but "her interest in the symphony does not interest me."

There was absolutely no tenderness or care in his discussion of his wife. She, during my brief meeting with her, betrayed no interest in the Grant Study or in her husband's fifty-year involvement in it. The way he put it later was that there was "nothing pulling our marriage apart, but we are not bound together."

People have intense reactions to a story like this. Like divorce, it evokes a lot of personal, and sometimes conflicted, associations. To the young and intense, it may sound like a tragic abdication of passion. To the old and tired, it may sound like a rueful reflection of all the dreams that never came true. To people living in marital war-zones, it may sound like a blessed relief. But personal reactions shouldn't obscure the important lessons of the Frosts' marriage.

One is that even in a population selected for health, as the College cohort was, pleasure in close relationship is not a predictable developmental achievement. It's not a clear correlate of one kind of childhood or another. It's not a black-or-white, yes-or-no kind of thing, either. The capacity for intimate marriage is near kin to the capacity for close friendship, and we tested that by asking the men to describe their closest friends. Virtually all women can describe close friends—in plural—but some men don't have any. Like musical talent, the talent for closeness exists on a continuum; there's a lot of room between a tin ear and perfect pitch. To some extent the skills and pleasures of closeness can be practiced and learned; John Adams learned them. But not everyone will become a virtuoso. And, as Eben Frost demonstrates, not everyone wants to.

Eben Frost mastered Eriksonian Intimacy in a way that Peter Penn and Algernon Young never did. But he had no interest in intimacy of other kinds. Godfrey Camille, in contrast, actively sought closeness; for him it was a life quest. John Adams kept trying to establish for himself a relationship in which closeness would be possible, and once he had one, he really worked at it—that was one of the adaptive aspects of his marriage. For Frost, as he made very clear, closeness was not a desideratum. And he enjoyed his life thoroughly.

A comfortable, productive, and enduring marriage like Frost's is no mean accomplishment, and for people without the desire for intense emotional closeness, it may be the best kind there is. It endured for fifty years without rancor or regret, and it gave its participants something closely approximating the modest levels of companionship they were looking for. If Patricia Frost complained sometimes that her husband's preferred level of intimacy did not precisely match her own, she too, by her daughter's report, appears to have been a person comfortable with more distance than with less.

Still, the Frosts' marriage makes me (and their daughters) a little sad. The capacity for emotional intimacy is not a moral virtue, and its lack is neither a sin nor a failure. But it is a blessing. A marriage like the Frosts' reminds us of how it feels to look for closeness and not find it.

Another provocative issue is the place of dependence in a successful marriage. Dependence is a dirty word these days; codependence is an even dirtier one. But *mutual* dependence is still allowed, and Study data suggest that some kinds of mutual dependence count among the apparently pathological adaptations that can in fact be healing. As I said in the last chapter, the most dependent adults came from the most unhappy childhoods. We've already seen (in Adam Newman and Sam Lovelace, for instance) that marriage, however imperfect, is an opportunity to assuage some of the loneliness of bleak early years. It turned out that happy marriages after eighty were not associated either with warm childhoods or with mature defenses in early adulthood—that is, you don't have to start out "all grown up" to end up solidly married.

Even when it isn't needed for repair, mutual dependence has its own pleasures. Study transcripts from the happily married are full of unapologetic paeans to mutual dependence. As Catherine Chipp said in one of her interviews, "It's very strengthening to have a person that you can always turn to." Eben Frost's story illustrates how the growth of closeness can be thwarted by too fierce a resistance to dependence. And it's worth noting that friendship and mutual dependence in marriage deepen late in life. There are three primary reasons for this. First, the so-called empty nest is often more of a blessing than a burden.

Second, the age-related hormonal changes that "feminize" husbands and "masculinize" wives make for a more level physical and emotional playing field. Third, physical infirmities make it ever plainer to both parties that mutual dependence is an advantage rather than a weakness. The deepening of dependence in marriage appears to correspond with a period of increasing subjective happiness as marriages age. More about this in a moment.

GOOD AND BAD MARRIAGES: SOME STATISTICAL DIFFERENCES

Table 6.2 depicts some of the relationships between Study variables and marital outcome. Most of them are self-evident. Divorce was most common among the alcoholic and the feckless, while bad marriages tended to endure among depressed men and men with bleak childhoods. The men with intact poor marriages were significantly more likely to have been depressed and to have used antidepressants or tranquilizers. As the lives of Newman and Adams suggest, while unhappy childhoods may lead to many ills in adult life, they do not preclude creating (or lucking into) a marriage good enough to help heal past miseries.

MARRIAGE AND SEX

There is one conspicuous missing piece in the Grant Study data set, and that is the sexual behavior of the Grant Study men as adults. We know a certain amount about the men's sexual attitudes as sophomores, because that was a major interest of the early investigators. Over half the men expressed serious worries about masturbation to the Study psychiatrists: "It caused my acne, and other people could tell"; "I feared I would go crazy, become impotent, or damage my penis"; "Greatest problem in my life." By modern standards, the men's training (or lack of it) in sexual matters seemed quite repressive. We

Table 6.2 Correlates of Good and Bad Marital Outcomes

Variable (n=159)*	Good marriages: n=50	Enduring bad marriages: n=48	Marriages that end in divorce: n=61
A. Good Marriages			
Warm father	38% (Significant)	21%	16%
Decathlon > 5	41% (Very Significant)	7%	5%
Mature defenses	42% (Very Significant)	16%	11%
Generative	74% (Very Significant)	46%	38%
B. Bad Marriages Or Divorce			
Lack of purpose and values	12%	11%	35% (Significant)
Poor social supports at 70	2%	47% (Very Significant)	50% (Very Significant)
Alcoholism in marriage	4%	46% (Significant)	57% (Very Significant)
Major depression	2%	27% (Significant)	20%
30+ days of tranquilizer use	12%	38% (Very Significant)	31% (Significant)
C. Surprisingly Not Significant			
Parents' marriage clearly stable	78%	54%	57%
Parents' marriage clearly unstable	14%	19%	24%
Deep religious involvement	20%	20%	24%
Bleak childhood	16%	38%	30%
Parents Catholic	14%	13%	10%

* The total N in Table 6.2 is 159 rather than the 242 in Table 6.1 due to the exclusion of the 73 so-so marriages, the 7 never married, and 3 missing data.

also know that in 1938 almost two-thirds of the men were strongly opposed to sexual intercourse before marriage: "I am disgusted with the idea of sexual relations"; "It is out of the question, nauseating"; "It would do something to my personality."

Of course we were very interested in adult sexual adjustment, but it didn't take long for the Study staff to notice that a questionnaire that delved too inquisitively into the men's sex lives was a questionnaire with a very low return rate. Over the years, therefore, the only information we systematically gathered on this subject were responses to a simplistic and rather ritualized question: Did a man perceive his sexual adjustment as "very satisfying," "satisfying," "not as good as wished," or "poor"?

What we found was that overt fear of sex was a far more powerful predictor of poor mental health than sexual dissatisfaction in marriage was. After all, marital sexual adjustment depends heavily upon the partner, but fear of sex is closely linked with a personal mistrust of the universe. The men who experienced lifelong poor marriages were six times as likely as men with excellent marriages, and twice as likely as men who divorced, to give evidence on questionnaires of being fearful or uncomfortable about sexual relations. Yet there were many good marriages in which sexual adjustment was less than ideal after the couple reached age sixty. We can't yet say much about why, or about what role biology may play.

Apart from that I don't have much of interest to report on the men's sex lives, except for one oddity of exactly the sort that makes lifetime studies so unpredictable and so fascinating.

At age eighty-five, we asked the men when they had last engaged in sexual intercourse. Only sixty-two (about two-thirds) of the men responded; of those 30 percent were still sexually active. In that small sample, the predictors of sustained sexual activity were: good health at sixty and seventy, good overall adjustment after (but not necessarily before) age sixty-five, and an absence of vascular risk factors. Surprisingly, the variables I thought would protect against early impotence—ancestral longevity, maturity of defenses, physical health at age eighty, and quality of marriage—showed no significant effect.

But I empirically defined a composite variable called *cultural and impractical*, summing five "touchy-feely" traits that had not been considered promising back in 1938—*Creative/intuitive, Cultural, Ideational, Introspective,* and *Sensitive affect*—and then subtracting from them the two practical traits that had largely been seen as laudable—*Pragmatic* and *Practical/organizing*. I chose these items because they correlated at least weakly with prolonged sexual activity. The result was not only very significantly correlated with a long sex life, but it also allowed a startling and dramatic separation of the Grant Study men by politics (more on this in Chapter 10, and on the curious association of the men's political beliefs with sustained sexual activity).

DIVORCE REVISITED

Thirty-five years ago I was largely wrong about the importance of divorce. It is clear to me now that divorce does not necessarily reflect an inability to achieve Eriksonian Intimacy or even to enjoy great relational closeness. Divorce is often a symptom of something else. Ultimately, a man's ability to cherish his parents, his siblings, his children, his friends, and at least one partner proved a far better predictor of his mental health and generativity than a mistaken early choice in his search for love.

There is unequivocal evidence that divorce occurs more frequently among people with every kind of mental illness.[12] But divorce does not imply mental illness. On the contrary, when a marriage is chronically unhappy, only divorce allows for the possibility of a new and better one. Most of us carry on an internal debate when a friend is considering ending a marriage. Divorce raises all kinds of personal anxieties about the safety of our relationships. It violates our sense of family stability and ruptures religious vows. It rarely makes for happy children. Yet it can also be a breakout from outworn social codes, chronic spousal abuse, or simply a bad decision. So it was interesting to compare the final remarriages of the divorced men with the marriages of men who remained unhappily with their first wives. The reasons for divorce had been one of Arlie Bock's founding questions when he first proposed the Study; he had appreciated all the way back in 1938 that it could only be answered in the context of lifetimes.

One man who stayed in a bad marriage wrote, "This marriage will stick, if for no other reason than we're a couple of stuffy, latter-day Victorians who just wouldn't face divorce anyway." Another wrote, "The marriage is stable if you will accept that it is held together as much by decision as by desire." "Divorce is pretty unthinkable," wrote a third, "so I grin and bear it. . . . Our marriage would have ended probably 15 years ago but for religion and the presence of children." Religion and children were the two reasons usually cited for staying together among the enduring bad marriages.

The characteristic adaptive styles of the two groups of men were different. The men who remained in unhappy long marriages were more likely to be passive in other aspects of their lives than those who sought divorce, and less likely to use humor as a coping mechanism than the men who made the happiest remarriages. Their mental health was less robust in that they had had significantly more recourse to psychotherapy and to mood-altering drugs in an attempt to self-medicate. They were also less likely to have had warm childhoods in their pasts, or to enjoy rich social supports in old age.

The endurance of some bad marriages, however, seemed to involve loyalty to a depressed or alcoholic partner. Of the forty-nine men with lifelong excellent marriages, no men and only two wives abused alcohol, and only one man was mentally ill. But in forty-eight lifelong bad marriages, eleven men were alcoholic, nine had alcoholic wives, and seven were depressed. Some of these marriages endured out of codependency (as in the case of Lovelace's first marriage), and somewhere one of the spouses was caretaker to the other, as in Albert Paine's last one. And even unhappiness in marriage isn't an all-or-nothing affair; three of these husbands reported at least one period when the marriage was excellent. Perhaps these are cases where hope died hard; as we've seen, the capacity to hope has a lot to be said for it.

CONCLUSION

So what do the seventy-five years of the Grant Study have to teach us about marriage, intimacy, and mental health? For one thing, they make clear that Lewellen Howland was right—the important thing is that "loving people for a long time is good." Why? Well, for one reason, as I said at the outset, it feels good. Most of us enjoy love when we can get it. But in developmental terms, both intimacy and positive mental health reflect the process of replacing narcissism with empathy, a progressive amalgam of love and social intelligence that is essential to the development of mature defense mechanisms and optimum adaptational skills (Chapter 8).

The Study also makes abundantly clear that people don't drink because they're in bad marriages, but that drink makes marriages bad. For twenty-eight men, happy marriages became unhappy following the onset of alcoholism in a partner; in only seven cases did alcoholism first become obvious following a failing marriage, and in some of them the "failing marriage" was clearly less a cause than a rationalization for the loss of control over alcohol. Popular belief notwithstanding, alcohol is a very bad tranquilizer.

Third, we learn something about why the percentage of happy marriages increases after seventy. It is known from studies of the general population that the divorce rate declines sharply with increasing age and length of marriage.[13] Explanations for this include the selective weeding out over time of vulnerable marriages, increasing commitment and resistance to change with increasing age, and perhaps the fact that greater joint assets make divorce more expensive. A decline in divorce rate is not necessarily the same thing as increased marital happiness, but increased happiness is what our data showed. During the period from age twenty to seventy, only 18 percent of both partners from the entire sample reported their marriages as happy for at least twenty years. By age seventy-five half of the surviving men did. And by eighty-five, the proportion of happy marriages had risen to 76 percent. Some of this improvement no doubt has to do with Laura Carstensen's theory of socioemotional selectivity, which suggests that as people get older they tend to remember the good over the bad.[14] Some seems to relate to the men's increased tolerance for mutual dependence as they aged; as George Bancroft said, contemplating the loss of his once-cherished driver's license, "You let your wife learn about you. . . ." Certainly the more the men became able to appreciate shared dependence as an opportunity rather than a threat, the more positive feelings they expressed about their marriages. Successful remarriage after divorce or widowhood was another contributing factor.

But the Grant Study offers an unexpected finding that even first marriages improve late in life. This is something that we would have been very unlikely to discover from a shorter or less comprehensive study. After age seventy, the men just seemed to find their marriages more precious. "Jane and I are at the age where what life we have left together is like the last few days of a great vacation," one seventy-eight-year-old Study member said. "You want to get the most out of them, so we want to get the most out of our togetherness."

Finally, the Study illuminates some subtle but vital distinctions that can't be perceived at all except through the long lens of lifetime study: the difference between facile optimism like Alfred Paine's (see Chapter 7), for example, and the lifelong faith in the universe of a Boatwright or an Adams. One is mere glitter; the other is true gold. Codependency may be a very false friend, but it was the capacity for *mutual* dependency that allowed the Chipps to share a warmth and comfort that the competent but chilly Frosts couldn't offer each other.

And the lives of men like John Adams and the others who turned my thinking about marriage upside down remind us again of the remarkable reality that people continue to change, and people continue to grow. An interviewer once asked Margaret Mead to what she attributed the failure of her three marriages. "I don't know what you mean," she answered. "I had three successful marriages, all for different developmental phases of my life."[15]

7 | LIVING TO NINETY

> It is a popular belief, but a fallacy, that we can learn about what contributes to extreme old age by studying very old survivors. This certainly needs to be done, but it is also necessary to know about the characteristics of other members of their birth cohort, in early childhood, adulthood and early old age.
>
> —M. BURY AND A. HOLME, *Life after Ninety*

IN MY PREOCCUPATION with love, joy, and relationship in old age, I tend to forget the more mundane but absolutely crucial importance of staying alive. Physical health is just as important to successful aging as social and emotional health. In 2011, the Grant Study and the Lothian Birth Cohort 1921 became, as far as I know, the world's first prospective studies of the physical health of nonagenarians.[1] (The Lothian Birth Cohort was formed in Edinburgh in 1932. It included more than 80,000 eleven-year-olds, many of whom have been followed over the years to the age of ninety.)

In this chapter I will focus on the sixty-eight surviving members of the Grant Study who are still active as of March 2012, but I will also review what happened to the men who died along the way. This exploration will take me through several important revisions of the received wisdom of earlier years, including some once convincingly backed up by hard data. There were surprises involved, and some shocks, too, for argument and theory can never be enough. We need documentation and proof. In the study of lifetimes this means prolonged follow-up and systematic retesting of serious hypotheses—not over a year or two, but over decades. That's the very first and most fundamental lesson of the Harvard Study of Adult Development, and the one on which all the other lessons depend: While life continues, so does development. The study of very old age is the ultimate case in point.

In 1980, Stanford University internist and epidemiologist James Fries recognized that modern medicine was not extending the human lifespan, and yet survival curves were changing. More people were living vitally until eighty-five or ninety, and then dying quickly, like the wonderful one-hoss shay in Oliver Wendell Holmes's poem, which ran perfectly for a hundred years and then fell apart all at once.[2]

Fries called this phenomenon "compression of morbidity."[3] In 1900, because most deaths were premature, the human survival "curve" was a diagonal line; now it is more of a rectangle—especially if you have no risk factors (Figure 7.1). In 2040 there will be ten times as many eighty-five-year-olds as there were in 1990. This is not because the normal human lifespan is any longer than it was, but because fewer people will die before eighty. *After* eighty the lifespan will reflect little increase. Medical advances like antibiotics, new cancer treatments, and kidney transplants all serve to decrease premature death. But they do not alter the fact that the bodies of most of us, like the one-hoss shay, have not evolved to live past one hundred.

In 2012 the longevity of the Grant Study men, who were selected for physical health, is clearly ahead of the historical curve. While only about 3 percent of the white American males born in 1920 are expected to celebrate ninetieth birthdays, 77 (28 percent) of the 268 Study men already have, and 7 (3 percent) more are eighty-nine and likely to join them. Their projected age of death is already four years longer than that for white males born in 2009.[4] (By way of comparison, only about 18 percent of the gifted Terman men made it to

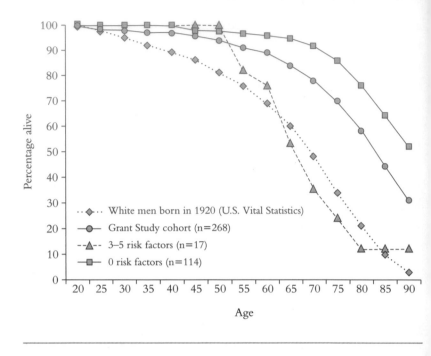

Figure 7.1   Percentage of men alive at each age.

ninety.[5]) One purpose of this chapter, therefore, is to give readers a view into the future, to a time when living to ninety will have become more commonplace.

Some may ask: How long do we *want* to live? Our society stereotypes the very old as frail, ill, and unhappy. No one likes the prospect of long years confined to a nursing home, helpless, mindless, or in pain. And it can't be denied that aging means loss. Our bodies begin their long, slow process of failure at thirty; by age seventy we can identify only fifty percent of the smells we recognized at forty. Our vision declines until by eighty few of us can safely drive at night.[6]

Even among the surviving Grant Study nonagenarians, 25 percent are cognitively impaired.

But that means that three out of four Grant Study ninety-year-olds are still cognitively intact. Although we often hear that our brains begin to shrink at twenty and that we will have lost fully 10 percent of our brain cells by seventy, modern brain imaging techniques suggest a less grim picture. Normal brain shrinkage is less than we feared, and estimates so far may also reflect the practice of averaging individuals with specific brain-destructive pathologies—Alzheimer's, trauma, alcoholism—in with everyone else.[7] In addition, we can speculate that some lost brain matter is the result of judicious "pruning" of cells that are no longer in active use; after all, due to such pruning, our brains contain fewer than half the synapses at twenty-one than they do at five.[8] There may not be as much cause for despair as we fear.

Furthermore, the mental life of the elderly is not as bleak as we are sometimes given to believe. After forty, the fear of death declines steadily and belief in an afterlife becomes more common. Careful epidemiological studies from multiple centers reveal that depression among the elderly does not increase.[9] Indeed, recent studies show that older adults report less depression, lower levels of negative affect, and more assertiveness and positive emotion.[10] As one Terman great-grandmother put it, "I hate my waist, but I love my psyche!" Most of the Grant Study men who reached their eighty-fifth birthdays were glad they did.

The most authoritative study of old age available is the MacArthur Foundation study, which is summarized in Rowe and Kahn's *Successful Aging*.[11] That study refutes the specter of dismal years in a nursing home, finding that the average eighty-five-year-old man will spend only about six months in an institution before he dies, and the average eighty-five-year-old woman about a year and a half. People who live to be one hundred, those data show, are usually pretty active at ninety-five.

In the Grant Study men, in fact, for a long time there seemed to be so little cognitive decline that we did not even begin testing them for cognitive function until they were eighty. At that time, 91 percent of them tested at normal levels. Of the seventy men who have so far survived to reach ninety, fifty-eight were cognitively intact at that age, and had showed over the preceding ten years an average decline of only one point on the 41-point Telephone Interview for Cognitive Status (or TICS).[12] (I'll say more about the TICS shortly.) Almost three-quarters of the Grant Study ninety-year-olds were as sharp as ever, if just a bit slower. It's true that a majority could no longer give the full name of the vice president: "I know he was a senator from Delaware. . . ." Yes, finding names can be a problem after seventy. So can finding your car in a parking lot!

The MacArthur Study supports Fries's view that the years of disability are growing shorter even as we live longer. Bury and Holmes report that chronic pain declines from a high of 28 percent at eighty-five to only 19 percent after ninety-five.[13] And while dementia is ghastly for afflicted individuals and their families, in the Grant Study, only ten of the forty-one men who became demented before eighty-nine lived for longer than five years after their dementia was recognized. Moreover, some, although not all, epidemiologists believe that the incidence of Alzheimer's declines after ninety-five.[14]

Progressive diminution of physical reserves is an inevitable part of aging, but the rate at which this happens varies. Forty percent of the general population remains "fully functional" after eighty-five. That is a hard criterion to define, but as Table 7.1 illustrates, about 55 percent of the Grant Study men between eighty-five and ninety appeared to meet it. And only 9 percent of them were confined to home or wheel-chair.

Table 7.1   Instrumental Activities of Daily Living at Age 88 ± 1 (N=60*)

No cutbacks	12%
Can walk 2 miles without resting	41%
Still engages in hard physical activities: jogging, hiking, chopping wood, moving sofas	46%
Has not reduced a major activity	61%
Has not stopped a major activity	52%
Can still move light furniture or carry a suitcase through airports	74%
Needs no help going downtown	67%
Still drives and/or uses public transportation	75%
Can climb 2 flights of stairs without resting	79%
Needs no help with dressing or personal care	90%

* Sixty people returned this questionnaire.

The questionnaires that the most energetic of the men returned at ninety (or almost) convey some of the flavor of their lives. One man who had been divorced four times described the happiness of his fifth marriage—which had lasted for twenty years—as 7 out of 7. He reported no cutbacks in any of his physical activities, except that he had given up night driving. For his energy level he checked "vigorous"; he continues to play tennis. Another man was still earning $60,000 a year at eighty-seven, and also engaging in thirty hours a week of volunteer activity. He, too, described his marriage as 7 out of 7, and at ninety he was sexually active. A third man exercised fifteen or more hours a week, was sexually active, and gave his marriage 6 out of 7 points. He engaged in four volunteer jobs.

TWO LIVES

The life of Alfred Paine.   The realities of the men's health and their subjective perceptions of it did not always coincide. Good self-care, high morale, intimate friends, and robust mental health often make the difference between having an illness and feeling sick. Let me now therefore give a more nuanced portrait of both ends of the aging spectrum, starting with Alfred Paine (previously discussed in *Aging Well*), who was both ill and sick, as per my differentiation above. His greatest strength was that he did not complain; his greatest weakness was that he knew himself very little. He could not acknowledge either his alcoholism or his depression. Paine had one of the best scores in the study on a subjective inventory of depressive symptoms. He had never sought psychotherapy; none of his doctors had ever called him mentally ill. On questionnaires, he described himself as close to his children and in quite good physical health. It was only by interviewing him personally, talking with his wife, examining his medical record, recognizing the disappointment in his children's questionnaires, and—finally—reading his obituary that Alfred Paine's misery could be fully appreciated.

His ancestors had been successful New England clipper-ship captains. One grandfather was a merchant banker; the other was a president of the New York Stock Exchange. His father had graduated from Harvard and his mother from a fashionable boarding school. Paine himself arrived at Harvard with a handsome trust fund. But his childhood received the lowest rating for warmth of any man's in the Study, and in the Grant Study, at least, parental social class does not guarantee successful aging. Paine's story illustrates that money doesn't, either.

At college, Paine was often in love. But to the Study staff it appeared that to him, "being in love" meant having someone to care for him. His multiple marriages were all unhappy, especially according to his wives. This was partly because of his alcoholism (which he denied), and partly because he was afraid of intimacy.

Paine described his physical health as "excellent" at sixty-eight; but objectively it wasn't. He was seriously overweight, and he had obstructive pulmonary disease as a result of lifelong smoking. By age seventy, gallstones and an ileostomy for diverticulitis of the colon were adding to his medical woes. When I interviewed him at seventy-three, he seemed ten years older than his age—like an old man in a nursing home. He had lost all his teeth; his kidneys and his liver were failing; he had a mild dementia as the result of a drunken automobile accident. He took two medicines for high blood pressure, two for diabetes, one for gout, and one for depression. There was no question that he was physically disabled.

Paine's Decathlon score was 0. Alone of the men who returned the Study questionnaire at seventy-five, he refused to fill out the Life Satisfaction Chart that we included with it. But his joylessness was evident. I always found it exhilarating to listen to men who were zestfully mastering life's slings and arrows, but interviewing Paine made me feel very sad. In his questionnaires, he said nice things about his children. But when I asked him what he had learned from them, he responded irritably, "Nothing. I hardly see my children." One daughter said in her questionnaire that she saw her father only every three years, while the other saw him once a year and viewed him as having lived "an emotionally starved life." Alfred Paine's only son said that he had never been close to his father.

Paine's third wife was protective and loving toward him, but he was disrespectful to her, and not very caring. I asked him once how he and his wife collaborated, and he replied, "We don't. We lead parallel lives." It doesn't help to be loved if you can't let yourself feel it.

When I asked Paine at seventy-three to describe his oldest friend, he growled, "I don't have any." He could climb stairs only with difficulty and he had great difficulty walking even 100 yards; he couldn't drive at night, and he had had to give up golf because of his gout. Both his wife and his doctor saw him as seriously impaired. Yet on his last questionnaire, in a shaky hand, the seventy-five-year-old Paine called his general health "very good" and reported that he had no difficulty with physical activities. The very next year he was in a nursing home; a year after that his multiple illnesses had killed him at an age when two-thirds of his Study-mates were still alive. Paine's way of dealing with trouble was to wear blinders all the time—a striking contrast with the always-game Boatwright and Adams, who relied on rose-colored spectacles, but mostly to help themselves keep up the good fight.

The life of Daniel Garrick. I never met Daniel Garrick, although I tested his cognitive function by telephone three times—when he was eighty-one, eighty-six, and ninety. I was impressed that he did better at eighty-six than at eighty-one. When I called that middle time, Garrick told me that I was lucky to have gotten him on his day off; he worked the other six. He was very excited about having just taught his first course at the San Francisco Art Museum—on American painters, complete with his own slides. I chose him as a counterpoint to Alfred Paine's early decline and miserable last years on the basis of those phone calls, and because I knew that at ninety-five he was the oldest living man in the Study. However, I didn't know much about his life before he turned eighty, so I had to read his record from the beginning.

Garrick was the oldest of six children. He grew up during the Depression in a family whose yearly income per person was $400— that was $3,200 per year total for a family of eight. His father, an accountant with a high school education, was often unemployed, and couldn't understand his son's decision to go first to acting school and then, at twenty-five, to Harvard. Garrick's father was a very conservative man who read only the *Reader's Digest,* but his mother was a sensitive, artistically inclined liberal, and Daniel modeled himself on her. The Study raters considered his childhood average for warmth.

Daniel was bottle-fed. He was fully toilet-trained by one. From the beginning, he was a "good-natured" and "very friendly" child. In school he got bullied by older children, was completely incompetent at all sports, and failed algebra. But in the ninth grade he had his first success in a school play. He loved the applause, and from then on he devoted himself to school theatricals, at which he excelled. Upon graduation from high school, he enrolled in drama school. He rejected New York's bohemian life and at that time didn't drink or smoke, but he took part in summer stock productions every July. He only wanted to act in plays by serious playwrights like Shakespeare and Ibsen, and he had an idealistic wish to invite the public (perhaps primarily his philistine father) into the aesthetic realm that he and his mother inhabited with such pleasure. He wasn't able to support himself as an actor in Depression-era New York City, however, and he came reluctantly to the conclusion that he was too emotionally inhibited to be a successful actor, despite his passion for the theater.

Garrick resigned himself to a future of stage managing and teaching drama in some small college. To this end he put himself through Harvard without family help or scholarship support. A private mental hospital gave him room and board in return for working the graveyard shift as a psychiatric attendant. He would sleep from 6:00 P.M. to midnight, cover his shift (which gave him some time to do his homework), and then bicycle the five miles to Cambridge at 8 A.M. to save the carfare. Reading this tale of an iron constitution, I found myself wondering about the truism that exercise leads to long life. Maybe it's the other way around. Maybe some people are just born with the stamina to live to ninety-five, and that stamina makes them good exercisers, too.

Even with his great physical toughness, Garrick's college life was difficult. He had only five dollars a year to spend on clothes. There was no time for a social life—no dances, no dates, no sports. He was chronically tired and unable to earn better than C's until he married, just before his senior year. Once his wife, a young summer stock actress, started helping with expenses, he received honor grades.

The Study staff did not all relate to Garrick. He received one of the top ratings in the Study for being "well-integrated" and "self-starting," and one observer appreciatively called him an "intense fellow." But when he was twenty-six, the Study psychiatrist summed him up this way: "He was chosen for the Study for outward manifestations of good health and stability. He has actually attained only very modest success in the field he has sought. . . . Although he has some insight and is making progress in understanding himself, he does not yet, and probably never will, reach the understanding necessary to be a happy person." The Grant Study was not always good at understanding artists—it rejected Norman Mailer and Leonard Bernstein!—and despite his manifold successes, Garrick's score on the Decathlon was only 3—average.

A decade after he joined the Study, Garrick was given a "D" for personality stability, a "D" for mental and vascular health, and a "B-" for projected longevity. Even Clark Heath called his health only "fair." The case discussant dismissed Garrick's wish to share his cultural pleasures as the "prestige motivation" of a show-off. Another staff member predicted "character disorder." The Study staff of the time expected people to aim at business, law, or medicine, and there definitely seems to have been a blind spot for artistic types.

Garrick was thirty-three when his local repertory theater went bankrupt. "My nerves were torn to shreds," he wrote. Still too emotionally constrained to act, he gave up theater work as too stressful, and returned to a safer goal. At forty he saw himself as "mediocre and without imagination," but he acquired the necessary Ph.D. and set out to make his way teaching drama and theater history in the "small college" of his imagination.

When I got to this point in my reading, I began to fear that my memory was playing tricks on me. Was I erroneously recalling the delightful telephone conversation of ten years before? Was my exemplar nonagenarian really a depressed narcissist? I didn't yet know that his doctoral dissertation had been published and won prizes, or that, when he was eighty-eight, a grateful former student gave his college a substantial gift in Professor Daniel Garrick's name. All I could see in Garrick's future was years of disappointment. Like Garrick himself, I viewed him as a failure.

My anxiety drove me to cheat. I opened his latest Study folder, where the first thing I saw was a newspaper clipping. Actually it was more than a clipping; it was the whole front page of the *San Francisco Chronicle*'s Sunday Arts section. The headline read: "For Daniel Garrick, 89, It's Always a Full House." I heaved a sigh of relief and went back to the chronological story. As Freud once said, "Before the problems of the creative artist, analysis must, alas, lay down its arms."[15] Clearly I had more in common with those early Study investigators than I wanted to acknowledge. I needed a reminder that artists need time to mature. Another reason for longitudinal studies. . . .

Garrick went through a lot of occupational and marital unhappiness in midlife, and both the Study staff and I dismissed him as a failure on the strength of it. He had given up his career of choice and settled for second best. At fifty-three he separated from his first wife, and his sex life for a while was so uninspired that he turned to archery in sublimation. But as it turned out, those setbacks were temporary. Garrick spent the second half of a very long working life in one distinguished repertory company after another. At eighty, he had long stopped needing archery; many fifty-year-olds would have envied him his passionate marriage. Daniel Garrick was not the only Grant Study man who rediscovered in middle age a life of feeling that had been left behind in childhood. If women, at least until recently, have been socialized into giving up vaulting ambition at menarche, boys in grammar school are encouraged to abandon right-brain heart for left-brain reason. If they're lucky, the tide may shift in maturity, so that they can develop conscious emotional lives. I can never resist quoting E. M. Forster in this context: "Only connect the prose and the passion and both will be exalted."[16]

As he got older, Garrick finally outgrew the emotional inhibitions that had so plagued him as a young actor. Neuroscientist Francine Benes has demonstrated that with maturity the frontal cortex of the brain, the executive center, becomes more securely attached to the brain's emotional center, the limbic system.[17] In his early fifties, Garrick quit his "day job," giving up his tenured drama professorship to become a paid actor in a Shakespearean repertory theater out West. That is the point at which I'd say he achieved true Career Consolidation, although even before that he had already received standing ovations in an off-Broadway *King Lear*. About that performance, one reviewer wrote: "Garrick is the greatest Lear I can imagine. He appears to be the living man, in all his arrogance, petulance, fear, and confusion. A truly unforgettable experience."

Daniel Garrick had matured emotionally to the point where he could now support himself by sharing not just his ideas, but his feelings, with an audience. He no longer needed to separate them, as young people, defensively, often do (see Chapter 8). He had also learned how to play; I think of Marlon Brando as the semi-retired Godfather, chasing his grandson joyfully around his garden. At fifty-five Garrick took his first vacation ever. After the long sedentary years of teaching, he began jogging and returned to long bicycle rides. And there's the chicken-egg question again. Was he unusually healthy because of all the bicycle touring at fifty-five? Or was the bicycle touring possible thanks to an unusual endowment of good health?

At seventy-six, Garrick had a stint of alternating performances of *King Lear* with a more famous actor. A reviewer wrote, "I found Garrick's interpretation the stronger in many ways. . . . his fury, his self-pity, his pain, and his madness more vital and deeper reaching. I feel very fortunate to have seen him." Garrick would eventually play opposite Gwyneth Paltrow, Lynn Redgrave, and Olympia Dukakis.

Garrick got married again at seventy-eight, to his longtime companion. When he was eighty-six, Maren Batalden wrote, "Garrick. . . . is thoughtful, intelligent, warm, self-aware, humble, curious, engaged, emotionally present, enthusiastic, honest. . . . He is extraordinarily successful in both love and work. His marriage to Rachel is clearly a deep, joyful connection that sustains him, surprises him, and makes him weep with gratitude. The marriage is remarkable for the freedom each has given the other to be independent—they each have separate spheres, their own friends and interests and social styles, but nevertheless enjoy each other's company and love each other's bodies." It's wonderful to hear that bodily love can arise *de novo* so late in life, and exist so passionately after eighty. Discoveries like that require longitudinal study.

Batalden continued, "In many ways, these seem to be the best of times for Daniel Garrick—he now relishes the fruits of his discipline. He loves loosely and well, works passionately, keeps his mind and heart open and engaged, makes new friends, relishes his hobbies, takes real nourishment from the past, permits himself a wide range of deep sorrow and great joy, adapts stoically to the inconveniences of aging in his body by staying focused on the things that matter to him."

Since age eighty, Garrick has enjoyed the highest possible mood scores. He did give up biking at seventy-six due to multiple aches and pains; nevertheless, at eighty-six his only medicine was Viagra "as needed"—twice a week, he boasted. Yes, he had begun to feel his age, but for the next five years he kept up with the demands of a major Shakespearean theater, memorizing his lines and rehearsing, and all the while serving as docent and teacher at the local art museum. Garrick's son noted changes when his father reached ninety-five. "My father has declined physically and cognitively to a far greater degree than he acknowledges in the questionnaire: gradually over the past 8–10 years, but most precipitously since retiring in 2006 from his acting career. He had worked less for a few years due to his arthritis pain and difficulty remembering lines. When he finally stopped, he has sat in his chair, read books, watched TV, listened to music, ordered more books and music and exercised hardly at all. His work at the theatre had provided a regular dose of social contact and the necessity of pushing himself to analyze the play and character in order to deliver a performance. Both the social interaction and the need to push himself [has now] dwindled to a trickle."

It's hard to say with certainty that the decline was really the result of a lack of social and physical activity. Surely the reverse is also possible: that it was more like the collapse of the one-hoss shay after its very long life, and that Garrick, as he approached his own hundredth birthday, gave up his earlier social and physical pleasures as a result. Reality probably lies somewhere in between. Until his death Garrick took only two medications: Aleve for arthritis and Ambien for sleep. At ninety his TICS score was showing some borderline decline, and at age ninety-one, he had a brief trial on Aricept (a drug designed to slow memory loss). When he was ninety-four, his wife, who had become very frail, had to move to assisted living. For a while Garrick remained full of beans, but at ninety-five he began to display some trouble orienting himself. That's when his son began to worry and came to live with him, doing his shopping, figuring out his taxes, and paying his bills. Garrick himself recognized that he had slowed down—he listed his activities on his last questionnaire as: "Read, watch TV and nap." But he didn't seem to be unduly troubled by the change. Daniel Garrick died at ninety-six.

Remember, Garrick was only the first of the men to reach ninety-five, and a gang of mentally alert Study-mates are waiting in the wings to surpass him. Remember, too, that for all of us the play has to end sometime. From sixty to ninety, Garrick had a long and successful run.

Can we learn anything from Daniel Garrick's life about the keys to graceful aging? Or about what might account for his extraordinary longevity? He certainly didn't follow most of the conventional "rules" for long life. His parents were not long-lived. (In fact our data indicate that parents' longevity is only modestly reflected in their children.) He didn't really start to exercise until he was almost sixty. He smoked a pack of cigarettes a day for twenty years, and nine pipes a day for decades after that. During his sixties, he drank enough that he and his future wife worried about it. Conventional wisdom tells us that we keep young by doing good for others, but he engaged in no community service activities except to become a docent in an art museum—at eighty-six. So when Garrick suddenly began riding his bicycle up to 100 miles a day, was that the cause of his great energy, or was it the effect of some innate natural vigor? Remember, way back in college Garrick had bicycled 5 miles back and forth to class after being up all night, and had gotten top marks for being "well-integrated" and "self-starting," both traits associated with longevity in the Study.

From the very beginning, Garrick had a quality of indomitability, and maybe that was really what enabled him to live so long. Nobody in the Grant Study waited longer or worked harder than he did to get a college education. Most students who lacked financial support received scholarships, but not Garrick. Most of the men who had explicit career ambitions were able to realize them long before they turned fifty. But even during the period when Garrick had resigned himself to being a teacher, he used his spare time to act. He never gave up, and he never gave up hope.

PATHS TO HEALTHY AGING

Sometimes it looks as though cancer and heart attacks are visitations from malicious gods, and that old age is ruled by cruel fate—or at least by cruel genes. So out of control does the whole process of aging often feel that I was relieved when our wealth of prospectively gathered data revealed that some aspects of successful aging—or lack of it—are in fact negotiable.

Twenty years ago Paul Baltes, a leading gerontologist, acknowledged that research has not yet reached a stage where there is good causal (as opposed to correlational) evidence for predicting healthy aging.[18] True, several distinguished ten- to twenty-year prospective studies of physical aging have contributed valuable understanding about the course of old age.[19] But few of the investigators in these studies knew anything about their members before they turned fifty, and only Warner Schaie has followed his subjects for more than twenty-five years.[20] The early facilitating factors in longevity remain unclear. But that doesn't stop people from having strong convictions about them.

For example, we are constantly urged to follow diet, exercise, or neutraceutical regimens that supposedly guarantee us sustained good health. But the formulas are rarely subjected to long-term verification, and the rules keep changing. The cod liver oil of the forties was replaced by the antioxidants of the seventies, which gave way to fish oil again ten years ago. For a while eggs were out, red meat was on everybody's no-no list, and fruit was on the side of the angels. But one of the few thorough studies of nonagenarians has found that 80 percent of them ate red meat regularly all their lives, and only 50 percent ate fruit weekly.[21] Perhaps fruit-eating vegetarians do not survive to ninety. Or perhaps, I, like the diet advocates, am merely revealing my own personal prejudices. Either way, survival is not as simple as the wellness gurus would have us believe. The risk factors that cure-all diets are supposed to counter aren't discrete entities; they are reflections of complex interactions.

My own confrontation with the arbitrariness of longevity prescriptions happened twelve years ago at the Boston Museum of Science. If visitors would punch in all their bad habits, a computer there promised, it would tell them how much longer they had to live. I told the computer that I consumed a half-pound of butter daily and three packages of Camels. I only rarely let a green vegetable touch my lips, I emptied a fifth of Jack Daniels every evening, and I got off the sofa only to change the batteries in the remote. The computer thought for a while, calculated the years that each risk factor was supposedly subtracting from my life, and then informed me that I was dead already, and had been since two years before I was born!

Now I'll grant that this is not a nice way for a person to behave who has given much of his life to collecting, fostering, and analyzing an invaluable hoard of data based on interviews and questionnaires. If the men of the Grant Study had been one-tenth as mendacious as I was. . . . Still, simply summing multiple risks together is very different from watching their complex interactions over real lifetimes, and here's a serious illustration of that.

Harvard sponsored a symposium on aging for its own 350th anniversary. Four experts were asked to describe their research on risk factors. One spoke on exercise, another on obesity, a third on smoking. I spoke on alcoholism. None of us addressed the importance of any factor other than our own, let alone the mischief that the four of them could get into together. Like the computer, we treated the different factors as if each were the only coffin nail in town.

In this chapter, I've tried not to do that. I'm going to recount the history of four separate investigations in which we examined the antecedents of successful aging and longevity. Each led me to a different set of conclusions, and thereby hangs a tale.

FOUR STUDIES OF AGING

Study I, 1978: Mental health and physical health.   In 1978, when I was in my early forties and the men were turning fifty-five, I viewed them as over the hill and therefore ripe for a study of successful old age. That's not much of an exaggeration. I thought that they were nearing their apogee, and that I'd better learn what I could before the long decline began. For my first study—the causes of health decline between age forty and age fifty-five—I began with the 189 active members of the Study who had been entirely healthy at forty. Eighty-eight of them had remained that way until fifty-five. What were the differences, I wondered, between them and the 101 men who by that age had either acquired minor chronic illnesses (66) or suffered serious chronic illness or died (35)? In 1979 I believed that Table 7.2 provided the answer.[22]

Remember, I was a psychiatrist, not an internist and not an epidemiologist. I wanted an answer to a straightforward question: What was the influence of mental health on physical health? I tested out a number of previously assessed psychological variables: bleak childhood, recourse to mood-altering drugs, major depression, number of psychiatric visits, poor relational skills at midlife, and especially a lack of good adult adjustment between the ages of thirty and forty-seven (as defined by the capacity to work, love, and play, and require no treatment for mental illness). As Table 7.2 shows, all six of these psycho-

Table 7.2 Factors Associated with Irreversible Physical Health Decline Before Age 55

	Important Before Age 55 (n = 189)
A. Mental Health Variables	
High use of mood-altering drugs	Very Significant
Major depression	Significant
Bleak childhood	Very Significant
Poor adjustment 30–47	Significant
Poor relationships at 47	Very Significant
Immature defenses	Significant
B. Physical Health Variables	
Heavy smoking	NS
Alcohol abuse	Very Significant
Ancestral longevity	NS
Education	NS

Very Significant – $p < .001$; Significant = $p < .01$; NS = Not Significant.

logical variables were significantly associated with subsequent poor physical health at fifty-five. By then, of the forty-nine men with the bleakest childhoods, seventeen (35 percent) were dead or chronically ill, contrasted with only five (11 percent) of the forty-four men from the warmest childhoods—a very significant difference. On the other hand, by age fifty-five, six (9 percent) of the sixty-six non-smokers and eight (17 percent) of the forty-eight heavy smokers had become chronically ill—a difference of no statistical significance at all.

I had made my case; mental illness was very important in hastening the aging process. Every five years I'd trot out these six variables and run them again, looking to see if they were still statistically important factors in looking at physical health. For fifteen years, they were. So far, so good. But alas, like the Science Museum computer, I had not fully controlled for other important risk factors. It took several decades before I statistically separated decline in health after age fifty-five from decline in health before age fifty-five. In retrospect, that should have been a no-brainer. Yet another of the many advantages of longitudinal studies is that they uncover sloppy thinking.

Study II, 2000: Decline in health between fifty-five and eighty. As the men began passing eighty, I was starting to wonder what factors might be associated with the development of dementia. The fact that there is frequently a vascular component to this dread condition led me to establish a five-point scale of vascular risk factors: smoking, alcohol abuse, hypertension, obesity, and Type II diabetes. For the first time I examined the five factors not only individually but as a group; summing risk factors increases their predictive power. When these factors proved to be strongly associated with dementia, I turned them loose on the larger question of physical health in general. And that led to some big surprises.

This time I followed until they were eighty the 177 men who had still been healthy or had had only minor illnesses at fifty-five. And I found with astonishment that the relationship between mental and physical health that had seemed so clear when the men were fifty-five no longer held. As Table 7.3 illustrates, the deterioration of the men's health between ages fifty-five and eighty was very significantly associated with the vascular risk factors, and not at all with the mental health risk factors. Of the 189 healthy men selected at age forty, 103 had no risk factors, and 86 had one or more. Sixty-five (that is, 76 percent) of the latter group were chronically ill or dead by their eightieth year. Of the men with no risk factors, only 45 (44 percent) were dead or disabled. This was a very significant difference.

But the early mental health risk factors predicted nothing about health at eighty. It appears that they did their damage early. The effects of past damage endured, and remained statistically significant even when the effects of alcohol, tobacco use, obesity, and ancestral longevity were controlled. The river of time flowed on, but poor mental

Table 7.5 Factors Associated with Irreversible Physical Health Decline at Different Age Periods

	Important Before Age 55 (n = 189)	Important Between 55 and 80 (n = 177)	Important Between 80 and 90 (n = 155)
A. Mental Health Variables			
High use of mood-altering drugs	Very Significant	NS	NS
Major depression	Significant	NS	NS
Bleak childhood	Very Significant	NS	NS
Poor adjustment 30–47	Significant	NS	NS
Poor relationships at 47	Very Significant	NS	NS
Immature defenses	Significant	NS	NS
B. Physical Health Variables			
Vascular Risk Factors:	Very Significant	Very Significant	Very Significant
a. Smoking	NS	Significant	NS
b. Alcohol abuse	Very Significant	Very Significant	NS
c. High diastolic blood pressure at 50	Very Significant	Very Significant	Significant
d. Obesity	Very Significant	Very Significant	NS
e. Type II diabetes	Very Significant	Very Significant	NS
Ancestral Longevity	NS	NS	NS
Education	NS	Significant	NS
Cognitive Competence at 80	NS	Very Significant	Very Significant
Diagnosis of Cancer	NS	Very Significant	Very Significant

Very Significant = $p < .001$; Significant = $p < .01$; NS = Not Significant.

health did no further damage. The vascular risk factors, as I documented subsequently (see Table 7.3), were important before fifty-five as well as after, a fact at first unsuspected by this young research psychiatrist. Maturity improves the discernment of researchers, too!

Study III, 2011: Getting to ninety.   The men got older, and I continued to try to identify the variables that allowed some of them to age successfully to ninety. I kept my eye on the vascular risk factors, and again I saw a pair of once-important variables lose their statistical significance as time went by. When the men were between eighty and ninety, the vascular factors as a summed total were still significant. Because so few heavy smokers and alcoholics survived into the ninth decade, however, the significance of those two individual factors faded after eighty. Of the forty-five heavy Study smokers, only sixteen were still alive at eighty and only four at ninety. Only three of the alcohol-dependent men lived until eighty and none until ninety. But the other vascular risk factors told a different story. They continued to affect the men's ability to remain cognitively sharp after eighty; and after eighty dementia was a major source of health decline. Of the men with no vascular risk factors, 50 percent were still alive at ninety, instead of the expected 30 percent. Of the seventeen men with three or four risk factors, all but one was dead ninety years after his birth.

Table 7.3 sums up the findings of all three of these studies; the first column refers not to the original analysis of the age fifty-five data, but to my later reappraisal, which coded individually for vascular and genetic risks. N in each column refers to living men; one of the reasons that predictive associations change is that certain risk factors, like badly adapted parasites, kill off their hosts, leaving themselves with nothing left to live on.

Study IV, 2012: How to live forever.   In 2012, as I write this, sixty-eight of the active Grant Study men are still alive; fifty are still cognitively intact. I wanted to learn what I could about what makes for good health and good mentation over so many years. Ideally this study will be done again ten years from now, by which time we will know exactly when each of the men died. But to bring it all together in a provisional way, I made the best estimate I could of how much longer the men would live, based on past experience. I added seven years to a man's current age if he still had only minor health problems, five years if he was chronically ill, and three years if he was disabled. Obviously there's speculation here, but after the age of ninety there's a limit to how wrong the guess can be. Then I plotted all the men's lifespans against many factors that had been significant in other contexts to see what I could learn about their relationship to longevity. Table 7.4 is the result.

Men with no vascular risk factors lived to an average age of eighty-six. Men with three or more lived to an average age of only sixty-eight. This complex of factors subtracted eighteen years from a man's expected life. The difference in longevity between men with very long- and very short-lived ancestors was seven years—eighty-four years as opposed to seventy-seven—a significant difference, but nothing like that predicted by the vascular risk factors. Surprisingly, parental social class made less than a year's difference. The men with the very best and worst adjustments at forty-seven lived to eighty-five and seventy-seven respectively—an eight-year difference in longevity that was very significant, but again far less than that produced by the vascular risk factors.

A curiosity: the sixteen men with the most combat exposure in World War II died significantly younger than the rest of the sample. They were six times as likely to die from unnatural causes (e.g., murder, suicide) and they were twice as likely to die before eighty. This is another of those unanticipated findings that in the coming years may turn out to be a clue to something unexpected and important.

An interesting independent predictor of longevity, which I recog-

Table 7.4   Factors Associated with Longevity (Mean Age at Death 81.5; N = 237)

	N	Mean Age	Statistical Significance
A. Significantly Shortened Longevity			
Alcohol dependence	18	67.6	Very Significant
3 + vascular risk factors	17	68.3	Very Significant
Heavy smoker	40	73.0	Very Significant
High combat	16	75.3	Significant
Neither well integrated nor self-driving	83	78.8	Significant
Only a college degree	81	79.0	Very Significant
Bottom sixth in ancestral longevity	36	77.1	Significant
B. Significantly Lengthened Longevity			
No vascular risk factors	113	86.4	Very Significant
Never smoked	77	84.7	Very Significant
Good adjustment age 30–47	58	85.1	Very Significant
Low diastolic blood pressure at 50	66	85.9	Very Significant
Top sixth in ancestral longevity	44	84.3	Significant
Both well integrated and self-driving	23	89.3	Significant
Graduate school	153	83.3	Very Significant
C. Did Not Significantly Affect Longevity			
Warmest childhoods	57	83.0	NS
Bleakest childhoods	61	80.0	NS
Cholesterol above 253 mg/dl	57	81.1	NS
Cholesterol below 206 mg/dl	57	82.9	NS
Low exercise (age 20–45)	79	80.4	NS
High exercise (age 20–45)	25	85.0	NS
Parental social class: Upper	80	82.4	NS
Parental social class: Blue collar	14	81.6	NS
No psychosomatic problems	50	82.0	NS
2 + psychosomatic problems	48	83.7	NS
No obesity	212	82.2	NS
Obese (BMI 28+)	23	79.3	NS
Mature defenses	39	84.0	NS
Immature defenses	34	78.4	NS
Highest social supports	54	84.1	NS
Lowest social supports	59	80.1	NS

Very Significant = p<.001; Significant = p<.01; NS = Not Significant.

nized only after Friedman and Martin published *The Longevity Project* in 2011, was the presence or absence in college of personality characteristics of perseverance and self-motivation (in the early Study terminology, *Well Integrated* and *Self-Driving*).[23] These traits were significantly associated with longevity; the twenty-three men who manifested both in college were likely to live ten years longer than the men with neither trait (see Table 7.4).

HOW *NOT* TO LIVE FOREVER

Most of the deaths before fifty-five were associated with factors over which the men had no control. We cannot pick our parents, or control the genes that lead to major depression. The vascular factors that were so strongly implicated in deaths after fifty-five, however, are a different story. They are all now widely accepted as causes of death before eighty, and we were able to confirm this. Using multiple regression analysis (unlike the Boston Museum of Science computer!) to investigate each of the five factors in the context of the other four, we found that each one made an independent contribution to health decline and death.[24] As Table 7.3 illustrates, there was a very significant association between these factors and irreversible decline in health in the Grant Study men. The risk factors were of equal importance in the Glueck Inner City cohort and the Terman cohort of gifted women.[25] But unlike the previously cited mental health factors, they are causes of early death over which we have considerable control. At the end of the day, good self-care before age fifty—stopping smoking, joining AA, watching the weight, and controlling the blood pressure—made all the difference in how healthy the men were at eighty and ninety.

Education is another factor to be reckoned with when it comes to health; it turned out to be critically important to the health of the Inner City sample (and modestly important for the Grant Study men, whose educational baseline was higher). The average College man who never sought post-graduate education had a life expectancy of seventy-nine years, while for the men with post-graduate educations, life expectancy was eighty-three years—a very significant difference. Years of education were powerfully associated with reduction of all of the five factors affecting vascular health. This was especially true of the Inner City men. This issue will be dealt with more fully in Chapter 10.

OUT OF OUR HANDS?

However, one interesting piece of conflicting evidence suggests that we still have to reckon with our stars. A recent paper by Thomas Perls and his colleagues on a cohort of 800-plus centenarians suggests that genes are just as important to survival past one hundred as lifestyle.[26] In *Aging Well* I could assert that ancestral longevity was not an important predictor of survival from fifty-five to eighty. But to live past eighty, it appears, ancestral longevity again assumes significance. Another reminder that follow-up must continue as long as life does.

Cancer.   Cancer is an increasingly important cause of death after seventy, and this was another factor that often seemed to lie in the men's stars—or, if you prefer, in their genes. With the exception of lung cancer, which is closely tied to smoking, cancer in the Grant Study seemed surprisingly independent of both mental health and the vascular risk factors. In larger samples, however, and in studies focused on specific cancers, environment, dietary and sexual habits, and alcohol abuse can also be demonstrated to be important.

Age and cognition.   We used the Telephone Interview for Cognitive Status, or TICS, to measure cognitive skills in men without an existing diagnosis of dementia.[27] In the Grant Study, the factors most associated with a high TICS score appeared to be the absence of vascular risk factors, good vision, college IQ and class rank, exercise at sixty, and, surprisingly, a warm relationship with one's mother when young. This is another curiosity. As we've gone along, I've pointed out that a warm childhood relationship with his mother—not maternal education—was significantly related to a man's verbal test scores, to high salary, to class rank at Harvard, and to military rank at the end of World War II. At the men's twenty-fifth reunion, it looked, to my surprise, as though the quality of a man's relationship with his mother had little effect on overall midlife adjustment. However, forty-five years later, to my surprise again, the data suggested that there *was* a significant positive correlation between the quality of one's maternal relationship and the absence of cognitive decline. At age ninety, 33 percent of the men with poor maternal relationships, and only 13 percent of men with warm relationships, suffered from dementia.

Dementia, like arthritis, is a curse of longevity. Vital aging at ninety is closely dependent upon the preservation of cognitive faculties, and the best way to achieve this is by minimizing the vascular risk factors. Alzheimer's disease, however, is a special case. It is a major source of health decline after eighty, but, unlike the purely vascular dementias, it seems to be surprisingly independent of the factors that I have been enumerating.

Death before eighty can be avoided to some degree by wise lifestyle choices, but so far the Study has provided no clues as to how to prevent the two greatest and most dreaded sources of total disability after eighty—most cancers and Alzheimer's disease.

WHAT WAS NOT IMPORTANT

Ancestral longevity.   Lacking lifetime studies of humans, scientists have studied aging in fruit flies. You can breed and study many generations of fruit flies in a year, and their longevity, it appears, depends heavily upon genes. Therefore, one of the first variables the Study looked at was ancestral longevity.

The myth that ancestral longevity is passed on is hard to test well; this is because most people are either so old that they no longer remember the exact ages of their grandparents' deaths, or so young that they don't yet know when their parents will die. It takes at least a two-generation study to learn from the subjects' parents the age that the grandparents died, and then to follow the subjects until the last of their parents have died as well. (In the Grant Study, the last parent died in 2002, sixty-five years and three generations after the Study began. The Study men themselves were very unreliable informants about the age of death of their grandparents.)

For the College men, we calculated ancestral longevity by averaging the ages of death of the oldest of the first-degree ancestors (parents and grandparents) on the maternal and paternal sides. As I've noted, the longevity of the forty-four men with the longest-lived first-degree ancestors was seven years longer than the longevity of the thirty-six men with the shortest-lived ancestors. That difference was significant, but just barely. Moreover, to my surprise, the average lifespans of the ancestors of the men with the best and worst mental and physical health at eighty were identical.[28] In the Inner City sample as well, the longevity of the men's parents seemed irrelevant to the quality of their own aging at seventy.[29] Fruit flies are clearly not always good models for human aging. Obviously, specific genes are very important in predicting the specific illnesses that shorten life, and soon the precise genes that facilitate longevity may be discovered. In the meantime, however, the McArthur twin studies and investigators using the Swedish Twin Registry confirm our finding that variance in longevity cannot simply be attributed to genetic inheritance.[30]

Psychosomatic stress.   When the Study began, psychosomatic medicine was in high fashion. Hans Selye had shown the world that stress could kill, and psychoanalytic theories about the role of emotions in medical illness were all the rage, attributing peptic ulcers to repressed anger or longing for love.[31] It would be decades before the medical profession brought itself to accept that the most common cause of duodenal ulcers is a gram-negative bacterium of the genus *Helicobacter.*

Besides, many people *do* feel physically sick under stress. They have headaches; they can't sleep; their stomachs ache; they get itchy; they're running to the bathroom all the time. This observation suggests an attractive hypothesis, to which in the 1960s I earnestly subscribed: that individuals who experience stress psychosomatically in midlife would suffer poor physical health in old age.[32] When I joined the Grant Study in 1966, I was thrilled to be in possession of prospective Study data that I suspected would prove this hypothesis.

A secondary hypothesis in psychosomatics, to which I also subscribed, held that individuals have characteristic "target organs" in which they experience stress.[33] This proposition was based on the observation that one person's place of experiencing somatic symptoms under stress may be quite different from another's, and it was much emphasized in my residency training. Theoreticians anticipated that the organ affected by stress would be the same one in which signs of psychosomatic illness appear.

A third related hypothesis, one that I felt certain of proving, was the implicit assumption that the development of "psychosomatic" illnesses (for example, colitis, asthma, and hypertension) reflect more psychopathology than the development of "real" illnesses (such as diabetes, myocardial infarction, and osteoarthritis). Indeed, many of the mental health screening tests of the 1950s and '60s, some of which, like the Minnesota Multiphasic Personality Inventory, are still in use today, use the presence of multiple physiological symptoms under stress as indicators of emotional illness.

In the 1970s I finally analyzed the data to test these three psychosomatic hypotheses so favored by me and other armchair speculators.

My intention was to demonstrate 1) that psychosomatic illness leads to accelerated aging; 2) that "target organs" for stress (the stomach, the lungs, and so on) are real and remain stable over time; and 3) that psychosomatic illness is a reliable hallmark of mental illness. Longitudinal study proved them all wrong.

(Note that the evidence for this belief system had been derived over the years from retrospective data. Furthermore, the source of the data was patients who presented themselves repeatedly for medical attention—a group likely to have mental health issues.[34] The Grant Study, on the other hand, was a prospective study of men who explicitly were *not* patients.)

Over the years, the Study men were systematically asked where in their bodies they experienced emotional stress. Decades of follow-up revealed that this site varied considerably over time.[35] The idea of stable target organs (hypothesis II) was not supported.

Nor did the number of physical symptoms under stress before age fifty predict physical health at age seventy-five or at ninety. If you wait a few decades, people often recover from psychosomatic illnesses. And at age eighty, the physical health of the men with multiple apparently psychosomatic illnesses was actually, if insignificantly, better than that of men with none. Hypothesis I was not confirmed either.

Hypothesis III, the supposition that there's a connection between psychosomatic illness and mental health, didn't fare any better than the other two. The Study has carefully followed the men's objective physical health from the age of forty, when it was still very good, up through the present or until their deaths. By fifty, over half of the Study men had required medical treatment for conditions thought by some physicians to be psychosomatic: hypertension, respiratory allergies, ulcer, colitis, and chronic musculoskeletal complaints. Admittedly, these five disorders do not define psychosomatic illness conclusively, but they make a good start. At age forty-seven, after almost thirty years of observation, the men were ranked for emotional health. There was no correlation with the number of psychosomatic complaints. The number of psychosomatic illnesses that the men suffered didn't predict mental health at age sixty or eighty, either. As a faithful believer in psychosomatic medicine, I was profoundly disappointed.

We also found that the childhoods of the men who experienced relatively few physical symptoms under stress and no psychosomatic illnesses were no warmer in blind ratings than those of men cursed with a great deal of psychosomatic symptomatology.[36] As I've noted, in our first study of aging, premorbid psychopathology did correlate with early onset of chronic (real) physical illness, but only up to the age of fifty-five. We had some middle-aged Study men who frequently visited medical doctors and who took more than five or more sick days a year. They displayed our distinguishing markers for mental health issues—drinking, psychiatric intervention, use of tranquilizers and antidepressants, and so on—and in this they resembled the self-selected populations of the old retrospective studies. In other words, people with mental health issues brought more illnesses of all kinds—real and imagined—to the attention of the medical community.

When in 1970 I first presented these findings at a national psychosomatic meeting, I was met with outrage and disbelief as a traitor to dynamic psychiatry. Today, my findings would be regarded as unremarkable. Over the last forty years, a medical model of psychosomatics has displaced the earlier one, thanks in part to the *Helicobacter* story and the fading influence of psychoanalysis in academic psychiatry. My research life, and the Study's, spanned both eras.

Cholesterol. Magazines would lose valuable advertising revenues, and probably readers, if they talked too much about really significant risks to health—like Virginia Slims and wine coolers. But it's OK to fuss about cholesterol, because the butter and egg lobbies don't advertise in *Cosmopolitan*. Moreover, as the TV ads tell us, your cholesterol can be magically lowered by statins, even if you don't eat less or exercise more. So the war on cholesterol would appear to be a win-win-win situation—for patient, for doctor, and for the pharmaceutical industry.

Countless studies show that the ratio between high- and low-density lipoproteins (HDLs and LDLs) is important and that reducing LDL levels is good for the heart. But in neither the College nor the Inner City sample did average cholesterol levels at age fifty distinguish the men who lived to ninety from those who died before eighty. The estimated age of death of the fifty-eight Grant Study men with cholesterol levels below 206 mg/dl was eighty-three. The estimated age of death of the fifty-seven men with cholesterol levels over 254 mg/dl was eighty-one, which was not a statistically significant difference. This finding has been confirmed by much larger and more representative studies.[37]

It's situations like these that persuaded me to include my experiment with the Boston Museum of Science computer—not as entertainment, but as an alert. To understand longevity we need longitudinal images, not snapshots. The more sacred our cows, the more they need longitudinal testing. This is one of the ones we were able to test.

Bleak childhood.   Alas, we cannot choose our families. Without asking permission they endow us with their genes. They also bathe us in their warmth and riches—or parch us with emotional and financial poverty. Many aspects of childhood are important to aging, as I've described in Chapter 4. But most of them do not predict length of life. Parental social class, stability of parental marriage, parental death in childhood, and IQ (at least in our samples with their two different but limited ranges) were not important to longevity. The men from the warmest childhoods lived only a year and a half longer than the men from the bleakest childhoods, a difference that was not statistically significant.

Vital affect and general ease in social relationships. These were the two personality traits most highly valued by the original Study investigators, and for the Harvard cohort, they correlated highly with good psychosocial adjustment in college and in early adulthood.[38] But they, too, failed to predict healthy aging.

CONCLUSION

After seventy-five years of collecting, analyzing, and reanalyzing our data on aging, what have we learned from all this? First, we've been reminded again and again that a strong association does not necessarily imply cause. Snow is associated with winter, but snow does not *cause* winter. Heavy smoking is strongly associated with fatal automobile accidents, but not because drivers take their eyes off the road to search for the cigarette lighter. The association is the result of a third factor—alcoholism, which very significantly increases the likelihood both of heavy smoking and of fatal accidents.[39] The seventy-year duration of the Grant Study have been invaluable in allowing us to draw lines between causes and associations.

For example, there is a strong association between exercise and physical health, which most of us understand to mean that exercise causes good health.[40] But might it not be the other way around? Healthy people enjoy exercise. Among the College sample, exercise at age sixty correlated more highly with health at age fifty-five than with health at eighty. Health at fifty very significantly predicted exercise at eighty, and health at sixty significantly predicted exercise at eighty. In other words, health predicted exercise at all ages, but exercise did not predict health in later years.

Admittedly, exercise at age thirty did significantly predict health at fifty-five and sixty, and exercise at sixty predicted health at age seventy through eighty-five, although not significantly. Presumably the exercise mavens must not be written off entirely. But it's important to keep in mind—as the Museum of Science computer didn't—that everything affects everything else, and that some things are horses and others are carts. When it comes to physical aging, alcohol abuse and education are looking more and more like horses. And, as Table 7.3 documents to my sorrow, after thirty years of betting the farm that maturity of defenses (the involuntary coping mechanisms that I'll discuss in the next chapter), would be the horses that pull us to late-life physical health, longitudinal study has proved me wrong. He who lives by the sword dies by the sword.

Social supports are often assigned a causal role in successful aging. In his classic review of the evidence, however, sociologist James House acknowledged that almost no attention has been paid to social supports as a dependent variable.[41] That is, social supports may be the result of the very variables that they are supposed to be causing. In the prolonged prospective view of the Grant Study, social supports at age seventy were strongly associated with the pre-age-fifty protective health factors identified in Table 7.2, yet only weakly associated with longevity (Table 7.3). In other words, good health predicts good social support better than good social support predicts future health. Indeed, good social supports in old age may be in large part a result of earlier habits that preserve physical health.[42]

There was a powerful association between the absence of cigarette and alcohol abuse before fifty and good social supports at seventy. Specifically, there were some with good supports among the

Grant Study men who abused alcohol or cigarettes. There were also men with poor social supports but no tobacco or alcohol abuse. In these asymmetric pairs, the risky habits of alcohol and cigarette use were the horse, and social supports the cart. Study men with good social supports but risky habits suffered health just as poor as men with poor social supports and risky habits. Men with poor social supports but good habits enjoyed good health almost as often as men with good social supports and good habits.

My point here is not that love and exercise aren't good for you. For most of us, the more social supports we have in our old age the happier we'll be, and the more exercise we get the better we'll feel and look. I only wish to underscore that the etiology of successful aging is multifactorial in ways that self-help books and cross-sectional studies do not necessarily take into account. Furthermore, the Grant Study demonstrates, if you follow lives long enough, the risk factors for healthy life adjustment change. There is an age to study the relation between mental and physical illness and an age to ignore it. There is an age to curse your arthritis and an age to appreciate it as the price of having survived to attend your granddaughter's wedding.

Please note too that I report these contrary findings as an investigator who claimed in Chapter 1 that love is the root of all blessings, and who believed, and for many years tried to prove, that mental health caused physical health. I've had to give up on that second claim, alas; my own worldview is as vulnerable to the upheavals of longitudinal follow-up as anyone else's. But I will retreat no farther than that. I am wiser than the Museum of Science computer, and I have grasped clearly that when it comes to healthy aging, everything really *is* connected to everything else. A happy old age requires both physical health and mental health. For mental health, love is a necessity. So is being alive. So is being able to think straight. We need physical and cognitive competence to build the social surrounds that give us love and support later on, and it is love and support that encourage us to care for ourselves well and keep ourselves healthy, even when the going gets rough. "Button Up Your Overcoat" isn't so far off the mark after all. The ninety-year-olds of the Grant Study took good care of themselves and of their important relationships. And for the most part, they've been very happy to be alive.

RESILIENCE AND UNCONSCIOUS COPING

The mechanisms of defense serve the purpose of keeping off dangers. It cannot be disputed that they are successful in this; and it is doubtful whether the ego could do without them altogether during its development. —SIGMUND FREUD

ONCE UPON A TIME at an amusement park in Florida, I watched some passengers (including my grandson) on a loop-the-loop roller-coaster. They gathered speed, swept up the curve, and hung suspended upside down at the top, waving their arms with excitement. I could see that for them the experience was one of ecstasy, exhilaration, and release. But it seemed to me that their elation, like the *Ode to Joy* of the angry and depressed Beethoven, reflected serious denial. For me, there would be nothing even remotely pleasurable in an experience like that. Just thinking about it I could feel the stress ulcering the lining of my stomach, upholstering the walls of my coronary arteries, and overwhelming my immune system with an avalanche of corticosteroids. A ride like that would take years off my life. But my grandson was over the moon.

By what alchemy had the brains of the laughing riders transformed into exaltation an experience that in me would evoke only misery and fear? It's not that we understood the risks any differently, at least not cognitively. I know, after all, that most of the time nothing really bad happens in amusement parks. It's not that the external physical stressors were any less for them than they'd be for me, dangling head down ten stories above the ground. Styles of conscious stress management don't account for it either; there isn't time for those. The difference is in the ways our individual minds work to convert a concrete situation into an experience of either excitement or terror. Who is sane and who is crazy—the excited teen or the cautious grandfather?

The Grant Study's second lesson, and perhaps the one dearest to my heart, is that any exploration of the links between positive mental health and psychopathology requires an understanding of adaptive coping. As cough, pus, and pain remind us with disconcerting regularity, the processes of illness and the processes of healing look startlingly alike.

In this chapter, I will use the terms *adaptation, resilience, coping,* and *defense* interchangeably; likewise *unconscious* and *involuntary.* The study of psychological adaptive mechanisms began for me years before I became involved with the Grant Study. I had learned in an earlier longitudinal study to admire the means by which some people manage to achieve lasting remission from schizophrenia and heroin addiction. But I was not interested only in the resilience that comes with seeking social supports and devising ingenious conscious coping strategies. We've all got a few of those, and we know about them. I was interested in *involuntary* coping, analogous to the ways we clot our blood and send white cells out to fight infection. Or transform terror, like my grandson did on that roller coaster.

What makes the study of defenses so fascinating is the ambiguity of the boundary between psychopathology and adaptation. Early nineteenth-century medical phenomenologists viewed pus, fever, pain, and cough as evidence of disease, but less than a century later their colleagues had learned to recognize these "symptoms" as involuntary efforts of the body to cope with mechanical or infectious insult. Similarly, psychological defense mechanisms produce behaviors that may appear pathological to others (or even at times to us), but in fact reflect efforts of the brain to cope with sudden changes in its internal or external environment without too much anxiety and depression.

We depend physiologically on multiple elaborate systems of homeostasis, the task of which is to buffer sudden change; we don't faint when we stand up quickly, for example, because our bodies adjust our blood pressure to the demands of the new position. The task of the psychological homeostatic system that I am calling *involuntary coping* is to buffer sudden change in the four sources of mental conflict: relationships, emotions, conscience, and external reality. Defenses are extremely important to comfortable and effective functioning, like our other homeostatic systems. But they are difficult to study. They resemble hypnotic trance in that their use alters the perception of both internal and external reality, and may compromise other aspects of cognition as well.

Even after I got to the Study and the investigation of unconscious coping became a central focus of its research activities, we never applied for an NIMH grant to inquire into this phenomenon. In fact, from 1970 to 2000 defenses took a backseat when we asked for money. Times had changed since the days when psychoanalysts were influencing the research agenda of the NIMH, and by 1970, defenses were too unfashionable for NIMH support. Yet at the Study, we were finding that defensive style predicted the future as well as, if not better than, any other variable we possessed.

The scientific community was trying to do its job when it dismissed defense mechanisms as a holdover from the (now outmoded) metaphysics of psychoanalysis. Yet defenses are real enough. They're elusive, yes, but not like fairies, yetis, and UFOs are elusive, and the other such will-o'-the-wisps that somehow always manage to elude our cameras. Defenses are more like rainbows and lightning and mirages—they're fleeting, but they can be photographed, replicated, and, above all, explained.

It was the many decades of detailed Grant Study recording, like sequential photographs or the transcripts of a neutral observer, that allowed us to identify in real behavior coping measures that are usually invisible to their users. This was necessary, because defenses are unconscious and involuntary. If I say, "You're projecting!" you will reply (probably angrily), "No, *you're* projecting!" and an argument is on that even an outsider wouldn't be able to resolve. A major contribution of the Grant Study, and one of its most appreciated results, has been to make the scientific study of defenses respectable.

WHAT ARE DEFENSES (INVOLUNTARY MENTAL COPING MECHANISMS)?

In 1856, Claude Bernard, a French physiologist and a founder of experimental medicine, started us on our way to understanding adaptation to stress when he wrote, "We shall never have a science of medicine as long as we separate the explanation of the pathological from the explanation of normal, vital phenomena."[1] Pus, cough, and fever are certainly unpleasant, and sometimes dangerous. But they can also be lifesaving; it is these superficially pathological homeostatic responses to physiological stress that in many cases permit us to survive it. In 1925, Adolph Meyer, a founder of modern American psychiatry and an early consultant to the Grant Study, believed that there were no mental diseases, only characteristic reactions to stress.[2] He thought that while patterns of mental reaction like denial, phobias, and even the projections of the paranoid character may look like illness, they may in fact be examples of Bernard's "normal, vital phenomena," promoting adaptation, healing, or at least psychological time-out. Just as fever, clotting, and inflammation use mechanisms that disrupt ordinary bodily equilibriums to do their healing work, so defense mecha-

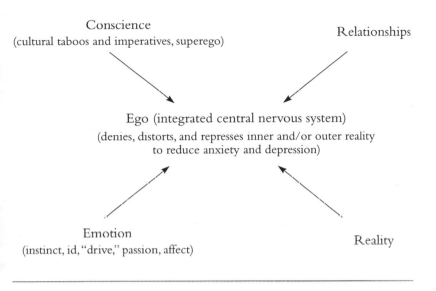

Figure 8.1 The four sources of intrapsychic conflict.

nisms heal through characteristic disruption of ordinary mental processes.

As outlined in Figure 8.1, defenses deflect or deny sudden increases in emotional or biological intensity, such as the heightened aggression and sexual awareness of adolescence. Psychoanalysts call this source of conflict *id*, fundamentalists call it *sin*, and cognitive psychologists call it *hot cognition*. Neuroanatomists locate it in the hypothalamic and limbic regions of the brain.

Defenses also enable individuals to mitigate sudden upsurges in guilt, such as might occur when a child puts a parent into a nursing home. Psychoanalysts call this source of conflict *superego*, anthropologists call it *taboo*, behaviorists call it *conditioning*, and the rest of us call it *conscience*. Neuroanatomists point to the frontal lobe and the amygdala. Conscience is not only the result of admonitions from our parents absorbed before age five, or even of cultural identifications; it is also formed by evolution, and sometimes by the irreversible learning that results from overwhelming trauma.

Defenses can moderate sudden conflicts with important people, living or dead, and protect us from the vulnerability and intensity aroused by sudden changes in intimacy. When a business partner walks out; when a marriage proposal is accepted; when a beloved child receives a fatal diagnosis—situations like these make for anxiety, excitement, or depression that can feel unbearable. In an adolescent striving for identity, even parental love that was once accepted without ambivalence may be unconsciously distorted for a while to make room for psychological separation to take place.

Finally, defenses allow us a period of respite, when necessary, to master inescapable realities that cannot be integrated immediately. They provide a mental time-out without which the individual would become acutely anxious and depressed. That is what would have happened to me had I been so imprudent as to join my grandson on the roller coaster. The events of 9/11 demonstrated this dynamic on a very large scale; the change in self-image resulting from an amputation is a smaller-scale example.

Over a period of forty years, Freud discovered most of the involuntary coping mechanisms that we recognize today, and identified five of their important properties.[3]

- Defenses are a major means of mitigating the distressing effects of both intense emotion and cognitive dissonance.

- They are unconscious.

- They are discrete from one another.

- Although they may look like mental illness—sometimes like severe mental illness—defenses are dynamic and reversible.

- They are potentially as adaptive, even creative, as they are pathological.

I will add one last property to Freud's list:

- To the user, defenses are invisible; to the observer, defenses usually appear as odd behavior.

In 1971 the Grant Study offered a hierarchy of defenses from the psychotic to the sublime.[4] In 1977 and 1993, we were able to show, not just tell, how defenses worked.[5] Over the last twenty years, Cramer, and Skodol and Perry, have reviewed several empirical studies investigating the clinical value of retrieving defenses from Freudian oblivion.[6] As a result, the fourth diagnostic manual for the American Psychiatric Association (DSM-IV) finally organized defenses into a hierarchy of relative psychopathology similar to ours, and formally included it as an optional diagnostic axis.[7]

A HIERARCHY OF DEFENSES

All defenses can effectively minimize the experience of conflict, stress, and change, but they differ greatly in their consequences for long-term psychosocial adaptation. Here they are organized into four levels, from least to most mature.

The psychotic defenses include *delusional projection, psychotic denial,* and *psychotic distortion.* They involve significant denial and distortion of external reality. They are common in young children and in dreamers, as well as in psychosis. To alter them requires altering the brain— either by maturation, by waking, or by the use of neuroleptic drugs.

The immature defenses include *acting out, autistic fantasy, dissociation, hypochondriasis, passive aggression,* and *projection.* Immature defenses externalize responsibility and are the building blocks of character disorders. They are familiar to most of us by observation. Immature defenses are like cigars in crowded elevators—they may feel innocent to the user, but observers often experience them as deliberately irritating and provocative. Defenses in this category rarely respond to verbal interpretation alone.

   The intermediate defenses include *displacement* (kicking the dog instead of the boss), *isolation of affect* or *intellectualization* (separation of an idea from the emotions that go with it), *reaction formation* (turning the other cheek), and *repression* (keeping the affect visible but forgetting the idea that gave rise to it). Intermediate defenses keep potentially threatening ideas, feelings, memories, wishes, or fears out of awareness. They are frequently associated with anxiety disorders, but they are also part of the familiar psychopathology of everyday life, and they may be seen clinically in amnesias and in displacement phenomena like phobias, compulsions, obsessions, and somatizations. Intermediate defenses tend to be uncomfortable for their users, who may seek psychological help for that reason, and they respond more consistently than the lower-level defenses to psychotherapeutic interpretation. Intermediate defenses are common in everyone from the age of five until death.

   The mature defenses include *altruism, anticipation, humor, sublimation,* and *suppression.* By allowing even anxiety-laden feelings and ideas to remain in awareness, they promote an optimum balance among conflicting motives and maximize the possibility of gratification in complicated situations. *Altruism* (doing as one would be done by), *anticipation* (keeping future pain in awareness), *humor* (managing not to take oneself too seriously), *sublimation* (finding gratifying alternatives), *suppression* (keeping a stiff upper lip) are the very stuff of which positive mental health is made. Although they may appear to be under conscious control, unfortunately they cannot be achieved by will-power alone; just try to be really funny on demand. Furthermore, their deployment must be facilitated by others who provide empathy, safety, and example. If Gandhi had lived under Hitler instead of Churchill, he would have been a victim, not a hero.

Even mature defenses alter feelings, conscience, relationship, and reality in the service of adaptation. But they achieve their distortions gracefully and flexibly. Observers tend to regard the adaptive defenses as virtues and to experience them as empathic; their success depends on sensitive connections with others, as can easily be seen in humor and altruism. The immature defenses are not considered virtuous at all, and are usually experienced as manifestations of narcissism.

ASSESSMENT OF DEFENSIVE STYLE (IDENTIFICATION OF DEFENSES)

We can't see the spinach caught in our own teeth, and it's difficult to identify our own defenses. This means that self-report measures to assess defenses have limited validity. Even when a person consistently and correctly recognizes when he is being defensive, he may still misidentify the kind of defensive behavior he is using (that is, he may label it incorrectly). Furthermore, "defense mechanisms," like "character traits," are abstractions. What does it mean to distinguish between ma ture or empathic coping ("good" denial) and immature or narcissistic/ unempathic coping ("bad" denial)?

An observer who wants to identify and appraise defenses objectively must triangulate between present behavior, subjective report, and past truth. Let's say, for example, that one woman founds a shelter for battered women and another breaks her toddler's arm in a tantrum. Defenses are unconscious; the first woman may attribute her altruistic behavior to a need to rent her house; the second may call the damage she did in her fury "an accident." It is only once we learn that social agency records from thirty years before reveal that both women had been taken at age two from the care of physically abusive alcoholic mothers that the defensive nature of both of their behaviors becomes apparent. Triangulation brings clarity to the defensive/adaptive nature of otherwise inexplicable behavior. You need the agency record and the mothers' explanations and the two tangible behaviors to label these unusual actions accurately.

The Grant Study's method of rating defenses can best be understood through a medical analogy. A symptom is a physical oddity. Perhaps the patient himself notices it, or perhaps a friend makes a comment. The patient may or may not understand what is going on. But when he goes to have it checked out, he'll be asked extensively about how it feels to him and what he notices about its context. He'll be examined; any relevant tests will be run. His self-report will be correlated with objective information from his medical history and recent workup. Eventually the symptom can be properly labeled and put in the context of its probable cause and mechanism.

Defensive behavior (including defensive symptoms or even creative products) is similarly called to our attention when something strikes us as odd or out of character. Once we've noted the oddity, we need to triangulate between what the subject says about it, what we know about his current circumstances, and the biological and biographical facts of his history. But objective documentation of mental health is scarcer and harder to come by than objective documentation of physical health. Like many other facets of mental health, therefore, the reliable identification of involuntary coping mechanisms requires longitudinal study. This is spelled out in much greater detail in my *Ego Mechanisms of Defense: A Guide for Clinicians and Researchers.*[8]

In the Grant Study, we had thousands upon thousands of pages of the men's self-reports, and also a great deal of observational, historical, and objective material recorded by others. We triangulated among these to assess defensive style. When the men were 47, we picked out of their reports examples of behavior that we (or others) thought odd, selecting the ones that seemed most characteristic of the subject's way of operating. We considered them in conjunction with the observations of the many members of the Study staff over the decades, and also in the context of the historical information available to us, and the vast background of objective test and other documentary evidence that had accrued over the years. The selections that appeared on this scrutiny to be defensive were excerpted as 100-word vignettes. We compiled an average of two dozen such vignettes for each man. Then we asked blind raters to identify and label the vignettes for defensive style, as per the hierarchy above. We made certain to check the reliability of the raters—that is, to make sure that raters came to similar conclusions—and then we scored the resulting defenses for their level of maturity. One of Anna Freud's great contributions to the understanding of adaptation was her recognition that you can learn more about children's defensive styles by watching them play than by listening (as her father did) to free association and dreams. And that's essentially what we did; many of the College men enjoyed day jobs that were more like play than work.

There were two weaknesses in our exploration. We analyzed the defenses of only 200 of the men; this was an immensely time- and labor-consuming process, and we ran out of money before we could complete the work on the other 68. And our rater reliability was shaky on the defense of dissociation. For almost four-fifths of the men, however, and for all but one of the defenses, we had extensive and reliable identifications of characteristic defensive styles.

Table 8.1 illustrates the maturational levels reflected in our vignettes of adaptation at different ages. The Study found that adolescents were twice as likely to use immature mechanisms as mature ones. Young adults were more than twice as likely to use mature

Table 8.1   Maturation of Defenses Over the Lifespan

| | Adaptive Category Assigned to Vignettes* | | |
	"Immature"	"Intermediate/Neurotic"	"Mature"
Adolescence	25%	61%	14%
19–35	12%	58%	30%
35–50	9%	55%	36%
50–75	6%	32%	62%

*The vignettes are the ministories excerpted from the men's histories as examples of adaptive coping behavior, and then identified and rated.

mechanisms as immature ones. Between ages thirty-five and fifty, individual men were four times likelier to choose mature mechanisms than they had been when they were adolescents.

We found that defensive style was relatively independent of nurture, and that was not a surprise. Almost by definition, adaptive mechanisms are the ego's antidotes to a poor environment; its adaptive capacity reflects the difficulties it has mastered at least as well as the assets with which it was originally blessed. Like Godfrey Camille, some individuals became progressively better adapted as they mature. Folk wisdom recognizes this: Pearls are what oysters do about irritation, and what doesn't kill you makes you stronger. Damaged children in their growing can find ingenious ways to compensate for earlier deprivation. Northwestern University psychology professor Dan McAdams recognizes this reality as the "redemptive self."[9]

Maturity of defensive style was not associated with parental education or social class, and bore no relationship to tested intelligence, body build, or physical responsiveness to stress. The use of mature mechanisms correlated very highly with variables that reflect psychiatric health and warm human relations.[10] And we found a very significant association between immature defenses and genetic vulnerability (that is, number of ancestors with major depression, alcoholism, or shortened longevity). The depressed and the alcoholic were most likely to use immature mechanisms later in life. Fifty percent of the ten men who had two mentally ill parents manifested immature patterns of defense, and only 9 percent of fifty-five men without mentally ill parents showed a predominantly immature pattern of defense. Of course, this finding could also be explained as identification (monkey see, monkey do).

DOES DEFENSIVE STYLE MATTER?

Once we had learned to rate maturity of defenses reliably, the next question was: Did choice of mechanism matter? Would the men's coping styles tell us any more than their handwriting had about their lives? To answer this question, for each of three cohorts, the 18 frequently used individual defenses were clustered into the four groups (psychotic, immature, intermediate, mature) I've outlined above. Correlations were calculated between the Study member's maturity of defenses and other indications of successful adult development. As Table 8.2 indicates, the associations were impressive. Maturity of defenses as measured from age twenty to forty-seven predicted the men's Decathlon scores at age seventy to eighty with a correlation of .43, which is very, very significant. Choice of involuntary defense mechanisms clearly does matter. Furthermore, the positive associations between maturity of defenses and mental health were, like our other maturity ratings, independent of the social class, education, and gender of the Study members.

Let me depart from abstraction now to show what the hierarchy of defensive styles looks like in real life. Here are the life stories of two men who have also appeared in *Adaptation to Life*. A few words first, though, about the chief adaptive mechanisms of the men you'll meet here.

Table 8.2 Correlations Between Maturity of Defenses and Measures of Successful Adult
Outcome

Variables	Terman women n = 37	College men n = 186	Inner City men n = 307
Life satisfaction, age 60–65	.44 Significant	.35 Very Significant	n.a.
Psychosocial maturity (Erikson stage)	.48 Very Significant	.44 Very Significant	.66 Very Significa
DSM-IV Axis V	.64 Very Significant	.57 Very Significant	.77 Very Significa
Job success, age 47	.53 Very Significant	.34 Very Significant	.45 Very Significa
Marital stability, age 47	.31 Significant	.37 Very Significant	.33 Very Significa
Job enjoyment, age 47	.51 Very Significant	.42 Very Significant	.39 Very Significa
% of life employed	.37 Significant	n.a.	.39 Very Significa

Very Significant = p<.001; Significant = p<. 01; NS = Not Significant.
n.a. = data not available.

Sublimation, suppression, intellectualization, and *repression* are four
distinct ways of dealing with desires and impulses that are experienced
as unacceptable or dangerous. In sublimation, the emotional energy
fueling these desires is invested in other goals that are acceptable (and
frequently also socially valuable). In suppression (sometimes called *stoicism*), the desires and emotional energy are held together in consciousness, and the anxiety and depression they evoke either tolerated
or ignored, until a way to pursue them safely and appropriately is discovered. In intellectualization (sometimes called *isolation of affect*), forbidden ideas remain in awareness but divorced from their emotional
intensity. In repression, unacceptable desires and impulses are excluded
from awareness along with the ideation associated with them, but the
emotion remains powerfully present, with no provision made for immediate or future gratification. Like all of the defense mechanisms I'll
be discussing in this chapter, these four are not employed deliberately
or even consciously; their use is automatic and involuntary, like the
mobilization of fever in the face of infection.

Repression is a short-term solution to the complexities and contradictions of emotional life. Our desires and impulses often have to be handled with care, but to dismiss them wholesale from awareness is to dismiss a great deal of emotional vitality as well. Repression serves us best, therefore, as a respite from conflict and confusion, a temporary battening down of the hatches until we are ready to look for more enduring ways to keep from being overwhelmed by our feelings. Intellectualization is much the same. It permits forbidden thoughts and desires to remain in awareness, but without their passionate emotional valence. Intellectualization allows for cerebral satisfaction and sometimes very valuable advances in understanding—one Grant Study man grew his mother's cancer in his lab as a tissue culture—but at the risk of alienation from emotional experience.

Sublimation and suppression are two more satisfactory long-term adaptive techniques. Sublimation—finding outlets for our passions—is the gift that allows successful artists to work themselves into the very marrow of our bones. In general we do not like being made to weep, but when Mahler, Verdi, Plath, Shakespeare, and Jesus do this, we bless them for transmuting the bitter poison of their own lives into an elixir of salvation.

In comparison, suppression is about as glamorous as a Mack truck. Perhaps less mysterious than the other three, it is an important workhorse of an adaptation, and a very successful one. Of all the defenses it was the one most associated with being *well integrated* in college, and with excellence at the events of the Decathlon three decades later. Sublimation, for all its elegance, was less associated with success or with joy. Artists are not known for freedom from mental illness, as Beethoven exemplifies—only for the joy they bring to others.

One way of coping with the hurly-burly of reality is to retreat into the pleasures of the mind. Several of the Grant Study men who went into teaching used this type of sublimation as a successful coping style. They tended to have productive and distinguished careers, like Daniel Garrick and Dylan Bright (below). The professors like Peter Penn who relied heavily on intellectualization as their major coping mechanism, however, tended to experience their jobs and their marriages as sterile, and they showed far more evidence of emotional problems.

THE LIFE OF DYLAN BRIGHT

Professor Dylan Bright was a vivid illustration of the coping potential of sublimation. Although he was less gifted intellectually than the average Study member, an exciting luminescence surrounded his life. As soon as I walked into his office, he put his feet up onto the desk and started to talk. He looked more like a prizefighter than an English professor, and I was moved by his affective richness. But initially I was not quite sure that I liked him. His first response to my request for an interview was, "Christ, that kills the afternoon!" His aggression was barely tamed and verged on the abrasive, and only his charm kept me from regarding the encounter as a pitched battle. He graphically described his worries and then growled, "If they get out, I'll kick your teeth in."

Bright was a football linesman and champion wrestler who, almost as an afterthought, became a professor of poetry. In high school he greatly preferred the excitement of athletics to the dreary world of the classroom; he was a rebellious D student, and at one point was nearly expelled. Nevertheless, his headmaster saw Bright as "vibrant and ardent in his beliefs," and the Study staff perceived him as "an eager, enthusiastic, attractive youngster with an outgoing personality." Bright's intensely competitive spirit, which he eventually harnessed in the service of his academic career—that is, sublimated—was never extinguished. His energy, his extraordinary capacity to win close friends, and his knack for exciting hedonistic activities made him one of the most dramatic members of the Study.

Unlike suppression and anticipation, the use of sublimation was not associated with particularly happy childhoods. Like Beethoven, Bright grew up in a family filled with turmoil. His father was an emotionally unstable alcoholic who was rarely at home and whose hobby was hunting. Dylan saw his parents' marriage destroyed by fighting, and early on he tasted both the triumphs and the dangers of (figuratively) taking the place of his father. His mother was an exuberant and energetic woman who was three inches taller than her son even after he reached adulthood. Her charm was appreciated by several Study observers, but from the beginning she taught Dylan to beware of instinctual pleasure. Before his first birthday he had been cured of thumb-sucking, bedwetting, and soiling himself. At two, his mother made him wear mittens to bed because of his "perfectly revolting" habit of masturbating. As a child, he conceived of God as "a person looking down on me, ready to conk me on the head with a thunderbolt."

In his lifelong quest to conquer fear, when Dylan Bright let go of his mother's skirts he became a daredevil quasi-delinquent. As a youth he incurred more cerebral concussions than anybody else in the Study—there's a statistic for you! But with the passage of time, his mastery became more graceful. After eighteen, he began concentrating on learning to do what he called "responsibly adventurous things," and there were no further injuries. First he was an all-state football linesman in high school, and then a fiercely competitive college wrestler. He played tennis for blood, shunning doubles for the joys of single combat. After college, once his devotion to tennis and wrestling had been replaced by an equally fierce devotion to poetry, he was still out to win. He raced through graduate school at Yale with top grades. He accepted an appointment at Princeton for the prestige, and a few years later exulted in his early acquisition of academic tenure.

Bright did not start out with creative and empathic ways of dealing with his feelings. As he and his ego matured, however, he replaced acting out (his delinquent rebellion) with reaction formation (containing emotional impulses by doing their opposite). For example, he suddenly found the first girl with whom he had slept "revolting"— unconsciously choosing the same word that his mother had used to condemn his sexual experimentation in infancy—and gave up intercourse with his next girlfriend out of an ascetic desire to "see if I could." In college, the once delinquent Bright seriously considered a career in law enforcement. As a young and lusty English instructor at Princeton, his devotion to enforcing parietal rules surprised the administration as much as it irritated the students. Even in middle life, Professor Bright conceived of his success in terms of rigid control. "If a person does not have self-discipline," he cautioned me, "he can go to rot so fast."

Bright didn't consider the possibility that his rigid self-control might not in fact have been protecting him, or that it might even have been steering him toward an empty disaster of a life. But it was only as the reaction formation of his youth gave way to sublimation that he caught fire. He risked his amateur standing by wrestling in exhibition matches, but "consecrated" his illegal fees by investing them in violin lessons. During his nineteen-year-old abstention from sexual intercourse, he substituted a close and exciting intellectual friendship and made his first discovery of poetry. He fought to be first in his graduate school class, but he gentled his naked ambition by writing his Ph.D. thesis on the poetry of Shelley.

When Bright was thirty-five, his wife broke up what had been a very close marriage. At the same time, he was becoming aware that his scholarship, although adequate to win him tenure at Princeton, would never win him national recognition. This was a critical period. Faced by two real defeats, for a while he lost himself in alcohol, careless affairs, and stock car racing—the poetry professor regressed to adolescent acting out. But he quickly replaced his temporary delinquencies with more acceptable and productive quests for excitement. He went scuba diving in the Aegean with a close friend, and made a discovery that allowed him to reinterpret a line of Homer's *Odyssey*. "Oh," he told me, "that was a heady experience."

Bright's adaptive responses were ingenious indeed. Sexual abstinence (this time following upon his divorce) once again brought him into relationship with a brilliant colleague who remained a friend for life. He withdrew from academic competition in which he anticipated only defeat, and involved himself instead in activities that permitted him to master danger with minimum risk and at the same time to anesthetize grief with real excitement. Sublimation not only facilitated the efficient expression of his instincts, it also permitted Bright to avoid the labels of "neurotic" and "mentally ill." He had once described himself as "a laughing man. I just let things slip off my back." But he didn't do that by permanently drowning his troubles in alcohol, or in self-destructive chance-taking, or in Scarlett O'Hara's ambiguous mantra of denial, "I'll think about it tomorrow." His capacity for sublimation enabled him to change the terms of his life. He went on the wagon to control his incipient alcoholism. He made a successful second marriage. He remained in touch with his feelings while softening them with excitement, laughter, and people. Asked if he had ever seen a psychiatrist, he spoke instead of his second wife and his best friend: "Professional assistance would be a pale shadow compared to these companions." Like art, love is an act of creation, but love far surpasses art as a cure for emotional suffering. Bright died at sixty-two. Eighteen years of heavy drinking and forty-five years of a two-pack-a-day habit had led to lung cancer. But like the Marlboro man, he died with his boots on.

THE LIFE OF FRANCIS DEMILLE

Francis DeMille grew up in suburban Hartford. He never knew his father, a businessman who left home before his birth and died shortly thereafter. His father's relatives played no part in his upbringing, and the DeMille household consisted only of Francis, his mother, and two maiden aunts. From one to ten, he lived in an entirely female ménage, with a playroom that he used as a theater. He was encouraged to play by himself on this stage; in fact his mother proudly reported to the Study that he "never played with other boys."

When he joined the Grant Study in 1940, Francis DeMille looked hardly old enough to be in college. His complexion was as fresh as a girl's, and despite his erect carriage, he struck several observers as rather effeminate. He was in the top 8 percent in "feminine" body build. But he also impressed the staff with his charm. His manner was open, winning, and direct, and he discussed his interest in the theater with a cultivated animation.

The staff psychiatrist marveled that at nineteen DeMille had "not yet begun to think of sexual experience." In fact, Francis as a college student "forgot" to an astonishing degree to think about sexual fantasy, aggressive impulses, or independence from his mother. He didn't remember his dreams well either, and he reported that "distressing emotional reactions fade quickly." He didn't date in college, he totally denied sexual tension, and he blandly observed, "I am anything but aggressive." He was a poster child for repression.

In retrospect it's a little hard to understand how DeMille came to be included in a study of normal development. It seems that dramatic skill got him in. The staff may have wondered at his sexual oblivion, but still they perceived him as "colorful, dynamic, amiable, and adjusted." Francis took an active and enjoyable part in college dramatics; the Study staff thought that his mother was pushing him into the theater, but Francis seemed unaware of that. Characteristic of people who use repression as a major defense, he reported that he preferred "emotional thinking to rational thought."

Like many actors, DeMille was also a master of dissociation, or neurotic denial. He found it "revitalizing" to free himself from inhibitions by becoming someone else in a play, and so "venting my emotions." Despite the fact that the staff worried about his inner unhappiness, during psychiatric interviews he seemed "constantly imbued with a cheerful affect." What, me worry? Alfred E. Neumann was good at dissociation, too.

When Lieutenant DeMille managed to remain at his mother's side both emotionally and geographically through the whole of World War II, the Study internist began to fear that he was going to be a lifelong neurotic. The Navy never took him farther from Hartford than the submarine base at Groton, Connecticut. But it was in the Navy, with continued maturation, that DeMille's repression at last began to fail. At first this was disconcerting and quite anxiety-provoking. He became aware of his peculiar lack of sexual interest, and was fearful about possible homosexuality. Discussing this problem in a Study questionnaire, he made, as many repressed people do, a revealing slip of the pen. "I don't know whether homosexuality is psychological or psychiological in origin." As it turned out, his unconscious was right; there was nothing physiologically wrong with his masculinity. Later in life he would father three children.

In manageable doses, anxiety promotes maturation, and over time DeMille began to replace repression and dissociation with sublimation. He wrote to the Study that he was always rebelling against the Navy, standing up for his own individuality and for that of his men.

His own reports of his behavior might have led us to call this passive aggression, an immature and destructive defense. But his military
efficiency records gave him his highest officer efficiency rating in
"moral courage" and "cooperation." This heretofore compliant man
had turned at least one rebellion into a veritable work of art that even
the military could appreciate.

By twenty-seven, DeMille's worries abut possible homosexuality
had been put to rest. "I enjoy working with girls!" he announced joyfully. He had found a job teaching dramatics at Vassar. He managed the
short move from Hartford to Poughkeepsie, thus gratifying the "great
necessity I feel for breaking away from home." Three years later, he
shattered maternal domination still further by marrying an actress
whom he had once directed in his amateur theatrical group. His marriage today may not be the best in the Study, but it is above average
and it has survived for more than fifty years.

DeMille was also becoming more aware of his use of repression as
a defense. In reply to a questionnaire that asked him about his marital
sexual adjustment, he wrote, "I must have a mental block on the questionnaire. My reluctance to return it seems to be much more than
ordinary procrastination." He was in conflict over his sexual adjustment, and he knew it. He was in conflict over work, too, mentioning
in that same questionnaire some concerns about the desire to work
aggressively for money. He could recognize now that on the one hand
he needed to feel that he wasn't greedily seeking money, but that on
the other, "I didn't tackle the career problem properly." However, once
involuntary coping styles have become conscious, they no longer
"work." And with insight comes resistance. That was the last the Study
was to hear from DeMille for seven years. He never saw a psychiatrist
during this time. But in keeping with his new capacity for sublimation, he wrote a successful comedy for his amateur theatrical group
entitled *Help Me, Carl Jung, I Am Drowning.* And after he returned to

the Study, he revealed the following ingenious way in which his ego had sublimated his issues with money and aggression.

If DeMille was more mercenary than he had wanted to think, his solution to the conflict was artistic. In his twenties, he had vowed that he would never associate himself with the "specter of American business," but in fact he left Vassar to return to his mother's city of Hartford, where Insurance is king. There, despite his theatrical interests, he became a very special corporate success story. In an industry not known for its openness to individual expression, he crafted a niche on the advertising side that gave him autonomy, high management perks, and a chance to exercise his dramatic flair. He took pains, however, to assure me that his success in the marketplace threatened no one and that he was not "overly aggressive." As he put it, "The ability to stay alive in a large corporation took all the craftiness I have." Only in his community theater group could he unashamedly enjoy playing aggressive roles.

In an interview at age 46, DeMille shared a vivid new recollection of a hyper-masculine uncle who had been an important, if previously unmentioned, role model during his adolescence. This was the recovery of a lost love, and another softening of his stern earlier repression of masculine role models, and all without psychotherapy. Five years after our interview, DeMille elaborated further on this uncle, whom he called "the only consistent male influence—very dominant—a male figure that earlier I had rejected." But not entirely, it seems, for with his pipe, his tweed jacket, his leather study furniture, and the bulldog at his feet, the middle-aged DeMille now rather resembled his uncle. The charming emotional outpourings of his adolescence were gone; he now hid emotions behind lists, order, and a gruff, hyper-masculine exterior. "In college," he said, "I was in a Bohemian fringe; but I've changed since twenty-five years ago. Maybe some clockwork ticked inside me and made me go down this route."

Perhaps this new identification explains why a few years earlier he had given up his mother's religion and was "suddenly smitten" with the Episcopalian tenets of the father he had never known. Certainly he hoped that his sons would never discover that *their* father had once worn long hair; by 1970 DeMille thoroughly disapproved of unshorn locks. Folks do change.

At sixty, after his mother died, DeMille took early retirement, delighted to be liberated from corporate life. But before he left his Hartford suburb for rural Vermont, he was honored as "first citizen" of the town to which he had given sixteen years of devotion, finally including a stint as chairman of the town historical board.

Once he was settled in Vermont, an acknowledged Guardian, his community service only increased. Over fifteen years of retirement he led one successful fundraiser after another. He rebuilt the small community church; he made possible the new library. He wrote, directed, and starred in plays for the town summer theater. He coped with life's kicks in the teeth with composure and admitted that he wasn't too much of a patriarch to let his wife take care of all the taxes and bills. But he still said of himself that "my mind is blank to things I don't want to remember."

"Sometimes make-believe and reality get mixed up," he had confessed at forty-seven. DeMille was a man who had always situated himself in places where he had access to a stage and did not have to play exactly by the rules. As a child he had preferred his playroom to the schoolyard; at the insurance agency he constructed for himself a special niche where he could do exactly what he wanted. Still, he had built a library and a church where neither had existed before. And in three different communities he was the town historian whose remembering of the past—admittedly, someone else's past—served the future. And in retirement his devotion to play and to plays was no longer a problem; in fact he was rewarded for it. He did not look for tragic roles like Lear; instead, he adapted realistically by playing (for pocket change) the leading role in *On Golden Pond*. As playwright and director, he could make life just the way he wanted it, and the results were real, not make believe. For Garrick, acting was a living and a passion; for DeMille it was a coping mechanism.

DeMille didn't begin "paring down" until after he turned seventy-five, when he was still taking seven-mile hikes and was still a community star. Until eighty, he saw life as a "pretty good ride," but after that he began to fail. He cut his walks back to just a mile. His wife died when he was eighty-five, and he developed dementia. He survived to ninety, but in a nursing home and unable to walk. Mature defenses affect how we feel about aging, but they can't, alas, guarantee that we'll live happily to a hundred and then collapse all at once like the one-hoss shay. It took the Grant Study twenty-five years to teach me that Daniel Garrick had been very lucky indeed.

PROSPECTIVE VALIDATION OF THE HIERARCHY

Do mature defenses make it easier for people to find joy in living? Or is it that joy in living allows us the luxury of mature defenses? I wanted to know whether maturity of defensive style has predictive validity as well as the uniformly positive correlations in Table 8.2. Predictive validity means that an association is not just a statistically significant coincidence, but that it can—as the term implies—predict the future reliably.

We approached the question this way. Raters unfamiliar with the lives of the College men before they turned fifty assessed their joy in working, their use of psychiatrists and tranquilizers between the ages of fifty and sixty-five, the stability of their marriages, and the course of their careers (that is, whether they had progressed or declined) since age forty-seven.[11] This assessment was then correlated with an assessment of defensive style. Mature defensive styles assessed between ages twenty and forty-seven were very significantly predictive of a superior adjustment at sixty-five for the College men (see Table 8.3). Only two men of the thirty in the bottom quartile of defenses (as assessed between twenty and forty-seven) were in the top quartile of adjustment at age sixty-five, and the mean Decathlon score for these thirty men was only 1.4 at age eighty. The men with the most mature defenses had Decathlon scores three times higher (4.6), a very significant difference, and of those thirty-seven men only one was in the bottom quartile of adjustment at sixty-five.

Then we contrasted the defenses of the twenty-three College men who at some point in their adult lives were clinically depressed with those of the seventy least distressed College men (that is, the men who over thirty years of observation eschewed both tranquilizers and psychiatrists, and never appeared to merit a psychiatric diagnosis).

Table 8.3   Late Life Consequences of Mature Defenses at Age 20–47 for the College Cohort

	n = 154*
I. Objective Evidence	Adaptiveness of Defenses
Income (50–55)	Very Significant
Psychosocial adjustment (50–65)	Very Significant
Social supports	Very Significant
Decathlon (60–80)	Very Significant
Mental health (64–80)	Very Significant
Good marital outcome (50–85)	Very Significant
II. Subjective Evidence	
Joy in living (75)	Very Significant

Very Significant = $p < .001$; Significant = $p < .01$; NS = Not Significant.

* Sample size reflects the fact that we did not have information on all the men on all the variables.

Table 8.4 Use of Defenses by the Most Depressed and Least Distressed Men in the College Sample

Defenses	Percent Using Each Defense as a Major Coping Style		
	Most Depressed (n = 17)	Least Distressed (n = 59)	Significance
Suppression	18%	63%	Very Significant
Altruism	0%	19%	NS
Reaction formation	28%	2%	Significant
Dissociation	59%	25%	Significant
Projection	24%	3%	Very Significant
Most mature	0%	31%	Very Significant
Least mature	53%	9%	Very Significant

Very Significant = p<.001; Significant = p<.01; NS = Not Significant.

The two groups showed a very significant difference in the overall maturity of their defenses.

As shown in Table 8.4, 61 percent of the least distressed and only 9 percent of the most depressed men exhibited generally mature defenses; whereas 53 percent of the depressed men and only 9 percent of the least distressed men consistently favored less mature defensive styles. Although altruism, a mature defense, is used by many people with unhappy childhoods to master adult life, sustained use of altruism was never noted as a major adaptive style among the depressed College men.[12] Instead, the most depressed men were more likely to use reaction formation and passive aggression, whether anger was directed against others or themselves.

Here, too, simple association does not prove cause. Immature defenses are associated with alcohol abuse and brain damage, but they do not cause them; rather, both alcohol abuse and brain damage can cause regression in maturity of defenses. Similarly, the association of immature defenses with depression is likely not a simple one. In some people, severe depression probably leads to a regression in adaptive style as less mature defenses, once developmentally surpassed, come back to the fore. In others, depression and immature defenses may both be responses to unmanageable stress, disordered brain chemistry, or both. Immature defenses probably predispose some people to depression. Much more evidence is needed to clarify the relationship between affective disorder and maturity of defenses.

To me, perhaps the most fascinating question of all was: Does the assessment of defenses in young adulthood predict future physical health? For the first twenty years that I worked with the Study, I fervently believed that because defenses mitigate stress, mature defenses would lead to better physical health than immature ones. In this belief I was wrong. For at least ten years after the men's defensive styles were recorded and rated at age forty-seven, the health of the men with mature defenses deteriorated less quickly. And, as I established in Chapter 7, immature defenses were one of the mental health variables that predicted decline in physical health in the period between forty and fifty-five. By age sixty-five, however, the association of mature defenses with continued good health could no longer be discerned. Once again, prolonged follow-up demolished both theory-based conviction and short-term evidence.

DEFENSIVE STYLE, GENDER, EDUCATION, AND PRIVILEGE

A major effort of the Grant Study on my watch has been to dissect the psychological mechanisms of homeostasis by which human beings achieve resilience in the face of sociocultural challenges. In biological medicine the task is easier. Blood-clotting is an elegant example of unconscious homeostasis, but the fact that the Romanovs died young from hemophilia and their peasants did not was not a class issue; on the contrary, it was a trumping of social class by genes. Clotting factors are distributed in an egalitarian fashion. It would be nice to think that biology is as democratic in distributing coping skills as it is about clotting factors and immune mechanisms, but there is room for doubt. Many aspects of mental health are a function of education, IQ, social class, and/or societal gender bias.

It is of great interest, then, that maturity of defensive style did not seem to be affected by socioeconomic status, intellectual ability, or gender. It is true that the College sample looked a little better than the Inner City men in this regard (11 percent and 25 percent respectively used predominantly immature defenses). But that can be explained by the original selection process; the College men were selected for mental health, while the Inner City men were not; in fact they had been deliberately matched with delinquents. Mental health and defenses are closely correlated.

A less distorted view of the effect of privilege on defensive style can be achieved through within-group comparison—that is, by comparing the members of one group to each other. In this way initial selection bias is circumvented. Table 8.5 examines the effect of social class, IQ, and education upon differences in defensive maturity within

Table 8.5 Correlations Between Maturity of Defenses and Biopsychosocial Antecedents

Background variable	Terman Women (n = 37)	College Men (n = 186)	Inner City Men (n = 277)*
Parental social class	NS	NS	NS
IQ	NS	NS	NS
Years of education	NS	—	Significant
Warm relations with father	NS	Significant	NS
Warm childhood environment	Significant	Significant	NS
Warm relations with mother	NS	Significant	NS

Very Significant = $p < .001$; Significant = $p < .01$; NS = Not Significant.
* Thirty men with IQs less than 80 were excluded.

three different groups. The associations are insignificant. Even the relationship of a warm childhood environment to maturity of defense is less than one might expect.

The effect of cultural diversity was tested in the following manner. Sixty-one percent of the Inner City men had parents who had been born in a foreign country, but the men themselves had all grown up in Boston, were fluent in English, and had been sampled and studied in the same way. Thus it was possible to vary ethnicity and culture of rearing while holding other demographic variables constant. In some facets of adult life, parental ethnic differences among the Inner City men were seen to exert a profound effect. For example, as I'll discuss in Chapter 9, men of white Anglo-Saxon Protestant (WASP) and Irish extractions had rates of alcohol abuse five times those of men of Italian extraction. But there was little cultural difference in defensive style. Dissociation was the only defensive style that appeared to be used significantly more by the WASPs than by Italians.[13] (Dissociation was also the defense with the lowest rater reliability.)

But if culture appears to have little effect on defensive style, this is not true of biology. The central nervous systems of some of the Inner City men had been impaired by chronic alcoholism. (By this I do not mean acute intoxication; most men were quite sober when interviewed.) In addition, some men had possible early cognitive impairment, as suggested by IQs less than 80. Both these groups exhibited significantly less mature defensive styles than the rest of the Inner City men. All immature defenses were two to four times as common in these two compromised groups as in the unimpaired sample.

CONCLUSION

Empirical investigation provided clear answers to the Study's three major questions about involuntary coping mechanisms. First, maturity of defenses can be rated reliably. Second, maturity of defenses demonstrated predictive validity toward future mental health. Third, maturity of defenses is independent of social class and gender, but is affected by biology.

Defense mechanisms are not just one more dogma of the psychoanalytic religion. On the contrary, the brain's mechanisms of involuntary adaptation are a fit subject for serious study by social and neurological scientists. But, as with all matters of lifetime development, long and deep access is needed to study them. It was the unusual longitudinal and naturalistic nature of the Grant Study that permitted the conclusions in this chapter. Fledgling efforts are now under way to image what the brain does while deploying defenses, but scientists will likely have to wait upon advances in brain imaging technology for further confirmation.[14]

9 | ALCOHOLISM

Remember that we deal with alcohol—cunning, baffling, powerful!

—ALCOHOLICS ANONYMOUS

ALCOHOLISM IS A DISORDER of great destructive power. Depending on how we define it, it afflicts between 6 and 20 percent of all Americans at some time in their lives. In the United States, alcoholism is involved in a quarter of all admissions to general hospitals, and it plays a major role in the four most common causes of death in twenty- to forty-year-old men: suicide, accidents, homicide, and cirrhosis of the liver.[1] The damage it causes falls not only on alcoholics themselves but on their families and friends as well—and this damage touches one American family in three. Life is not a cog railway that we step onto at birth and off at death, secure in the knowledge that we are safe from accidental derailments and the tug of gravity. No matter how blessed by good fortune we start out or how blighted by its lack, our circumstances can always change, and so can the conditions under which we meet them. This is the Study's sixth lesson, and much of what we learned about it came from our prospective longitudinal investigation of alcohol use and abuse.

The Grant Study's involvement in alcohol research was one of the silver linings of our perpetual anxiety about funding, and true sterling. Without it, I would not have been forced into this last ten years' reexamination of marriage, divorce, and the development of intimacy. Yet that reexamination called into question not only some cherished assumptions of my own, but also assumptions that predated my tenure at the Study, and the received wisdom of several generations. Lifetime studies are bread cast upon the waters. You can't know in advance everything you should be finding out. But on the other hand, some of what you find out and have no idea what to do with may turn out to be invaluable unforeseen years later. The work with alcohol was like that.

It isn't easy to identify who is and is not an alcoholic. Until now, most major longitudinal studies of health (for example, the Framingham Study in Massachusetts and the Alameda County Study in California) have taken into account only alcohol consumption, not alcohol abuse.[2] Unfortunately, as I've said before, reported alcohol consumption identifies alcohol abuse almost as poorly as reported food consumption reflects obesity. In contrast, the Grant Study has always focused on alcohol-related *problems*. Where alcohol is concerned, it is what people do, not what they say, that is important.

Our study of the College and Inner City cohorts is the longest and most thorough study of alcohol abuse in the world. It has established answers to seven major questions:

1. Is alcoholism a symptom or a disease?

2. Is alcoholism environmental or genetic?

3. Are alcoholics premorbidly different from nonalcoholics?

4. Should the goal of alcohol treatment be abstinence?

5. Can "real" alcoholics ever drink safely again?

6. How can relapse be prevented?

7. Is recovery through AA the exception or the rule?

In several instances, the Study's longitudinal findings differ from those of well-respected cross-sectional studies.

METHODS

The Study's unique structure gave it three advantages as we made our usual effort to replace opinion with science. First, the men were followed for their entire lifetimes—a rarity but a necessity, because alcoholism is a relapsing and evolving disease. Second, the Study quantified alcoholism not by reports of quantity or frequency of drinking, but by objective numbers of alcohol-related problems. Third, over its own lifespan the Study has had between thirty and fifty contacts with each of its members, which greatly facilitated the collection of this data.

A few notes on how the men were studied. The questionnaires they received every two years asked if they, their friends, their families, or their physicians had expressed concern about their drinking, and whether and for how long they had ever *stopped* drinking (evidence not of control, but of loss of control). At interview, alcohol abuse or its absence was always specifically recorded. When the men reached forty-seven, 87 percent of them participated in a two-hour semi-structured interview with a detailed twenty-three-item section on lifetime problem drinking.[3] Since forty-seven, they have undergone physical examinations every five years. Any man not previously classified as an alcohol abuser who answered yes twice in a row to two or more of the four concern questions, or who through interview or telephone contact acknowledged alcohol abuse, or whose physical exam revealed evidence of alcohol abuse, was classified an alcohol abuser. The age that each participant first met DSM-III criteria for alcohol abuse was estimated from all available data: questionnaires, any relevant court records, social service data, family interviews, etc.

(In 1962, before the Inner City cohort joined the Harvard Study of Adult Development, arrest records and records of psychiatric hospitalization, if any, had been searched for more than 95 percent of the men, who were at high risk for alcoholism, and for the preceding two generations of their families.[4] These are data that are well-nigh irreplaceable, as recent privacy legislation now precludes such searches.)

By mining interviews, clinical data, objective documents, and self-reports obtained from the participants by clinicians experienced in treating alcoholism, we were able to establish both a categorical and a dimensional scale of alcohol use for the men of both cohorts. The categorical scale was derived from DSM-III, the third edition of the *Diagnostic and Statistical Manual of the American Psychiatric Association;* that was the version current in 1977–1980, when this analysis was taking place.[5] It distinguished three types of alcohol use: *social drinking* (that is, no chronic problems related to alcohol use), *alcohol abuse* (chronic problems but no physiological dependence), and *alcohol dependence* (presence of withdrawal symptoms or hospitalization for detoxification). In this chapter, I will use the term *alcoholism* to refer to both of the latter two categories.

The dimensional scale, the Problem Drinking Scale, or PDS, assessed problem drinking on a continuum of severity by means of sixteen equally weighted questions (similar to those of the Michigan Alcoholism Screening Test).[6] The PDS inquired about social, legal, medical, and job problems caused by alcohol abuse. It also asked about blackouts, going on the wagon, seeking treatment, withdrawal symptoms, and problems with control. Scores of 4–7 on the PDS usually met the DSM-III criteria for "alcohol abuse," and scores of 8–12 usually met the criteria for "alcohol dependence." Men with fewer than 4 lifetime problem points on the dimensional scale were usually classified as social drinkers.

Observant readers will note that the numbers in this chapter vary sometimes from those in older reports. This is because in those earlier analyses we included all the Study men who had ever met DSM–III criteria for alcohol abuse (153 out of the 456 original Inner City men and 56 of the original 268 College men). In refining the original analysis for this book, however, men whose problem drinking scores were borderline (3 or 4) and who abused alcohol for less than five years and who returned to social drinking for the rest of their lives (13 Inner City and 2 College men) were reclassified as social drinkers.

We have excellent and inclusive death data, including death certificates for all Study members, including those who withdrew, except for two who died abroad. Survival or mortality has been ascertained and documented through the National Death Index or credit agencies, whichever was applicable, and death certificates obtained. Data from death certificates, and recent physical examinations from participating Study members, were used to infer major causes of death.

We assessed the alcohol status of all cooperating Study members, whether or not they had been alcoholic in the past, every year between the ages of twenty and seventy, using the biennial questionnaires and triangulating them with other material. (Since the Glueck men had not been personally followed between thirty-two and forty-five, for those years we had to depend on history and public records, including records of arrests.) We categorized the alcoholics as follows: *Abstinent:* less than one drink (0.5 ounces of ethanol) a month for a year. *Return-to-controlled-drinking:* a former alcohol abuser consuming more than one drink a month for at least three years with no reported problems. *Continued alcohol abuse:* Clear past history of sustained alcohol abuse and one or more acknowledged problems caused by drinking in the past three years. When data was missing for three years, the yearly status was rated as unknown. Data on alcohol abuse for every man was obtained between twenty and forty times over sixty (on average) years of observation. Nonresponders for two consecutive ques-

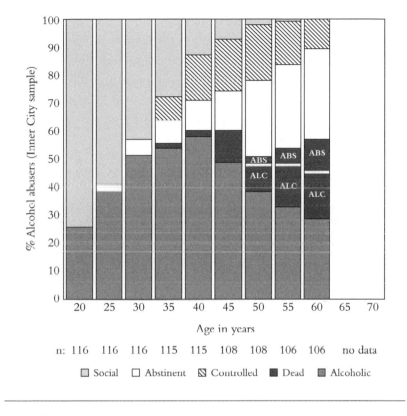

Figure 9.1   Final alcohol outcome status.

tionnaires were interviewed in person or by telephone. Men were classified as *Dropped* if they asked to withdraw or ignored questionnaires and follow-up telephone calls for ten years.

Figure 9.1 reflects the men's alcohol status at age sixty.[7] In the case of death or institutionalization, outcome status was based on the last three years of community residence prior to death or chronic institutional care.

By age seventy (approximately 1990 for the College men and

2000 for the Inner City men), 65 (46 percent) of the 140 Inner City alcoholics and 19 (35 percent) of the 54 College alcoholics were dead. (Recall that under "alcoholics" I am subsuming both alcohol-abusing and alcohol-dependent men.) In 2003, we found that all but 3 of the 18 alcohol-dependent College alcoholics had died by age eighty, and that their lifespans on average had been seventeen years shorter than those of their social-drinking Study peers.

Figure 9.1 illustrates vividly why the prevalence of alcoholics goes down with time. The problem isn't "burnout" (as AA members put it, getting "sick and tired of being sick and tired"); this is rare among alcoholics. Nor is it poor case-finding among the elderly. It's that over time, alcoholics become abstinent or die.

Active alcoholics die twice as fast as abstinent alcoholics, but even the latter die significantly sooner than social drinkers—often because cigarette abuse continues after alcohol consumption stops. As we will see, it is only for the first ten years of abusive drinking that the diagnosis of alcoholism is ambiguous. Over decades, alcoholism becomes a black-or-white disease.

THE SEVEN QUESTIONS

1. Is alcoholism an evanescent symptom or a chronic disease?   Social scientists generally consider matters such as intelligence, drinking behavior, and even eyesight as continua; physicians in the trenches have little patience for such nuances. Who is right? Both. The definition of retardation, alcoholism, or blindness may depend upon a host of independent contextual, interpersonal, and motivational factors. But when an alcoholic actually goes to the expense and inconvenience of seeking help for a self-acknowledged disease, there tends to be a certain crispness of definition as to what the problem is.

The medical model of alcoholism is one of inexorable progression. It was made famous by William Hogarth in the series of paintings known as *The Rake's Progress.*[8] It has been retrospectively documented by E. M. Jellinek.[9] It is taken as an article of faith by Alcoholics Anonymous.[10] Yet how can this model be reconciled with the unpredictable oscillations observed in fine-grained prospective studies of alcoholics?[11] Short-term prospective investigations indicate that during any given month a majority of alcoholics are either abstinent or drinking asymptomatically. This cannot be said of either cigarette or heroin addicts. Is alcoholic progression therefore a myth? When does the state of drunkenness (which is often voluntary) become the trait of chronic problem drinking (which is largely involuntary)? The Study helped to clarify.

Virtually every alcohol abuser in the Study, no matter how chronic, had periods of abstinence lasting a month or longer. In fact, a history of going "on the wagon" is a commonly accepted criterion for the diagnosis of alcoholism. The more physiologically dependent and symptomatic the alcoholic, the more likely he or she is to have attempted abstinence, usually more than once. As I've said elsewhere, Mark Twain found it so easy to stop smoking that he did it twenty times.

That is why it is only the number and frequency of alcohol-related *problems* that can truly define the clinical phenomenon known as alcoholism. Alcoholism is a construct of a higher order of complexity than mumps or retardation. It does not reflect a specific pathogen, like mumps. It is like retardation in being another diagnosis that depends on where a clinician draws the line. But more than anything else, like Type II diabetes, hypertension, and coronary artery disease, the disease of alcoholism is the endstage effect of bad habits on facilitating genes.

Although its symptoms may come and go, alcoholism acts like a chronic disease, and chronic diseases are forever. Without specific treatment, diabetes will plague you until you die—young. The Harvard Study of Adult Development's seventy-five years have enabled us to document that without sustained abstinence, the vast majority of problem drinkers continue to have alcohol-related problems until they, too, die young. Seventy-two percent of the College social drinkers lived to be eighty, but only 47 percent of the College alcohol abusers and a minuscule 14 percent of the alcohol-dependent College men—a very significant decline.

On the one hand, the Study demonstrated that alcohol abuse is inexorably progressive only in its initial stages. Once a drinker's alcohol consumption has gotten "out of control" and become a source of problems, it can remain so for a lifetime without necessarily progressing to morning drinking, job loss, or even severe withdrawal symptoms. Seven of the eighty-year-old College alcohol abusers had misused alcohol for decades (about thirty years, on average), but without evidence of worsening symptomatology. Similar courses can be seen in cigarette dependency.

On the other hand, alcoholism does not get better. Those same seven men over those same thirty years continued to report alcohol-related problems affecting their self-esteem, their health, and their families. In this, alcoholism does conform to the conventional disease model, and here too there is a resemblance to cigarette dependency. Two-pack-a-day smokers rarely revert to half-a-pack-a-day social smoking, and once an alcoholic has developed the problems typical of dependence, he or she is unlikely to revert to social drinking or even abuse. More on this in a moment.

Thus I think it is both appropriate and useful to consider alcoholism a disease. The diagnosis is not made so lightly that social drinkers are likely to suffer from incorrect labeling. Nor are the manifestations of alcohol abuse so varied as to render a unifying diagnosis meaningless. There is a benefit in calling severe problem drinking a disease; alcoholics who label themselves "ill"—as opposed to "bad"—will feel less helpless; they will have higher self-esteem; and they will likely try harder to change and to let others help change them.

A final reason to consider alcoholism a disease is that it kills people—tens of thousands of people a year. By age eighty, a College alcoholic was twice as likely to be dead as a nonalcoholic. By age seventy, almost half (46 percent) of the alcohol-abusing Inner City men were dead, as opposed to 29 percent of the nonabusers. Admittedly, much of this increased mortality does not reflect direct physical effects of alcohol itself, but it does point up graphically that cigarette consumption among problem drinkers was vastly greater than that of social drinkers. The average nonalcoholic smoked for fourteen-pack years (half a pack a day for twenty-eight years or two packs a day for seven years would both equal fourteen-pack years). But the average alcohol abuser smoked for twenty-seven-pack years, and the average alcohol-dependent individual for *fifty*—more than three times the cigarette consumption of social smokers. There is no evidence that heavy smoking increases drinking. However, as of 2010, one College alcoholic in four had died of heart disease. Of the social smokers, only one in eight had. Three percent of the College social drinkers died of lung cancer; 15 percent, five times as many, of the alcoholics did. Similarly, Inner City alcoholics were twice as likely to die from lung cancer as social drinkers.

The overlap between drinking and smoking does not indicate the presence of an "addictive" or "oral" personality or any other such abstraction, only the concrete reality that conscience and judgment are soluble in ethanol. Alcoholics are fearless in barrooms and automobiles. They're not careful about safe sex, and they don't worry about cigarettes, either. Deaths from cirrhosis, accident, suicide, and pharyngeal cancer were also far more common among alcohol abusers. Study findings in this regard are very similar to those from eight other longitudinal studies of premature mortality in alcoholics reviewed elsewhere.[12]

2. Is alcoholism environmental or genetic? In 1938, the year the Grant Study began, Karl Menninger, America's most famous psychiatrist, made a bold statement: "The older psychiatrists . . . considered alcoholism to be a hereditary trait. Of course, scarcely any scientist believes so today. Although it's still a popular theory, alcoholism cannot possibly be a hereditary trait, but for a father to be an alcoholic is an easy way for a son to learn how to effect the retaliation he later feels compelled to inflict."[13]

Menninger was wrong. When it comes to alcoholism, genes trump environment. Our data revealed that a family history of alcoholism more than doubled the chance that a Study member would become alcoholic, even when all other suspected etiologies (such as ethnicity, social class, and multiple family problems) were statistically controlled. But children with alcoholic stepparents were *not* more likely to become alcoholic. Here the data from the Inner City men was even more convincing than that from the Grant men.

Until very recently, most social scientists believed unhappy childhoods to be a cause of alcoholism. Our data showed that the association of childhood environmental weaknesses with future risk of alcohol abuse exactly paralleled the extent of parental alcohol abuse: that is, the more severe a parent's alcoholism, the more it will be reflected in his child's environment, and in the severity of alcoholism in the child. Contrary to Menninger's belief, however, alcoholism does not arise in children in response to an unhappy childhood or even to life with an alcoholic stepparent. It is the *fact* of an alcoholic biological parent, whether or not the child lives under the same roof, that increases the child's risk. Of the fifty-one men who had few childhood environmental weaknesses and an alcoholic parent, 27 percent became alcohol-dependent. Of the fifty-six men with many environmental weaknesses and no alcoholic parent, only 5 percent became alcohol-dependent. Alcoholic parents do not have to live with their children to pass on the disease.

The presence of a genetic component does not free us from issues of nature and nurture. While the number of alcoholics in one's ancestry increases the likelihood of alcohol abuse for genetic reasons, it also increases the likelihood of lifelong teetotalling, presumably for environmental reasons. Almost half of the forty-eight teetotalers of Anglo-Irish-American descent in the Glueck Study had an alcoholic parent.

3. Are alcoholics premorbidly different? This question essentially explores whether alcoholism is a symptom of mental illness or a cause. It has long and widely been considered the former. In the year the Grant Study was conceived, Robert Knight, a prominent psychoanalyst at the Austen Riggs Center, said it directly: "Alcohol addiction is a symptom rather than a disease. . . . There is always an underlying personality disorder evidenced by obvious maladjustment, neurotic character traits, emotional immaturity or infantilism."[14] In 1940 Paul Schilder, an Austrian psychiatric researcher who has given his name to four diseases, concurred. "The chronic alcoholic person is one who from his earliest childhood on has lived in a state of insecurity."[15] Two decades later, E. M. Jellinek, Yale's great alcoholism scholar, wrote: "In spite of a great diversity in personality structure among alcoholics, there appears in a large proportion of them a low tolerance for tension coupled with an inability to cope with psychological stresses."[16] And in 1980, psychiatrist Michael Selzer wrote more generally in the leading textbook of psychiatry: "Despite occasional disclaimers, alcoholics do not resemble a randomly chosen population."[17] None of these world-renowned experts, however, possessed any prospectively gathered data as to what alcoholics were like *before* they became alcoholic.

Three premorbid personality types have repeatedly been proposed as causal contributors to alcoholism: the dependent, the depressed, and the sociopathic.[18] The Grant Study confirmed none of these hypotheses.

The College alcohol abusers did not exhibit more premorbid evidence of dependent personality disorder than men who remained social drinkers all their lives, and 58 percent of the College men who became alcohol abusers did not lose control of their alcohol use until after age forty-five.

There were men who displayed many dependent traits as young adults and who showed lifelong difficulties with loving, with perseverance, and with postponement of gratification. Such so-called oral-dependent men were also more anxious and more inhibited about expressing aggression. Yet these traits in young adulthood were not significantly more common in future alcohol abusers than in everyone else. Oral-dependent traits did become very common in the College men, however, once they began to abuse alcohol.

It is also true that alcohol-abusing College men were five times more likely to report severe depression than men who did not abuse alcohol. Furthermore, of the thirty-one men who ever appeared to manifest major depressive disorder, fourteen (44 percent) also manifested alcoholism. Following these men for twenty-five years, I received the impression that many of the fourteen had turned to alcohol to relieve their depression. In 1990, however, the longitudinal data were subjected to blind analysis, and that impression proved to be illusion.[19] One psychiatrist, blind to age of onset of depression, reviewed each man's entire record and estimated the year that he first manifested evidence of alcohol abuse. A second psychiatrist, blind to age of onset of alcoholism, reviewed each man's record and determined the age of onset of major depressive disorder (or probable major depressive disorder). In four of the fourteen cases, the psychiatrist looking for evidence of primary depression believed that the depressive symptoms could be explained entirely by alcohol abuse. In six more cases, the rater noted that the first episode of major depressive disorder had occurred (on average twelve) years after the patient met the criteria of alcohol abuse. In only four cases had a man's depression actually preceded his alcoholism. Given the prevalence of alcoholism and affective disorder among the 268 men in the College sample, chance alone could account for primary alcoholism and primary depression occurring independently in four cases or even more.

As for the sociopathic personality, alcoholics are somewhat more likely to be premorbidly antisocial than asymptomatic drinkers.[20] And some antisocial adolescents initiate alcohol abuse as their antisocial behavior develops. But most alcoholics are not premorbidly antisocial; they become antisocial only after developing alcoholism. The role of sociopathy in alcoholism remains murky and I've discussed it extensively elsewhere.[21]

Two of the three supposed personality antecedents of alcoholism having been dispatched, and the third on hold for the time being, we looked for more general premorbid factors that might predict later alcohol abuse. However, data from the College sample revealed that neither bleak childhoods, childhood psychological problems, nor (more positively) psychological stability in college differentiated future social drinkers from future alcohol abusers.

Far more surprising, most future alcoholics do not appear to differ from future asymptomatic drinkers even in terms of premorbid psychological stability. A hypothesis like that could not even be seriously entertained until prospective studies were available, so compelling was the illusion that unhappy, nervous people turn to alcohol for self-medication, and so unlikely did it seem that depression and the

Table 9.1   Correlation of Alcoholism and Dependent Traits with Premorbid
              Variables

	Dependent Traits $N = 95$	Alcohol Abuse $N = 185$
Family history of alcoholism	NS	Significant
Poor childhood environment	Very significant	NS
Poor father–child relationship	Significant	NS
Poor high school social adjustment	Significant	NS
Poor college "psychological soundness"	Significant	NS

Very significant = p < .001; Significant = p < .01; NS = Not Significant.

so-called alcoholic personality might be secondary to the disorder of alcoholism. But alcohol in high doses is neither a stimulant nor a tranquilizer; it is the very opposite of a successful drug for these conditions. And alcohol ingestion makes both insomnia and depression worse.[22]

In the Inner City sample, the three childhood variables that most powerfully predicted positive adult mental health—warmth of childhood, freedom from childhood emotional problems, and boyhood competence—did *not* predict freedom from alcoholism; similarly, the three variables that most powerfully *did* predict alcoholism—family history of alcoholism, ethnicity, and adolescent behavior problems—did not predict poor future mental health. Alcoholism and poor mental health cannot be yoked invariably together.

In dismissing these factors (unhappy childhoods, multiproblem families, depression, and anxiety) as major causes of alcoholism, I do not mean to say that they are not important. They are always important, and they will make any chronic disease worse. I simply wish to underscore that in a prospective design, when more salient variables like culture and familial alcoholism can be controlled by the choice of sample, premorbid personal and family instability do not make a statistical contribution to the risk of alcoholism. Let me also reiterate that the Inner City men with an alcoholic biological parent but an otherwise stable family were five times as likely to develop alcoholism as men from clearly multiproblem families but without an alcoholic biological parent.

Our data suggested only two areas in which alcoholics appeared to be premorbidly different from asymptomatic drinkers. Before they become abusers, future alcoholics can tolerate relatively larger amounts of alcohol than controls can without hangovers, vomiting, or staggering.[23] This difference appears at least in part to be another reflection of the hereditary component of alcoholism.

The second difference is that future alcoholics are more likely to come from environments that tolerate adult drunkenness and discourage children and adolescents from learning safe drinking practices—Ireland and the United States are prime examples. Alcoholics were less likely to come from cultures that permit adolescents to consume low-proof alcoholic beverages at ceremonies and with meals, but condemn drunkenness—Italy, for example. Thus, the parents and grandparents of the alcoholics in our samples were several times more likely to have been born in English-speaking countries than in Mediterranean ones. Controlling for genetic risk, the Irish Study members had very significantly higher rates of alcohol dependence than the Italians. One highly controversial conclusion of the Study is that alcoholism is more affected by parental modeling of *how* children should drink than by legislation of *when* children should drink.

1 Should the goal of treatment be abstinence? When the Grant Study began, we knew far more about the natural history of most cancers than we did of alcoholism. Even the clinical course of alcoholism was poorly understood. What actually happens to alcoholics over time—not just those who attend clinics, but the whole constellation of treated and untreated alcoholics? Why does the prevalence of alcohol abuse decline sharply with age?[24] Is the explanation for this decline "burnout" (no), or return to asymptomatic drinking (no), or stable abstinence (yes), or high mortality (yes)? Another question is how long abstinence or return-to-controlled-drinking must persist before an individual's recovery from alcoholism can be considered secure. In smoking and even in cancer, remission must usually endure for five years before relapse is considered unlikely. In alcohol treatment studies, however, investigators often speak of "recovery" after the drinker has been symptom-free for six months or a year. The Harvard Study of Adult Development demonstrates that such a short time frame is not realistic. Only after five years of abstinence can remission from alcoholism be regarded as secure.[25] And as with smoking, the remitted alcoholic can almost never return to social drinking.

Table 9.2 summarizes the outcome status for the 148 alcoholics from the College and Inner City cohorts successfully followed until death or age seventy. It clarifies why the numbers of alcoholics seem to decline with age, and why at age seventy only a quarter of the men were still abusing alcohol. Some had been stably abstinent (for a mean of nineteen years), and a few had returned to social drinking, less stably. But almost half of them were dead.

An interesting finding: the alcoholic Inner City men were more likely to become abstinent than the alcoholic College men. Only nine of the College men meeting criteria for alcohol abuse achieved three or more years of abstinence, while 51 alcoholic Inner City men achieved this status.

5. *Can "real" alcoholics ever drink safely again?* The answer to this question is "Yes, but . . ." My hesitation is based on four caveats: one from the general literature, and three based on Study findings. In fifty years of alcoholism literature, every study that reported the successful

Table 9.2 Outcome Status for Study Alcoholics at Age 70 or at Time of Death

	College Cohort				Inner City Cohort			
	In Community at Age 70		Dead at Age 70		In Community at Age 70		Dead at Age 70	
	N	%	N	%	N	%	N	%
Stable abstinence (3+ years)	4	14%	5	26%	34	64%	17	36%
Return-to-controlled-drinking (3+ years)	5	17%	2	11%	3	6%	8	17%
Chronic alcohol abuse	20	69%	12	63%	16	30%	22	47%
Total	29	100%	19	100%	53	100%	47	100%

return of its clients to social drinking (often making the evening news in the process) has been found to have been in error after ten years' follow-up.[26] One highly publicized cautionary tale was the case of Audrey Kishline. Kishline founded Moderation Management in 1994, but in March of 2000, she swerved out of her lane and killed two occupants of an oncoming car in a drunken driving accident.[27]

Second, our data revealed that of the men who returned successfully to social drinking, most had only barely met the criteria for a diagnosis of alcoholism. This was true in both the Inner City and College samples. Third, of those who had maintained social drinking for three years or more, half relapsed or ended up seeking abstinence. Fourth, even the successes (the men who at last follow-up have had no further problems due to alcohol) often found it impossible to return to the carefree use of alcohol manifested by most social drinkers.

The year 1977 was the end of the age twenty-to-forty-seven follow-up period for the Inner City men. Twenty-one of them had achieved stable abstinence of three years or more, and twenty-two had returned to controlled drinking for three years or more.[28] Continued follow-up until 1992 revealed that eighteen (86 percent) of the twenty-one abstinent men maintained their abstinence until age sixty or until death.[29] The range of known abstinence for these eighteen men was three to thirty-seven years; the mean length of abstinence was twenty years. In contrast, over the same fifteen-year follow-up period, seven of the twenty-two men with three or more years of controlled drinking—almost a third—relapsed to sustained alcohol abuse, with a mean duration of alleged controlled drinking before relapse of twelve years. Three of the twenty-two sought sustained abstinence, four withdrew from the Study, and three, because of the brevity of their abuse, were reclassified as non-alcoholic. Thus, in 1992 or at death, only five of the twenty-two men were still believed to be

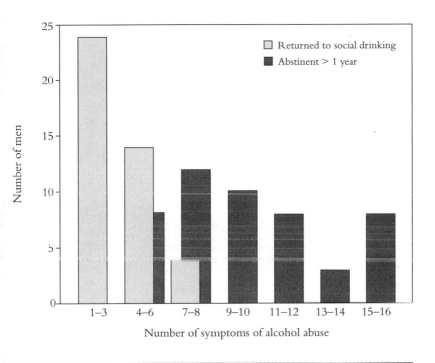

Figure 9.2   Relation between number of alcohol-related problems and likelihood of a return to social drinking.

true alcoholics drinking in a controlled fashion. Figure 9.2 illustrates the relation of the number of alcohol-related problems to the likelihood of a return to social drinking.

6. How do we prevent relapse?   Observers of human nature have long puzzled over the phenomenon of individuals who become "converted" to a new religion or otherwise abruptly alter their life course. The Study sought to discover whether in the case of alcoholism this usually occurred through clinical treatment (no), through willpower (no), through becoming sick and tired of being sick and tired (no), or through nonclinical factors known to affect relapse prevention in other addictions (yes). Alas, going on the wagon is more often evidence of alcoholism than of recovery, just as dieting is more often evidence of obesity than of slimness. Nevertheless, prolonged relapse prevention is key, both in alcoholism recovery and in obesity. Thus, our continued study of two cohorts of alcoholics over fifty years provided a naturalistic way of uncovering who did manage to prevent relapse.[30]

In the vast majority of alcoholics, we found that while counseling, detoxification, and even hospitalization could be temporary lifesavers, they did not change the natural history of the disease. As with diabetes and obesity, permanent changes in self-care are the only way to extend life over the long run. Table 9.3 covers four well-established means of relapse prevention in any addiction. Stably abstinent Study members on average employed two out of four of them during their first year of abstinence. About half of the ever-abstinent men found an alternative for alcohol, and some found more than one. Substitute dependencies varied from candy binges (five men) to benzodiazepines like Valium or Librium (five men); from compulsively helping others (two men) to returning to dependence upon parents (two men); from marijuana (two men) to mystical belief, prayer, and meditation (five

Table 9.3   Factors Associated with Abstinence for a Year or More for 49 Untreated and 29 Clinic-Treated Inner City Men

	Untreated Abstinent Alcoholics n = 49	Treated Abstinent Alcoholics n = 2⟨
Compulsory supervision	49%	34%
Substitute dependence	53%	55%
New relationships	32%	31%
Inspirational group membership (usually AA)	49%	62%

men); from compulsive work or hobbies (nine men) to compulsive gambling (two men); from compulsive eating (three men) to chain smoking (seven men).

Another half of the men achieved relapse prevention through compulsory supervision or behavior modification. By that I mean the presence of factors independent of willpower that systematically altered the consequences of alcohol abuse. This was often probation, or a painful medical consequence that, as AA members say, "keeps the memory green."

Two other means of relapse prevention were employed by almost half the men. One was to get involved in an inspirational group (for our sample, usually AA). The second was to find new relationships; sometimes these were love relationships and sometimes opportunities to help needy others. But they were always relationships with people toward whom the alcoholic did not feel guilty for past misdeeds. In their review of the literature on remission from abuse of tobacco, food, opiates, and alcohol, Stall and Biernacki identified these same four factors.[31] The four factors in Table 9.3 appeared to be the most important keys to relapse prevention; only 30 percent of the men resorted to clinic attendance or hospitalization during their first year of abstinence.

Surprisingly, absence of risk factors for alcohol abuse did not predict successful remission. The skills that get you out of a hole are likely independent of the forces that got you in. The remitted alcoholics abused alcohol for an average of two decades at least, and the severity of their alcoholism and their genetic vulnerability had been if anything greater than that of their nonremitting counterparts. Limited education, which is also a risk factor for alcohol abuse, did not inhibit stable remission. Indeed, the less-educated Inner City men were significantly more likely than their Grant Study counterparts to become abstinent. Although the per capita cigarette consumption of the alcoholics was almost twice that of the nonalcoholics, the severity of cigarette abuse among alcoholics was not statistically associated with eventual abstinence.

7. Is recovery through AA the exception or the rule? For both cohorts, regular AA attendance was strongly associated with sustained abstinence. All four of the factors in Table 9.3 are embodied in the AA program and in many other incorrectly named "self-help" recovery programs organized along similar lines. I put *self-help* in quotes because AA is as much about self-help as a barn-raising. In both of these community activities, success is at least as much about helping other people as about helping yourself. Four variables were associated with Study men joining AA: severity of alcoholism, Irish ethnicity, absence of maternal neglect, and a warm childhood environment.

Of the nine alcohol-dependent College men who achieved stable abstinence, five (56 percent) attended AA for between 30 and 2,000 meetings. Two other alcohol-dependent College men attended about 50 meetings but relapsed. Of the thirty-nine alcohol-dependent Inner City men with stable abstinence, at least fourteen (36 percent) attended AA for 50 to 2,000 meetings. The Problem Drinking Score assessment of the alcohol-dependent Inner City men with fewer than 30 AA visits was about 9. Men who attended more than 29 AA visits (the mean was 400) had a mean problem drinking score of 12—a very significant difference. One doesn't usually seek painful hip replacement before one's arthritis has become quite severe, and "hitting bottom" increases the alcoholic's willingness to sit on hard church chairs, drink bad coffee, and take "the cotton out of one's ears and put it in one's mouth" several times a week. In both cohorts, the men who were stably abstinent attended about twenty times as many AA meetings as the chronically alcoholic (Table 9.4).

Table 9.4 Question: How Does AA Work? Answer: AA Works Fine!

	COLLEGE COHORT		INNER CITY COHORT	
	Abstinent n = 9	Chronic n = 32	Abstinent n = 57	Chronic n = 44
Average years of abstinence	15 years	1 year	16 years	1 year
Years of active alcoholism	20 years	23 years	18 years	22 years
Problem drinking score	9	6 (VS)	10	8
AA meetings	137	2 (VS)	143	8 (VS)

Very significant = p<.001; Significant = p<.01; NS = Not Significant.

THE STORY OF JAMES O'NEILL

The story of James O'Neill illustrates how alcoholism reverses what are commonly thought to be cause and effect; it demonstrates that alcoholism is the horse in life's troubles, not the cart. O'Neill behaved very badly while drinking, but—however difficult this may be to believe—in 1950, before his alcohol abuse began, he had been assessed by the Study staff as a man of "unqualified" ethical character, and the rather prim director of the Health Services had described him as a "straightforward, decent, honest fellow, should be a good bet in any community."

James O'Neill did not come to psychiatric attention until 1957, thirteen years after his Harvard graduation, when he was first admitted to a psychiatric ward at a VA hospital. A thirty-six-year-old father of four and former assistant professor of economics, O'Neill described himself as a "failure at his marital and professional responsibilities because of drinking and missing teaching appointments." His admission note stated, "Present symptoms include excessive drinking, insomnia, guilt and anxiety feeling." The diagnosis was "behavior disorder, inadequate personality."

O'Neill provided the following history, paraphrased from his hospital record. He began drinking and gambling in the summer of 1948, while depressed over a poor performance on his Ph.D. generals. He was drinking during the daytime, and missing teaching appointments. However, he did continue to teach and to keep his family together. He finished his Ph.D. without difficulty, and in 1955 he left his large West Coast school for a research university in the South.

At the time of his admission, O'Neill expressed suspicion and anger toward the important people in his life, all of whom, he alleged, had treated him badly. Otherwise he showed little emotion, and the interviewer commented: "His pattern of drinking, sexual infidelity, gambling and irresponsible borrowing led him to recognize from his reading that it adds up to diagnosis of psychopathic personality, especially since he has experienced no real remorse about it." It was known that he had given his son some books to sell, and that four books from the university library were among them; he was accused of stealing university property, and fired for moral turpitude. He assured the hospital staff that he did not sell university books knowingly.

The psychiatric record continues:

> During all of the time that O'Neill was frequenting bars, contacting bookies and registering in hotels to philander, he always used his own name. It's interesting that when he was carrying on his nefarious pursuits, he got considerable satisfaction out of it being known that he was a professor. . . . When his mother died in 1949, he felt no remorse [sic] at her death. He did not remember the year of his mother's death. In view of the fact that he dates his extracurricular activities as beginning in 1948, this confusion is probably significant.
>
> During his eight-month hospital stay, the patient . . . was able to work out a great deal of feelings toward his family, in

particular toward his mother and also toward his wife. The patient felt quite hostile and anxious about the fact ... that his parents were always very cold. ... He harbored many feelings of hostility toward his wife whom he feels does not appreciate the fact that she's married to such an intelligent college professor. All she wants is to have money and bigger homes.

The discharge diagnosis was "anxiety reaction manifested by feelings of ambivalence about his family, his parents and his work." The precipitating stress was considered to be "death of the patient's mother and a long history of drinking and gambling and going into debt." His predisposition was considered to be "an emotionally unstable personality for the past 20 years." At one point the VA even called him "schizophrenic." A diagnosis of alcoholism was never even considered.

But the Grant Study record told a completely different story. In college, James O'Neill had been the embodiment of the Grant Study's ideals of optimal health and achievement. He was one of the brightest men in the Study, and after three years of observation he received an A in psychological soundness. A child psychiatrist blind to his life after age eighteen was asked to compare his childhood environment with those of his Grant Study peers. She placed it in the top third, and summarized the raw data on his childhood as follows:

O'Neill was born in a difficult delivery. The mother was told not to have more children. His parents were reliable, consistent, obsessive, devoted parents. They were relatively understanding, and their expectations appear to have been more non-verbal than explicit. The father was characterized as easy to meet, the mother was seen as more quiet; no alcoholism

was reported. Warmth, thoughtfulness and devotion to the
home were some of the comments. The subject spoke of going to his father first with any problems, but of being closer
to his mother than to his father. His peer relations were reported to have been good, and little or no conflict with his
parents was reported.

She went on to predict that O'Neill would develop into "an obsessional, hard-working, non-alcoholic citizen, whose work would be
related to law, diplomacy and possibly teaching. He would rely on his
intellect and verbal abilities to help in his work. He would probably
marry and be relatively straight with his children. He would probably
expect high standards from them."

Other observers summed O'Neill up equally favorably in the
years before he turned thirty. The dean's office ranked his stability as
"A" while he was in college; the Study internist described him as "enthusiastic, whimsical, direct, confidant, no grudges or chips, impressed
me as an outstanding fellow." The staff psychiatrist was impressed by
his "combination of warmth, vitality and personality," and also put
him in the "A" group. When he was twenty-one, he married his childhood sweetheart, with whom he had been in love since he was sixteen; in 1950, six years after they married, the marriage still seemed
solid. When O'Neill was twenty-three, his commanding officer wrote
that he gave "superior" attention to duty and was a particularly desirable officer.

From the prospective record it was also possible to record a more
accurate picture of O'Neill's feelings about his mother's death. The
child psychiatrist who assessed the prospective record saw his as among
the best mother-child relationships in the Study. His mother's physician commented that O'Neill had been "devoted and helpful during
the illness," and in 1950, six months after her death, a Study observer

noted that O'Neill felt her loss deeply. It was only seven years later, on his admission to the VA, that O'Neill reported having no feelings toward his mother and blamed her alleged coldness for his current unhappiness. Over time, alcoholics develop excellent collections of "resentments."

O'Neill was one of the lostest of the Study's lost sheep. He had stopped returning questionnaires long before his hospitalization. It was not until 1972 that he finally brought the Study up to date on the progression of his life and his alcoholism. He had begun drinking heavily in 1948 while still in graduate school, and by 1950 he was drinking in the morning. In 1951 O'Neill's wife's uncle, an early member of AA, had suggested the possibility of alcoholism. But his wife insisted to the Grant Study, with whom she had maintained connection even while O'Neill did not, that her husband was not abusing alcohol. Furthermore, in 1952, at his first admission, the health services at his university whitewashed his drinking as due to "combat fatigue." His prospective 1946 military record revealed, however, that O'Neill had experienced no combat in World War II.

In 1972 I interviewed O'Neill, and he filled in some long-standing gaps. We met in his apartment. He was balding and sported a distinguished mustache; his clothes were worn but elegant. He came across as an energetic man who kept a tight rein on his feelings. At first during the interview he had a lot of trouble looking at me and seemed very restless. He chain-smoked, walked back and forth, lay down first on one bed and then on the other. Although he avoided eye contact, there was a serious awareness of me as a person, and I always felt he was talking to me. He behaved like a cross between a diffident professor and a newly released prisoner of war. As he put it to me, "I'm hyper-emotional; I'm a very oversexed guy. The feelings are there, but it's getting them out that's hard. The cauldron is always bubbling. In Alcoholics Anonymous, I'm known as Dr. Anti-Serenity."

He admitted that he had been chronically intoxicated between 1952 and 1955 while writing his Ph.D. thesis, and that he had regularly sold books from the university library to buy alcohol. By 1954 his wife had begun to complain about his drinking; by 1955 it was campus gossip. But no diagnosis of alcoholism was made during his 1957 VA hospital admission or the subsequent one in 1962. In our 1972 interview, I felt that O'Neill himself still did not understand the cause-and-effect relationship between his drinking and his misery.

In 1970, O'Neill became sober in Alcoholics Anonymous. By our 1972 interview, AA was clearly the most important force in his life, besides his wife. He made frequent reference to it; when I asked him what his dominant mood was, he replied, "Incredulity. . . . I consider myself lucky. Most people in Alcoholics Anonymous do."

Even after two years of complete sobriety, O'Neill described himself to me as "a classical psychopath, totally incapable of commitment to any man alive." To me, though, he felt like a lonely but kindly man. I never had the feeling that he was cold or self-absorbed. If anything, he suffered from a hypertrophy of conscience, not a lack thereof. Remember that, although alcohol does not help insomnia, chronic anxiety, or depression, it is the best antidote for guilt that we have.

As I was leaving, I noticed several books related to gambling on the bookshelves. Aha, I thought. Were these the lingering remnants of the sociopathy he talked about? No, as it turned out. Once sober, he had sublimated his interest in gambling. He had consulted to the governor of Louisiana while the state lottery there was being set up—a considerably more profitable occupation for an economist than frequenting racetracks. In other words, with the remission of his alcoholism, O'Neill's ego functioning had matured; instead of compulsively acting out his interest in gambling, he had yoked his passion to his Ph.D. in economics, and harnessed them both in a socially and personally constructive way.

In closing, O'Neill told me that he could not agree with AA in calling alcoholism a disease. "I think that I *will* the taking up of a drink," he said. "I have a great deal of shame and guilt and remorse and think that's healthy." I heartily disagreed; I suspect that his shame had facilitated his denial of his alcoholism for twenty years, and that by reframing it as a disease, AA had rescued him. Sadly, O'Neill died two years after our interview from coronary heart disease, a fate undoubtedly hastened by twenty-five years of chain smoking.

THE STORY OF FRANCIS LOWELL

The history of Francis Lowell, aka Bill Loman, illustrates how different alcoholism looks to sociologists and to physicians. In this case, the two viewers were myself! Without realizing it, I narrated one man's story twice, never recognizing him the second time as a person I had studied fifteen years before.

Francis Lowell was an effective and well-paid upper-class New York lawyer. In 1995 I used his life as evidence that the misuse of alcohol, like heavy smoking, is not a disease, but a lifestyle choice.[32] Given enough education, willpower, social support, and a forgiving occupation, a fortunate drinker could drink as long and as much as he wanted. In college Lowell had been a heavy social user of alcohol, and very guarded about answering Grant Study questions related to his alcohol use. By age twenty-five, this gregarious man had established a pattern of heavy drinking Friday through Sunday, and none during the rest of the week. He continued this pattern over the next forty years. His heavy weekend drinking sometimes expanded into five-day binges, with a loss of one or two days of work, but Lowell abused alcohol from age thirty to age seventy without any noticeable decline in his physical health or serious damage to his legal career (most of his clients were rich family members).

Lowell was aware that he had a problem with his drinking by the time he was thirty. He felt guilty about how much he drank; his friends criticized his drinking; he failed to keep his promises to cut down; and when he was drinking he avoided his relatives. At age thirty-nine he had his first drunk-driving arrest; there was a second one at age forty-seven. He had his only detoxification at age fifty-two, but his physical exam and liver chemistries were normal. At age fifty-six, Francis Lowell said of himself, "No doubt about it, I do drink heavily at times," but he never stopped drinking, except for giving it up for Lent. Many weeks he drank within social limits, and usually he did not drink during the week. He attributed his successful steady pattern of alcohol abuse to the fact that his stomach would not tolerate more than five days of drinking. And, he added, "I don't want to sound pompous, but a sense of duty drilled into me from family and from St. Paul's School contributes to my control. . . . You just can't let everything go."

By fifty-nine, Francis Lowell was making $200,000. His heavy alcohol intake did not interfere with his work, although his career did not advance after sixty. And it did not (much) interfere with his relationships; it had contributed to his loss of the woman who had most touched his heart, but by remaining single he limited further damage. After he turned sixty-two, his doctor began encouraging him to cut down on his drinking, and at age sixty-six he had a seizure "possibly related to alcohol." Nevertheless, at seventy Francis Lowell was still working forty hours a week and earning his handsome salary. Compared to his college classmates he was still very active physically, and his liver chemistries were still normal. I wrote in my first biography of him, "At no time in his life has he described a wish to become abstinent and he continued to drink ten drinks a day on the weekend." In short, I believed that Francis Lowell had a lifelong problem with alcohol, but not a "progressive disease."

THE STORY OF BILL LOMAN

But alcoholism has an unstable, chameleon-like quality. After I had forgotten my original description of Lowell, I wrote the life history of a chronic alcoholic whom I called Bill Loman. It wasn't until after the fact that I discovered that I had written about the same man in 1983, seeing his life from a very different point of view, and making a point very different from the one I was aiming at the second time. Although there had been only four more years of follow-up and a little bit more data, my view of Lowell/Loman had shifted from the sociological to the medical one. Light can act as both wave and particle, and alcoholism can present as both habit and disease. Only years of observation allow us to identify both of these patterns in the same person. Bill Loman was one of the Study's great illustrations that the genes for alcoholism can derail anyone, no matter how promising his beginnings.

Bill Loman was a man destined for great things; he became a tragic figure not because he deserves our scorn, but because he had the disease of alcoholism as his implacable enemy. At St. Paul's School, Loman had been a senior prefect and captain of the football team. He was elected to the most prestigious club at Harvard, and he graduated *magna cum laude.* College descriptions of Loman included "unspoiled by his wealth," "well-poised and very attractive," "rather mature."

His World War II record was exemplary too. He won three battle stars for active participation in the Battle of the Bulge and the crossing of the Roer and the Rhine Rivers. His commanding officer described him as "intensely loyal, collected and cool under most trying conditions Sense of humor never deserts him." He was promoted to first lieutenant and then to captain. The Study director, summing up the twenty-five-year-old Loman's military record, remarked, "This boy could go quite far." After the war, Loman went to Harvard Law School and finished in the top tenth of his class. He returned to New York to practice corporate law at a prestigious firm. He was elected to the very best clubs in the city, and spent his weekends playing golf and bridge with other members. At thirty, he was an upper-class football captain poised to be a superstar when he grew up.

But he didn't. There was another thread running through Bill Loman's life, and it could be seen even while he was still in college. He spent his weekends drinking. In college he experienced three- and four-day episodes of "depression"—probably alcohol-related—when he would see the world as a "very sorry place." The college psychiatrist saw the hard-drinking Loman as "undependable, careless, self-centered and evasive." In the military, too, Loman recalled, "I spent most of my spare time drinking and chasing women."

In law school, Loman dared to drink only on weekends—he already recognized that he could not have even one drink without going on a binge. By the age of thirty he had established a pattern of drinking heavily from lunch on Friday through Sunday evening. He was abstinent during the week, but he took frequent sick days, especially on Monday when he was recovering from weekend hangovers.

Intimacy, Career Consolidation, and Generativity were not in Bill Loman's future. He became emotionally involved with only one woman, in his twenties. At thirty he proposed to her, but she turned him down. Pressed, he admitted that her reluctance to marry him might have been due to his binge drinking. They remained closely involved for the next twenty-five years, both living with their mothers on the weekends. When he was fifty-three, her "dominating" mother died, and she married someone else. Loman continued to live with his mother until she died. He was discontented throughout his legal career, feeling undercommitted and undercompensated.

I interviewed Bill Loman when he was fifty-nine. He was insecure and unable to make eye contact. Instead of enjoying the interview as most Study members did, he acted like an unhappy adolescent being grilled. "I don't think I'd have joined the Study if I knew it was going to last so long," he grumbled.

Loman manifested a pervasive melancholy. When he was fifty, his brother, another Study man who *did* grow up to be a superstar, sadly revealed that Bill had stopped making new friends. Despite his high income, he took no vacations, involved himself in no civic activities, and had no exciting relationships with the opposite sex. There had never been anyone in whom he had confided. Asked who he turned to in unhappiness, Loman replied, "I turn to me."

At age sixty-five when asked, "What is your most satisfying activity?" Loman answered, "None." Unlike many unhappy Study men and women who could rely on strong religious affiliations even when socially isolated, Loman went to church just once a year—on Christmas, with his mother. He had been raised in a church school. But alcohol abuse interferes with spiritual solace as well as the more mundane kinds.

After our interview, Loman's alcoholism continued to progress. By sixty-five, despite reasonable health, the once-sociable Bill no longer attended any of his clubs. He could not learn new things. Unlike most of his Harvard cohort, he never even began to master the computer. The personal losses that accrue in life had never been replaced; among his friends and relatives there was no longer a single person with whom he could say he had an intimate relationship. Not surprisingly, he saw the present as the unhappiest period of his life. The happiest time, he said, had been the war years. It was hard not to be disconcerted by that. The Battle of the Bulge is not everyone's idea of a day at the beach.

Loman's alcoholism never affected his liver, but it destroyed his life. At thirty, both he and the girl he loved were already worrying about his drinking. At age forty, after three DWI's with damage to others, his mother, his brother, and the police were worrying too. At age fifty-three he sought his first detoxification. But at sixty he was still going on binges where he'd drink a quart of whiskey a day for five days. The only thing that stopped him was that he always eventually became too ill to drink anymore. He had his first seizure due to alcohol withdrawal at sixty-five.

After that Loman made repeated unsuccessful efforts to go on the wagon until he died from an alcohol-related fall at seventy-four. Seventy percent of his Study mates were still alive. He did not begin his life among the Loveless, but he ended it that way.

My readers will have noted, I am certain, the—unconscious—shift in focus, and even in detail, between these two versions of Lowell/Loman's life. As I've said before, biography is more vivid than statistics, but it is much more vulnerable to vicissitudes in the writer's intent than numbers are. The fact is that both Lowell and Loman had a problem drinking score (PDS) of 11—in the top 10 percent of both the College and Inner City samples. For many years I struggled to make sense out of conflicting views of alcoholism, both of which had fierce partisans: was it a disease or a career path? For most of his life, Lowell/Loman's lab tests and physical exams were clean. For most of his life his mates drank as much as he did. His career, in its own way, prospered. I can see now that he was the perfect example of why it's so hard to say what alcoholism "really is." It wasn't until his last years that the evidence of what alcohol had done to him physically and psychologically really began to show. Another reason for lifetime studies. Alcoholism is a crafty foe, and even under the intense scrutiny of the Grant Study, in the case of Lowell/Loman it managed to keep itself hidden for a very long time.

We can all consider for ourselves whether in my first effort to tell this story I was in my own kind of denial, missing the disease that is alcoholism, or whether in my second I was bending the facts, forcing a heavy habitual drinker into the mold of an alcoholic. But however we explain it, both Lowell and Loman suffered from something that was indeed cunning, baffling, and powerful. It was also ultimately fatal.

CONCLUSION

Prospective study has consistently shown alcoholism to be the cause, not the result, of dependent, sociopathic, neurotic, or aggressive personality disorders. Alcoholism is the cause, not the result, of unhappy marriages. Alcoholism is the cause of many deaths, too, and not only through liver cirrhosis and motor vehicle accidents—suicides, homicides, cancers, heart disease, and depressed immune systems can all be chalked up to this serial killer. The critical factors predicting recovery from alcohol dependence, besides its severity, appear to be finding a (preferably) nonpharmacological substitute for alcohol, compulsory supervision (with immediate negative consequences for relapse), new loving relationships, and involvement in inspirational programs.

Prolonged follow-up reveals two fundamental paradoxes in predicting the life course of alcoholism. Socially disadvantaged men, men with strong family histories of alcoholism, and men with early onset of severe alcohol dependence were more likely than other men to become stably abstinent. In contrast, alcohol abusers with excellent social supports, high education, good health habits, and late onset of minimal alcohol abuse—epitomized by the College sample—were more likely to remain chronic alcohol abusers. If their alcohol-related problems were truly minimal, they also had an excellent chance of returning to lifelong social (controlled) drinking. In short, it appears to be the most and the least severe alcoholics who enjoy the best chance of long-term remission.

10 | SURPRISING FINDINGS

> There are more things in heaven and earth, Horatio,
> than are dreamt of in your philosophy. —WILLIAM SHAKESPEARE

As I'VE TRIED TO CONVEY throughout this book, longitudinal studies are constructions of intrinsic contradiction and paradox. They require investment in massive information collection before there's any way to know for certain whether the information being compiled is the kind that can answer the questions the Study is posing. Much of this vast accumulation will almost assuredly never answer any questions at all. Yet in the huge heaps of data, little glints may sometimes be perceived. Some will turn out to be fool's gold. But some, suddenly—with a different cast of light or a sudden shift in context or a new analytical technology—identify themselves as twenty-four karat. There are unexpected, unexplainable, curious, and just plain odd findings in every large longitudinal study, and these are the very ones that beg for review every so often, just in case.

Here are some of the tantalizing glints from the Harvard Study of Adult Development. In time they may, as my final lesson from Chapter 2 suggests, help to elucidate one or two of life's enduring mysteries.

I. WHY DO THE RICH LIVE LONGER THAN THE POOR?

The last fifty years of epidemiology are making it ever clearer that the human lifespan is finite; after a point, the greatest riches in the world can't help you live longer. Nonetheless, the poor die sooner. In the

United States it is socioeconomic status (income, education, occupation, access to health care) that is thought to account for this disparity, not the malnutrition and infection that shorten lifespans in third world countries.

Some people blame society for discriminatory access to health care and for toxic neighborhoods, poor nutrition, poor schools, and high unemployment; some blame the victim for dropping out of school, delinquency, bad habits, and poor self-care. While there is danger that the current emphasis on individual health promotion can be used in the service of victim-blaming, I've shown in Chapter 7 that the role of health-related behaviors cannot be carelessly dismissed out of political correctness.[1]

Nevertheless, as Marcia Angell, former editor-in-chief of the *New England Journal of Medicine,* has pointed out, "Despite the importance of socioeconomic status to health, no one quite knows how it operates. It is, perhaps, the most mysterious of the determinants of health."[2]

Nowhere is the Harvard Study of Adult Development as powerful as when it addresses this mysterious relationship between health and social class. The College and Inner City groups were matched for several important confounders: gender, race, geography, absence of delinquency, and a 1920–1930 birth cohort. But the two samples were clearly dichotomized by social class and intelligence (at least according to standard IQ tests), since the Inner City men were matched to a low-scoring group of delinquent youths. As Figure 10.1 illustrates, the Inner City cohort has been becoming disabled ten years earlier on average than the College sample, and dying ten years sooner; the estimated average longevity of the Inner City men is seventy years, of the College men, seventy-nine years.

The morbidity of the two samples is similar with regard to illnesses that are independent of self-care: that is, cancer (excluding

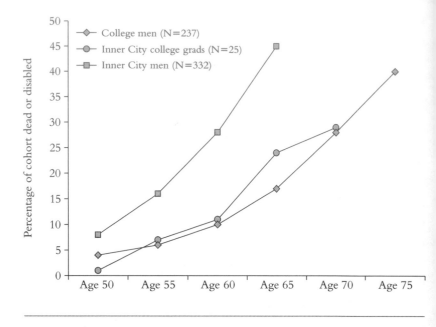

Figure 10.1   Death after fifty for College men, Inner City men, and Inner City college graduates.

lung), arthritis, heart disease, and brain disease. But there was twice as much lung cancer, emphysema, and cirrhosis, and three times as much Type II diabetes, among the Inner City men as in the College sample. The Inner City men were also more than three times as likely to be overweight. *All these differences, however, disappeared for Inner City men who graduated from college.*

The average Inner City man was much less educated than the Harvard graduates, and he also led a far less healthy lifestyle. But the more education an Inner City man obtained, the more likely he was to stop smoking, avoid obesity, and be circumspect in his use of alcohol. The estimated average age of death for the Grant Study men who did not go to graduate school and the college-educated Inner City men was identical—seventy-nine years, if the World War II deaths are excluded.

So we have to ask: Does education really predict healthy aging independent of social class and intelligence? The college-educated Inner City men were neither more intelligent nor more privileged socially than their peers who did not attend college, so those two factors do not explain the nine-year difference in their estimated lifespans. The college-educated Inner City men had carefully tested IQs that were on average 30 points lower than their Harvard counterparts', and they attended lesser colleges. They were a full inch shorter, suggesting inferior childhood nutrition, and none of them had the upper-middle- or upper-class advantages typical of two-thirds of the Harvard men. In middle age, only half as many of the Inner City men as College men had made it into the ranks of the upper class, and they made only half as much money. Thus, neither intelligence nor status nor wealth can account for the disappearance of the nine-year shortfall in lifespan once an Inner City man had graduated from college. Parity of education alone was enough to produce parity in physical health.

Well-trained medical sociologists will scoff at this assertion. Do not the Whitehall Studies by Sir Michael Marmot in England appear to show that social class is one of the leading causes, if not the leading cause, of premature death in England? Did not the health of Whitehall civil servants improve with every step in their pay grade?[3] Yes. But Marmot's early Whitehall studies did not control for education or alcoholism. Our study of the Inner City men showed that education is very significantly associated with income and job promotion, and alcoholism is very significantly deleterious to both.[4] That is, with every step in pay grade, the chances rose of a man being nonalcoholic and better educated. It is likely that it is these factors, not the job or the pay grade, that facilitate better health. This is yet another instance where the choices we make appear to influence how long we will live.

Most social science studies, including Marmot's own early ones, control only for self-reported alcohol consumption, which, as I have been at pains to point out, correlates very poorly with objective alcohol abuse. In our samples, however, the difference in health outcomes among occupational levels diminished sharply once we controlled for alcohol abuse, which tends to depress occupational level. Alcoholism is bad for career advancement as well as for health.

The question then becomes: If education so powerfully affects self-care (as well as the more obvious job status), what affects whether or not people stay in school? In general, pursuit of education is most successful in coherent communities that invest heavily in their families and their school systems in an atmosphere of gender and racial tolerance. People with no hope for the future don't pursue education effectively. Providing that hope is the responsibility of the community, not the individual.

That said, however, the pursuit of education also reflects individual personality traits of *perseverance* and *planfulness*—traits that Friedman and Martin have shown to be important to longevity.[5] An important postscript to these findings is a recent paper by David Baber and colleagues, who find that it is not only increased education per se that reduces mortality; of equal importance was what Baber et al. called *health reading fluency*—the capacity to read prescription bottles, understand preventive services, and so on.[6]

II. IS PTSD DUE TO COMBAT OR TO PERSONALITY DISORDER?

The frequency of posttraumatic stress disorder (PTSD) among returning Vietnam veterans has led many to wonder whether the principal causal factor is not severe combat stress after all, but pre-existing personality disorder. To address this question, the Grant Study took advantage of the fact that most of its members were World War II veterans. Not only had they all been extensively studied before the war, but they were all extensively debriefed after the war on their combat experiences, their physical symptoms during combat, and their persisting stress-related symptoms. John Monks, an internist particularly interested in combat experience, carried out these debriefings.[7] Forty years later, sociologist Glen Elder and I asked all the surviving College veterans (excluding the early Study dropouts) to fill out questionnaires regarding persisting symptoms of posttraumatic stress. (PTSD had not yet been "invented" in 1946, but its principal symptoms had been anticipated by the prescient Monks.) One hundred and seven men returned questionnaires. These men had also completed the NEO, which, as I've described in Chapter 4, is an extensive and popular multiple-choice scale that includes a scale for the trait *Neuroticism*.[8] We were particularly interested to learn if the men who developed symptoms of posttraumatic stress had already looked vulnerable before the war, or if combat really was the primary factor in symptom development. (Please note here, however, that since the Study men had been protected to some degree by education and rank from the circumstances in which frank PTSD develops, we were studying *symptoms* of posttraumatic stress, not the disorder per se.)

First, the men who experienced the most intense combat did not appear to have been more vulnerable by nature; in fact, they manifested superior psychosocial health in adolescence and also at age sixty-five. Second, it was only the men with high combat exposure who continued to report symptoms compatible with PTSD after forty years. Third, we found that symptoms of posttraumatic stress both in 1946 and in 1988 were predicted independently by two factors: combat exposure and number of physiological symptoms during combat stress (but not during civilian stress). As I've already noted, severe combat exposure also predicted early death. It was noteworthy that the symptoms of posttraumatic stress reported in 1946 were not correlated with evidence of subsequent major depressive disorder, alcohol abuse, or poor psychosocial adjustment. Only combat exposure made a significant statistical contribution to posttraumatic stress symptoms, and that contribution was very significant. And while the NEO trait *Neuroticism* was associated with bleak childhood, psychiatrist utilization, poor psychosocial outcome at age forty-seven, and physiological symptoms during civilian stress, it was *not* associated with PTSD.[9]

Sixteen men who endured severe combat reported no posttraumatic stress symptoms in 1946, and in 1988 still could not recall ever having had such symptoms. When we compared these sixteen resilient men to the eighteen with high combat exposure who did experience symptoms, their *Neuroticism* scores were the same. However, their defensive styles were not. The high combat veterans who manifested less mature defenses as young adults had very significantly more symptoms than those with more mature defenses (Chapter 8). Equally important, seven (39 percent) of the men with high combat experience and less mature defenses were dead by age sixty-five, while no man with mature defenses was. In the Grant Study, high combat exposure per se was not associated with a higher incidence of postwar alcoholism, but as I've said, the Grant Study men also did not develop full-blown PTSD.

The take-home messages here are: first, that combat exposure predicted symptoms of posttraumatic stress, but pre-existing psychopathology did not; and second, premorbid emotional vulnerability predicted subsequent psychopathology, but not symptoms of posttraumatic stress. Without the kind of prospective data and prolonged follow-up that the Grant Study made available, this finding would not have been possible.

III. POLITICS, MENTAL HEALTH, AND . . . SEX

As an optimist, a psychoanalyst, and a Democrat, I brought many prejudices to the Grant Study. I've always been delighted to have them confirmed, and this actually has happened occasionally. There *is* such a thing as a happy marriage, for example. And although colleagues derided for years my conviction that alcoholics could not return safely to controlled drinking, patience (as I detailed in Chapter 9) has seen me vindicated.[10]

Still, the greatest value of longitudinal study is its way of shattering prejudices and superstitions, and the Grant Study has done a real job on some of mine. I held a deep belief, for example, that Republicans are neither as loving nor as altruistic as Democrats. But after it occurred to me that Gandhi was a very bad father and John D. Rockefeller rather a good one, I thought that I had better test that notion out.

My first step was to identify political preferences from 1950 to 1999. Every biennial questionnaire had asked about politics, and we knew a lot about the men's political views. We knew, for instance, that in 1954 only 16 percent had sanctioned the McCarthy hearings. We knew that like many liberal members of their generation, the Grant men were "for" equal rights, at least after the fact. They applauded the Supreme Court decisions and civil rights legislation after they happened, but only very few took an active role in trying to bring about racial or gender equality. In 1967, 91 percent were for de-escalating our involvement in Vietnam; this was true of only 80 percent of their classmates. Had the Grant Study subjects had their way, Eugene Mc-

Carthy and Nelson Rockefeller would have been nominated in 1968 rather than Humphrey and Nixon; Gore would have easily beaten Bush in 2000, and the U.S. would never have invaded Iraq without further consultation or U.N. permission.

With all this information, it was not hard for independent raters to score the men along a political continuum from 1 (very liberal) to 20 (very conservative). Let me illustrate these abstractions briefly. Bert Hoover received a 20 on our scale (the highest score possible) for political conservatism. In 1964 he was a Goldwater supporter who thought that America should plan an invasion of Cuba. In 1967, he advocated "dropping Lyndon and Hubert on Hanoi" to resolve the problem of Vietnam. He saw student protests (and hippies and drug users as well) as "evidence of a sick society, a culture that is overripe and ready to be clobbered by a more vigorous one." In 1972 Hoover disapproved of long hair, premarital sex, anti-war demonstrations, and marijuana, and said that he would forbid his children to date a black person. Asked in 1981 what blessings he perceived in a Reagan presidency, he almost shouted, "The country may survive. Especially if we can rid the House of idiot liberals!" The public figure in the last two hundred years that he most admired was Herbert Hoover; and in 1985 when asked to check his current political stance—conservative or liberal—he wrote in, "just left of Genghis Khan." In 1988, with regard to "student protests over disarmament and university divestment in South Africa," he circled "Disapprove" and inked in, for emphasis, "The dumb bastards." In 1995, he was asked to check "What do you think of Newt Gingrich?" with choices ranging from "Delight" to "An abomination." Hoover circled "Delight" and again ad-libbed for emphasis, "Sorry. The abomination [that is, Bill Clinton] is in the White House!" In 1999, Hoover nominated Ronald Reagan for *Time*'s Man of the Century.

Oscar Weil, on the other hand, got a score of 3 on the political continuum. He had returned far fewer questionnaires than Hoover had, but of the Study black sheep Weil was one of my favorites. In 1964 he had supported Johnson over Goldwater. In 1967, he suggested that the way to end the crisis in Vietnam was to "Halt all offensive operations; start drawing up a time-table with South Vietnam for return of U.S. forces [to U.S.]."

Weil did not buy the whole Sixties package. In 1972, asked how he felt about his children and long hair, he responded insightfully, "I know it is irrelevant, but somehow I can't cheer it on." He added, "I know it's silly, but I have difficulty with premarital sex." But he had no problem with his daughters dating black people, and about the antiwar demonstrations he wrote, "On the general level of protest, I am most excited. I think that this is a really revolutionary time in this country and holds great promise for a real leap forward." He supported Walter Mondale for president in 1984, and Michael Dukakis in 1988. With regard to student protests for disarmament and against university investment in South Africa, he checked "approve" and inked in for emphasis, "I am more sympathetic to purely symbolic acts than I once was." In 1992 he voted for Clinton "with pleasure." He voted for Clinton again in 1996.

Most of the men in the Study clustered toward these two poles of the political continuum, and most maintained their positions for fifty years. The distribution of scores along the liberal-conservative axis was not a bell curve; it looked more like the twin humps of a Bactrian camel. And when I ran my test of how politics correlated with love, nobility, and so forth, it proved that my lifelong convictions about the virtues of liberals had no validity at all. Democrats had no superiority over Republicans with regard to the stability of their marriages, their mental health, their altruism, or their aging. Conservatives and liberals enjoyed identical scores on the Decathlon of Flourishing, and there was no difference in their relationships with their children.

That didn't mean that there were no real (and significant) differences between liberal and conservative. Liberals were more likely to be open to new ideas, and to approve of the younger generation's behavior. They were more likely to have had highly educated mothers, to have gone to graduate school, to display creativity, and to use sublimation as a defense. Conservatives were less open to novelty, but they made more money, played more sports, and were twice as likely to be religious.

I wondered what influences might inform political alignment, and went back to the men's college records. The two political poles could easily be identified in college, but—at least in this cohort—choice of one or the other had nothing to do with the quality of the men's childhoods, and little to do with parental social class. Remember, Identity in the Eriksonian sense means disentangling your values, your allegiances, and your politics from your parents'.

Surprisingly, however, the men's *personalities* in college had everything to do with their politics at eighty. As I've recounted, in 1942 each man was rated on twenty-six personality traits.[11] Two of them, *Pragmatic* and *Practical organizing*—both consistent with hardheaded common sense—were not significantly correlated with many facets of mental health. But they were very significantly associated with conservatism half a century later. Conversely, five traits quite *un*correlated with mental health and not well regarded by the early Study staff (but epitomizing many people's stereotype of Harvard students)—*Introspective, Creative and intuitive, Cultural, Ideational,* and *Sensitive affect*—were very significantly predictive fifty to seventy years later of being classified as a liberal.

Before he died at eighty-six, my very Republican, very intolerant, and as best I can tell, very mentally healthy grandfather used to chaff me with his take on the old epigram: If you're not a socialist before you're thirty, you have no heart; and if you're still a socialist afterward, you have no brains. But as a good developmentalist and champion of prospective study, I can now see that he was misled by retrospection. He'd been a McKinley Republican even as a young man. The years of Grant Study questionnaires confirmed that political convictions tend to endure, and that most young liberals and conservatives evolve smoothly into older ones. William James wasn't all wrong.

At the end of the day, those differences were not important. Even if "openness" were to be counted as a virtue—as my liberal prejudices would have it—that trait did not correlate with successful aging. For a while I was prepared to say that political conservatism predicts nothing about the future but one's politics. But, as it turned out, there was one big surprise hidden in the men's political profiles. As I noted in Chapter 6, the two early traits that predicted conservatism also predicted early sexual inactivity, and the five traits associated with being a liberal predicted sustained sexual activity fifty years later. The seven most liberal men (with scores of 0–5) did not on average cease sexual activity until after the age of eighty. The seven most conservative men ceased sexual activity at an average age of sixty-eight. That was a significant difference. I have consulted urologists about this; they have no idea why it might be so, but it is an interesting finding nevertheless. It was curious too that the variables I thought would protect against early impotence—ancestral longevity, maturity of defenses, physical health at age eighty, and quality of marriage—showed no effect, and the men who at eighty-five were still sexually active were eight times as likely to have been rated *Shy, Sensitive, Ideational,* and *Self-conscious and Introspective* as those who became sexually inactive before seventy-five.

IV. HEALTH AND RELIGION

The modern world seems to be of two minds with regard to religion. On one hand, in the last fifty years Sunday school attendance in the U.K. has dropped from 74 percent to 4 percent, and Oxford University evolutionary biologist Richard Dawkins can declare, "I think a case can be made that faith is one of the world's great evils, comparable to the smallpox virus but harder to eradicate."[12] On the other, Gallup polls point out that 85 percent of Americans believe in God, and an increasing number of researchers report a positive association between religious commitment and health.[13] Nevertheless, despite growing evidence that there is a relationship between health, religious beliefs, and participation in religious activities, investigators remain uncertain about what might account for it. Some critics suggest that the relationship between religious commitment and physical health may be illusory, due to inadequate control of confounding factors that undermine both physical health and involvement in religion.[14] Such confounders—such as, and especially, poorly matched control groups and inadequate assessment of alcohol and smoking history—have marred prior studies. A fifty-year-old church-going Mormon is almost certainly a lifelong abstainer from alcohol. That fact must be separated out from his religious beliefs and engagements as a factor in his physical health outcome. Here again, prospective study is essential.

The College cohort was ideally suited to explore such concerns. Participants were a well-studied, culturally and socioeconomically homogeneous sample of men who varied widely in religious involvement and in health habits, but who all had excellent access to education and health care. Starting in 1967, in the College men's twenty-fifth reunion questionnaires, I began investigating the depth of their reported religious faith and the extent of their participation in it. I wanted to test a hypothesis that religious interest would increase with age. Also, I suspected that religious involvement would make a positive contribution to aging, both from a medical and a psychosocial perspective.

I assessed religious involvement over a thirty-year period on a 5-point scale: 1 = no involvement; 3 = some involvement; 5 = deep involvement. By design, different questionnaires explored different aspects, including how frequently a man attended religious services, how important he considered a personal belief in God, how applicable he considered religious considerations to his own life, and so on. Ninety men (two-fifths) fell into category 1 (close to atheism), and forty-eight men (one-fifth) fell into category 5 (devoutly religious).

We also assessed interest in "private spiritual practices" and other markers of interest in individual spiritual development as separate from faith-based beliefs and institutions. In our sample, the two admittedly different concepts of religion and spirituality overlapped so markedly that it was not possible to find any significant outcome difference between their effects. Fewer than 3 percent of the men fell in the top third in the religion category and the bottom third in the spirituality category, or vice versa. In short, our measure of "religious involvement" appears to have good face validity in measuring interest in "spirituality" too.

Counter to my first hypothesis that religious involvement would increase with age, 58 percent of the men reported little religious involvement at sixty-five in contrast to only 28 percent in college. Nine (26 percent) of thirty-five men who had gone to church regularly when young had virtually no involvement after age sixty. However, fifteen (25 percent) of the sixty men with little religious involvement when young became very involved after age sixty. Why? One possible reason might be that those fifteen men manifested nine times as many symptoms of depression, and spent three times as many years disabled or dead before age eighty, as the nine men whose religious involvement declined over time. This doesn't mean that religion is bad for your mental state. But the ill are more likely to go to the hospital than the healthy, and the depressed and anxious are more likely than the psychologically robust to seek religious comfort.[15]

Table 10.1 contrasts the associations and consequences of religious involvement with the associations and consequences of mental health. Associations included parental social class, warmth of childhood, psychosocial soundness in college, smoking (in pack years), and alcohol abuse.[16] I expected to find a positive relationship between physical health and religious involvement, but Table 10.1 reveals none. There were 224 men for whom we had adequate data about their participation in religion over the course of their lives. By 2012, 63 (70 percent) of the 90 men with no religious involvement had died, and 36 (75 percent) of the 48 men with deep involvement had died. Notably, in these two well-matched groups of agnostic and religiously committed

Table 10.1   Religious Involvement and Adult Adjustment

	Religious Involvement, age 45–75 N = 224	Adult Adjustment, age 50–65 N = 224
Mental Health		
Childhood social class	Not significant	Not significant
Warm childhood	Not significant	Significant
Psychological soundness (age 21)	Not significant	Significant
Warm relationships (age 47)	Not significant	Very significant
Warm marriage (age 50–70)	Not significant	Very significant
Close to kids (age 50–70)	Not significant	Very significant
Social supports (age 55–75)	Not significant	Very significant
Adult adjustment (age 50–65)	Not significant	—
Neuroticism (age 60)	Not significant	Very significant
Smoking (age 20–60)	Not significant	Very significant
Alcohol abuse (age 20–65)	Not significant	Very significant
Physical Health		
Objective health (age 45)	Not significant	Very significant
Objective health (age 60)	Not significant	Very significant
Objective health (age 70)	Not significant	Very significant
Years disabled or dead <80	Not significant	Very significant

Very significant = $p<.001$; Significant = $p<.01$

men, not only was there no difference in mortality, there was also no difference in cigarette and alcohol abuse. Mormons, Seventh-Day Adventists, and evangelical Christians live longer than their agnostic neighbors; they abuse cigarettes and alcohol far less. This was not true, however, for the devout Grant Study Irish Catholics and Episcopalians. The likelihood is that abstinence from drinking and smoking play more of a role in increased longevity than religious devotion itself.

Depression and an increase in stressful life events correlated significantly with increased religious involvement and also with psychiatric visits, but psychiatric visits and religious involvement were not correlated with each other except that they both appear to reflect independent ways that the men coped with life's difficulties. And if religion was not important to the longevity of the men in general, it was useful to the forty-nine men with the highest "misery quotient," including eighteen men who suffered from major depressive disorder, twenty-five who had experienced multiple stressful life events not of their own making, and six men of whom both these conditions were true.[17] Among the ten "miserable" agnostics, only one survived until age eighty-five. But among the fifteen "miserable" but deeply religious men nine were still alive—an almost significant difference (p<.02). Looked at from a psychosocial perspective rather than a strictly medical one, religious devotion remains one of humanity's great sources of comfort, and for many denominations replaces the use (and abuse) of alcohol and cigarettes.

My last hypothesis—that religious involvement would be associated with enhanced social support (as a consequence of more loving behavior toward others)—was not confirmed. The items in Table 10.2 were selected from two personality questionnaires, one administered when the men were about fifty years of age, and one when they were

Table 10.2   Spirituality and Warm Relationships

	Religious Involvement Age 45–75 n = 91–98	Adult Adjustment Age 50–65 n = 91–98
Faith-based beliefs:		
I believe that each person has a soul	Very significant	Not significant
I believe that people are born to do good	Significant	Not significant
I believe in a universal power, a god	Very significant	Not significant
In the last 24 hours, I have spent 30 minutes in prayer, meditation, or contemplation	Very significant	Not significant
Pragmatic beliefs:		
I believe I am the most important person in someone else's life	Not significant	Very significant
I have cared deeply about someone for ten or more years	Not significant	Significant
I always feel the presence of love in my life	Not significant	Very significant

Very significant = p<.001; Significant = p<.01; NS = Not significant.

seventy-five.[18] The first four items were chosen as common reflections of religious faith, and the second three as reflections of real-life loving relationships. All items used a True = 1, False = 0 format.

The items chosen to reflect spirituality did correlate highly with religious involvement, but not with adult adjustment from age fifty to sixty-five. Conversely, the second group of items correlated significantly with adult adjustment, but not with religious involvement.[19] To embrace the last three items requires a happy childhood or a loving spouse. The first four items do not; they can be honored in the abstract, whatever one's life circumstances or relational capacities. More, for people whose circumstances or capacities do not facilitate loving real-life relationships, they promise other sources of contact. Even if a bleak childhood has taught you not to trust, or a major depression has distanced your friends, God continues to love you. Religion is a source of comfort for people whose concrete sources of love are limited. Beethoven saved himself in an angry depression by writing incomparable faith-inspired music, and he set to it Schiller's words: "Be embraced, all ye millions, with a kiss for all the world. Brothers, beyond the stars surely dwells a loving father."

There are some disparities between these Grant Study findings and those of other well-designed studies, and they deserve comment. An impressive body of research suggests that attendance at religious services protects against premature mortality.[20] But none of them show a direct causal effect, and too many rely on self-reports of alcohol abuse and physical health rather than on objective evidence.[21] As I've said, some deny the effects of alcohol and cigarettes entirely. One elegantly designed and analyzed study by leading social scientists notes that "church attendees were also less likely to smoke and drink, but those variables did not significantly affect mortality; thus they are not considered further."[22] Yet in virtually all studies conducted by physicians, alcohol abuse and smoking are among the most important causes of premature mortality. In the Grant and Glueck Studies, early mortality was four times higher among the alcohol and cigarette abusers than among nonsmoking social drinkers.[23] In our underprivileged Inner City sample, religious involvement at age forty-seven appeared to predict significantly fewer years of disability by age seventy. But the significance of this effect disappeared when smoking, alcohol abuse, and years of education were controlled for.

Another compelling explanation for our aberrant findings is that many American studies linking religious observance to physical health come out of the so-called Bible Belt, where agnostics are, at least statistically, also social outliers. In our sample of highly educated men centered in the northeastern United States, high religious involvement was not the cultural norm; more important, it was not deeply tied in to other sources of social supports.[24] In other words, in samples where healthy social adjustment usually includes clear religious involvement, such involvement is likely to correlate with warm relationships, social supports, and good physical health. Evidence of that correlation, however, does not reflect a direct causal relationship between religious involvement and health.

V. THE IMPORTANCE OF MATERNAL GRANDFATHERS

The mean age at death of the College men's grandparents (an 1860 birth cohort) was seventy-one years, and the mean age at death of their parents (an 1890 birth cohort) was seventy-six years. By historical standards, such ancestral longevity is remarkable, and comparable with that predicted for some contemporary European birth cohorts.[25] The likelihood of adventitious ancestral death from poor medical care, hazardous occupation, poor nutrition, or infection was less in the Grant Study sample than for less socioeconomically favored groups. This reduction in environmentally mediated mortality increased our chances to identify some genetically mediated effects on longevity.

While examining the effects of ancestral longevity on sustained good physical and mental health, we noted a marked and unexpected association between age at death of maternal grandfathers and the mental health of their grandsons.[26] The age of death of the other five first-degree ancestors (mother, father, maternal grandmother, and both paternal grandparents) was associated with virtually nothing in the men's lives except, modestly, with longevity. But the associations with the maternal grandfather's (MGF's) age of death were extraordinary. For example, while the average longevity of the other five ancestors was not relevant to the men's scores in the Decathlon of Flourishing, the maternal grandfathers of the men with the highest flourishing scores lived nine years longer than those of the men with the lowest scores—a significant difference. The average age of MGF death for 147 Study men who never saw a psychiatrist was seventy years. The average age of MGF death for the thirty-two men who made one hundred or more visits to a psychiatrist was sixty one—a very significant difference—but the age of death of the other five relatives made no difference as to psychiatric visits. The maternal grandfathers of upper-class men lived only three years longer than those of men from blue-collar families.

In 1990, two board-eligible psychiatrists blind to other ratings were given the complete records of the sixty-one men in the Study who, by age fifty, manifested objective evidence of sustained psychosocial impairment (that is, psychiatric hospitalization, scoring in the bottom quartile of psychosocial adjustment at age forty-seven, or having used tranquilizers or antidepressants for more than one month). Many of these men were dependent on alcohol. They were rated for eight correlates of depression, the DSM-III criteria for major depressive disorder not yet having been developed. These eight correlates were (1) serious depression for two weeks or more by self-report, (2) diagnosis of clinical depression at some point in the man's life by a non-Study clinician, (3) use of antidepressant medication, (4) psychiatric hospitalization for reasons other than alcohol abuse, (5) sustained anergia or anhedonia, (6) neurovegetative signs of depression (e.g., early morning awakening or weight loss when depressed), (7) suicidal preoccupations, attempts, or completions, and (8) evidence of mania.

Of the sixty-one, thirty-six men were categorized as alcoholic or personality disordered. The remaining twenty-five included twelve men categorized by only one rater as having major depressive disorder and thirteen who both raters agreed manifested major depressive disorder. These twenty-five men met at least three—and an average of five—criteria of major depressive disorder. The other thirty-six men met an average of only one-tenth as many criteria.

As a contrast, we identified fifty undistressed men who through the age of sixty had never reported evidence of alcohol abuse, had made no visits to a psychiatrist, and had not used psychotropic drugs more than one day a year on average over twenty years. (This stipulation covered the contingency, say, of a man's having had to take Librium briefly during a surgical admission, or Ambien to recover from jet lag before an important overseas conference.) In addition, these men were classified in college as having well-integrated personalities, and the Study had never assigned them a psychiatric diagnosis. They were the antithesis of the depressed men.

Table 10.3 shows our four diagnostic groups: undistressed, alcoholic/personality disordered, major depressive disorder, and "intermediate"—that is, the men who didn't fall into any of the first three. The mean age at death of MGF for the twenty clearly depressed men was sixty. The mean age at death of the fifty-eight undistressed men was seventy-five—a very significant difference. The age of death of the MGF of the ten men with the highest anxiety scores on the NEO was fifty-seven, and that of the MGF of the ten men with the lowest anxiety scores was eighty-three, an even greater distinction.

This unexpected discovery—that of the six ancestors, it was only the maternal grandfather's longevity that was associated with affective disorder in the grandsons—is consistent with a linkage mediated via the X-chromosome. X-linked disorders such as hemophilia, color blindness, and baldness come through the mother's father, when it is he who supplies the grandson's only X chromosome. Such illnesses tend to skip the mother, who has a second, usually unaffected X chromosome to protect her. For half a century, researchers have speculated that there might be an X chromosome linkage in the etiology of de-

Table 10.3   Mean Age at Death of Maternal Grandfather and of Other Primary
Ancestors in the Four Affective Disorder Categories

Affective Disorder Category	Maternal Grandfathers, Mean Age of Death	Other 5 Ancestors, Mean Age of Death
Major depressive disorder (n = 23)	60 Very significant	71 NS
Personality disorder/alcoholism (n = 35)	66 Significant	72 NS
Intermediate (n = 114)	69 NS	74 NS
"Undistressed" (n = 58)	75 Very significant	73 NS
Total sample (N = 230)	69	73

Very significant = p<.001; Significant = p<. 01; NS = Not significant.

pression.[27] But this hypothesis has been confirmed only inconsistently in the search for a specific gene for bipolar disorder.[28] It seems clear that bipolar disorder and major depressive disorder are genetically heterogeneous—that the genes for the two disorders are different.

To establish firm evidence for X-linked transmission of affective disorder would require full genealogical analysis of the affected families with knowledge of presence or absence of affective disorders in both male and female subjects over three or four generations—evidence that we do not have. Our findings do indicate, however, that for unknown—dare I say mysterious—reasons, early death of maternal grandfathers predicts an increased incidence of affective disorder in the grandsons. Still more exciting is that long-lived maternal grandfathers predict unusual psychological stability in their grandsons—evidence that positive mental health may be in part genetic. The association of very long lived maternal grandfathers with men scoring low on the NEO anxiety score is particularly intriguing.

My own guess is that someday soon an accomplished geneticist with a larger study, better characterized maternal grandfathers, and complete DNA analyses will win the Nobel Prize for explicating this phenomenon. At present it must be considered a preliminary finding, of interest only to the curious. Still, it is provocative, and would not have been discovered but for 60 years of follow-up. And it's a perfect example of how a seductive gleam can wink out unexpectedly from unruly heaps of unsorted longitudinal data, to be revealed eventually by Time's assay as either brass or true gold.

11 | SUMMING UP

All's well that ends well; still the fine's the crown;
Whate'er the course, the end is the renown.

—WILLIAM SHAKESPEARE

LEARNING FROM LIFETIME STUDIES does not stop until the lives have been fully lived—and not even then, because archives of prospective data are an invitation and an opportunity to go back and ask new questions time and time again, even after the people who so generously provided the answers are gone.

Each time the Study of Adult Development was threatened with extinction—in 1946, in 1954, in 1971, and in 1986—the grant-makers asked, "Hasn't the Study been milked dry?" There was a time when *I* thought that after the College men reached sixty-five and retired, there was nothing more to do but watch them die. Yet the Study always survived—to teach, to surprise, and to give. Granting agencies must be selective, it is true, but they must be selective the way foresters are. Longitudinal studies are the redwood forests of psychosocial studies. Fallen branches and felled trees are useful in the short run. Faithfully maintained and imaginatively harvested, the older the forest gets, the more it is worth. But once cut down it can never be restored.

Obviously, the point of the Grant Study's seventy-five years is not that it gratified the narcissism of a ten-year-old who wanted the most powerful telescope in the world, nor that it helped to resolve the grief of a boy whose father died too young. I am a real part of this Study, but a small one. It is very much worth noting that so many provocative findings have come out of it—findings that could have been discovered by no other means—that it has taken *ninety-five different authors* to elucidate them.

It is certainly appropriate to assess the cost-to-benefit ratio of a follow-up study that endured for more than seven decades at a cost of twenty million dollars. And in some ways that is easy to do. The bean-counter in me points out that granting agencies have paid only $10,000 in award, on average, per peer-reviewed—and sometimes highly cited—book or journal article. That's not an exorbitant cost as these things go. But the Study's three greatest contributions, which justify its cost and give meaning to the extraordinary generosity, patience, and candor of the men who exposed their entire lives in the interests of science, are less easily subject to financial valuation.

The first contribution is the absoluteness of the Study's demonstration that adult development continues long after adolescence, that character is not set in plaster, and that people do change. Even a hopeless midlife can blossom into a joyous old age. Such dramatic transformations are invisible to pencil-and-paper explorations or even ten-year studies of adult development.[1]

Second, in all the world literature there is no other study of lifetime alcohol abuse as long and as thorough as this one.

Third, the Study's identification and charting of involuntary coping mechanisms has given us at once a useful clinical tool, a route to empathy for initially unlikable people, and a powerful predictor of the future. Without this long Study of real lives, the importance of the maturation of defenses would still be out of fashion, dismissed as a relic of failed psychoanalytic metaphysics. In this final chapter I will briefly review each of these three contributions.

DEVELOPMENT IS LIFELONG

Adult development is a lifelong process. To investigate it properly means lifelong study—for which even seventy years is not really quite long enough. Many students of adult development, most brilliantly Warner Schaie in the Seattle Longitudinal Study, have tried to speed up the process by studying several groups of varying ages concurrently over ten or even twenty years.[2] This method worked in Schaie's study of the rise and fall of intelligence, which can be studied in populations. But it cannot work for the study of personality, which is unique to individuals. The study of old age in particular requires patience and empathy, not the restless intolerance of frightened fifty-year-olds such as I was in 1986. Without persistence, endurance, and restraint, we learn little about the delicate processes of growth that endure even as life is fading.

Piaget's and Spock's delineations of child development changed parenting forever. Erikson's appreciation of adult development as dynamic growth rather than decay was likewise a major paradigm shift.[3] But without empirical evidence to back it up, it gave rise to more theory and speculation than knowledge. The Grant Study has changed that. Four major empirical works have come out of the Study under my name—*Adaptation to Life* (1977), *Wisdom of the Ego* (1993), *Aging Well* (2002), and this final summary volume—which clarify the second half of life a decade at a time. As the men went on maturing, as hypotheses were tested and retested over time, the books have become progressively less theoretical and more evidence-based. Future studies and wiser authors will take us further still. But the optimum (to borrow a concept from Arlie Bock) study of lifetimes will always require a hundred years of toil, generations of dedicated investigators, and stories of real lives supported by statistical verification. As I keep reminding myself, what people say doesn't mean much. It's what they *do* that predicts the future. It was the facts of people's long-term love relationships, not their belief systems, that showed us what we needed to know first about their capacity to love, and then about their mental health.

The realities of adult development cannot be discerned through speculation, through biography (by definition retrospective), or even through diaries (by definition biased). Yet theorists of adult development have had to depend upon just such sorts of materials, so scarce have alternatives been. The very distinguished Terman and Berkeley longitudinal studies rarely published case studies. Before the Grant Study, accurate individual records of adult development were rare.

Sadly, the bellwether publication in the field, the *Journal of Adult Development,* while rich in theory and informative about given stages of adult development, pays almost no attention to real lives or to maturational changes, which are full of paradox. This can be seen in the quixotic truth that at the same time as the Grant Study was demonstrating incontrovertibly that adult development continues throughout the lifespan, it was offering support for the contradictory view of William James and other nay-sayers of whose work I once made light. Lifetime studies hoist us investigators with the very petards that we delight to deploy against our challengers.

Let me illustrate that apparent contradiction, which lies in the vulnerability of all belief systems. In sitting down to write this book, I believed that it was only the tracking of behavior over time that could predict the future, not inventories or questionnaires or any of psychology's other myriad "instruments." Paul Costa and Robert Mc-Crae believed that because a person's NEO has not changed over thirty years, his character has not changed either.[4]

I had no problem with the first half of that contention; indeed, we found that among the College men the NEO did not change very much over *forty-five* years (see Chapter 4).[5] I even found, in running the numbers, that the NEO traits of *Neuroticism* (negatively) and *Extraversion* (positively) were significantly predictive of some Decathlon outcomes. But I did have trouble with the second half—that the lack of change in the NEO meant a lack of change in personality. The NEO traits—uncomfortably like the elegant and empirically derived anthropologic measurements with which this book began—are static. They do not address processes of growth, any more than a masculine body build, which may well endure over a lifetime, predicts the kind of officer its owner will grow to become. The NEO deals with personality as reflected in multiple choice questions, and although I tried, I could not correlate any of its traits with the five mature defenses (altruism, sublimation, humor, suppression, and anticipation) that do change over time, and in their changing exert such an important influence on the outcomes of real lives.

But having hoisted Costa on *his* own petard—for failing to recognize the difference between static measurements and dynamic processes—I proceeded to hoist myself on mine. It occurred to me that if *Extraversion* (positively) and *Neuroticism* (negatively) could predict the Decathlon to a limited degree, it might be interesting to create a theoretical composite value: *Extraversion* minus *Neuroticism*. That is, what would happen if I removed from consideration the thwarting effects that *Neuroticism* had on Decathlon success? What happened was that the *Extraversion* minus *Neuroticism* values correlated at least significantly with every Decathlon event for which data was available, rivaling adaptive style and childhood in predictive power. As illustrated in Table 11.1, *Extraversion* minus *Neuroticism* estimated at age twenty-one could predict the Decathlon measured with data gathered forty to sixty years afterward, and it could predict it as well as adaptive style assessed twenty years later! My ship of fine theories had crashed upon the rocks, but truth was well served.

The moral of this story is that they were right and I was right. My rightness did not preclude theirs; theirs did not preclude mine. My rightness was not complete, and theirs wasn't either, and the whole is greater than the sum of its parts. There's more than one path to the top of a mountain, and there are prizes enough for everyone: Woods, Soldz, Bowlby/Vaillant, Costa/McCrae. But note that it was only in

Table 11.1   Statistical Strength of Association of Alternative Predictors of Flourishing*

Decathlon of Flourishing	Maturity of Defenses, Age 20–47	Warm Childhood Environment	Extraversion Minus Neuroticism, Age 21	Extraversion Minus Neuroticism, Age 68
Decathlon	Very Significant	Very Significant	Very Significant	Very Significant
Who's Who in America	Very Significant	NS	Very Significant	NS
Highest earned income	NS	NS	Very Significant	Significant
Low lifetime distress	Very Significant	NS	Significant	Very Significant
Success and enjoyment at work, love and play (age 65–80)	Very Significant	Very Significant	Significant	Very Significant
Generativity achieved	Significant	Significant	Significant	Very Significant
Subjective health (age 75)	Very Significant	NS	NS	Significant
Successful aging (age 80)	Significant	NS	NS	Very Significant
Social supports (other than wife and kids) (age 60–75)	Very Significant	NS	NS	Significant
Good marriage (age 60–85)	Significant	NS	NS	Very Significant
Close to kids (age 60–75)	NS	Significant	NS	NS

* 168 men were included in this table. Only the classes of 1942–1944 were rated for defenses, and some men have been lost to death.

Very Significant = p<.001; Significant = p<.01; NS = Not Significant.

the context of lifelong study that the question of changing adaptive style could even be tested. Context is everything, and lifetime studies provide it most generously.

The Grant and Glueck Studies make clear another important dynamic reality that personality inventories, or even retrospective and cross-sectional studies, cannot: childhood trauma becomes less important with time (even though recovery may take decades), while the good things that happen in childhood endure. A related finding was that environmental conditions that appear very important to outcome in ten-year studies—such as parental social class, loss of a parent, or membership in a multiproblem family—appear much less important over the entire course of a life. Some environmental circumstances, of course, were not tested in the Grant Study and are likely to remain important, such as racism and other forms of societal discrimination. In our admittedly small Inner City sample, however, confined as it was to 456 white urban males, a warm childhood environment appeared to be a far better predictor of future social class and of adult employment (or unemployment) than was either childhood intelligence, parental dependence upon welfare, or the presence of multiple problems within the family.[6] For the Inner City men, the percent of life employed (a behavioral item) was one of the best objective predictors of mental health that the Study possessed, and employment was uncorrelated with social disadvantage. Children of great privilege with personality disorders have greater difficulty with stable employment than the mentally stable uneducated poor.

Another point of interest: it is through lifetime studies that we discover the tools we need for future investigation. The best tool we have for predicting longevity was discovered only through the study of entire lifetimes. Summarizing findings from the Berkeley and Oakland Growth Studies, arguably the greatest study of human development in the world, the distinguished sociologist John Clausen observed that when his subjects were sixty-five to eighty-five years old, the childhood trait of *Planful competence and dependability* was the most important predictor of future mental and physical well-being.[7] Howard Friedman arrived at the same conclusion in summarizing the longevity of the Terman men; I confirmed his finding seeking predictors of estimated longevity among the Grant Study men with the traits *Well integrated* and *Self-driving*.[8] But until the Terman, Berkeley, and Grant Studies reached maturity, the longevity literature had never even considered the importance of what Howard Friedman called *Conscientiousness*.

As I've described at some length in Chapter 6, the Study has made some important discoveries about marriage: that 57 percent of the Grant Study divorces were associated with alcoholism; that divorce led to future marriages happier than those enjoyed by the couples in the lowest third of sustained marriages; that marriages become happier after seventy. These issues have gone largely unaddressed in the literature on marriage. Time will take care of that. The point I want to make here is that these findings were not perceived *even in the Grant Study itself* until sixty or seventy years had passed. To understand lives takes a lifetime.

ALCOHOLISM

Much of what we infer retrospectively about our lives is not true. Nowhere is this more evident than in lifetime studies of alcoholism, where the Study of Adult Development has made perhaps its greatest contribution. In 1980 the Study was already the longest study of alcoholism in history, and had been awarded the international Jellinek Prize, a biennial award for alcohol research.[9] If it had stopped there, though, two other important discoveries that were published many years later would never have been made.[10]

To disprove the illusion that securely diagnosed alcoholics can return to successful social drinking required that we follow Inner City alcoholics for thirteen more years after they had been rated as returned-to-social-drinkers in 1983. Disaster was only a matter of time, like driving a car without a spare tire. And proofs that are a matter of time take time.

It also took many years to spot alcoholism as an unrecognized confounder in many classic studies of psychosocial effects on physical health. It is often unacknowledged alcoholism that causes precisely those stressors—the bankruptcy, the job loss, the divorce—on which the victim's poor health is blamed. Not until the men were eighty did the Study discover data showing that alcoholism (entirely unmentioned in prior major studies of marriage and divorce) was by far the most important factor in Grant Study divorces. Prolonged prospective study is a gift that keeps on giving—and surprising.

INVOLUNTARY ADAPTATION

The elucidation of resilience is the third way that the Grant Study has earned its lifetime of support. Its time-lapse photographs of human development illustrate and validate what is perhaps Freud's greatest discovery—the existence of involuntary coping mechanisms and their profound effects on adult life.[11] Freud's discovery of defenses was analogous to the discovery of Pluto by his astronomical contemporaries. In the nineteenth century both Pluto and defenses were invisible; they could be appreciated only by the systematic distortions they caused in the otherwise predictable behavior of visible objects. But twentieth-century telescopes allowed Pluto to be seen, and the longitudinal telescope of the Grant Study has permitted the visualization of defenses and their transmutations.

Still, in 2010 a leading textbook of human development referencing 2,500 different investigators and 1,200 topics says not a word about defense mechanisms.[12] Why do involuntary coping styles receive so little attention in the resilience literature now that they are documented as such powerful predictive tools? One general reason is that the latest editions of major textbooks in the field, Kail and Cavanaugh's *Human Development* and William Crain's *Theories of Development,* make almost no mention of empirical longitudinal life studies at all.[13] But the more particular answer is that measuring defenses reliably is extremely costly in time, effort, and money; this unpalatable truth tends to obscure the proven validity, and the elegance, of assessing adaptation by defensive style. Paper-and-pencil tests are cheap and easy, and even scientists are sometimes tempted to look for their car keys where the light is good instead of where they dropped them.

Certainly the Grant Study has so far failed to make Freud's inductive discovery of defenses part of the developmental landscape, and until fMRI evidence becomes available, the study of defenses may retain its scent of unreliability. Human resilience may continue to be conceptualized according to a model very different from my model of involuntary coping, and the baby may continue to be thrown out with the bathwater. It's not too surprising that Kail and Cavanaugh don't mention psychoanalysis, although I think that Freud deserves a historical footnote, at least. But the omission of defense mechanisms is another matter altogether. They are real, they are visible, and they are very important. And it is the Grant Study that has documented all three of these assertions.

The good news is that involuntary defense mechanisms are now included in the American Psychiatric Association's new *Diagnostic and Statistical Manual* (DSM-IV), and Joshua Shenk's 2009 cover article for *The Atlantic* on the Grant Study's investigations of defenses and development won a fascinated national readership.[14] The next few decades of the history of science will decide if the conceptual scheme (that is, the paradigm) of defenses (that is, involuntary coping) is a lasting contribution to resilience and development or not. The jury is still out. Nevertheless, since the study of involuntary coping matters to me a great deal, I will offer a final illustration of its importance to human resilience and development.

THE LIFE OF ERNEST CLOVIS

Artists are people who can share their most private dreams with others through the mature defense of sublimation. Although all the College cohort's successful academicians were skilled sublimators, like Dylan Bright in Chapter 8 and the elusive Art Miller in Chapter 3, it was Professor Ernest Clovis who was the true artist in this mode of involuntary coping—for a while. What was so striking about Clovis was that he had suffered more personal tragedy than any other man in the Study, yet at midlife he could write, "Perhaps I have not struck any particularly rough spots." An understanding of defenses enables us not to dismiss such resilience as denial, but to understand and appreciate it as a technique of coping. In Ernest Clovis, the transformation of a healthy childhood coping strategy (narcissistic fantasy) into mature generative sublimation fostered his own resilience and enriched hundreds of grateful students.

Clovis was raised in an austere and religious farming family. He maintained that he had had an enjoyable relationship with both his parents, but he also felt them to be emotionally constricted. He grew up with little physical affection, and his mother revealed to the Study staff that "very early Ernest developed qualities of self-reliance." He was sent to boarding school at eight years old.

Ernest's Calvinist family did not tolerate "selfishness," even the temporary kind that supports separation and individuation, and as a young child of four and five, Ernest took comfort in fantasy. He had an imaginary playmate whose outstanding trait was his egotism, and by his family's report, this companion was far more real to Ernest for a while than his toys, and maybe even than his brother and sisters. Not only did Perhapsy (that was his name) provide friendship to a geographically isolated little boy, but he also had the marvelous knack of always getting one up on the boy's dominating father.

As his father recalled it when Ernest entered the Study, "Perhapsy was a most wonderful fellow and could do anything. No matter what we did, especially something out of the ordinary, Perhapsy always did us one better. Once we were taking a train, and while we were waiting for it to leave, Ernest saw other trains moving about the yards. Perhapsy was 'in the big train over there,' or he was 'riding on top of the engine' or some such courageous stunt. He could climb the highest buildings and mountains in the world, etc., etc. He never took a back seat for anybody."

But as Ernest grew up, real games took the place of fantasy. "One of us asked him," his father went on, "when we noticed he was no longer talking about his 'friend,' what had become of him. The answer was, 'Perhapsy is dead.'" Ernest was finding fresh ways to bring pleasure to a cerebral life. He learned to beat his father in tennis, and through this sublimated form of aggression they became close. In college, many of the inhibited scholarship boys like Peter Penn rationalized their lack of dates by pleading lack of money. But the equally impecunious Clovis was very successful with the opposite sex. He entertained his companions by taking them to art museums, which were free. Sublimation took the place of money as a gateway to romance.

In college, Clovis impressed Study observers as showing very little emotional color. Some men with childhoods similar to his grew up into emotionally frozen adults, like the lawyer Eben Frost from Chapter 6. Some chose work in the physical sciences, and led rigid lives devoid of observable pleasure. Not so with Clovis. He become a first-rate squash player, a distinguished medievalist, and a good father. After fifteen years of marriage, he and his second wife agreed that they still enjoyed "a very satisfactory" sex life.

During World War II, Clovis fought with Patton's army through France, and afterward he felt an emotional pull to the French and their culture. Several of his classmates, guilty over their participation in so much destruction overseas, got involved in tangible (and ambivalent) postwar efforts to rebuild Japan and Germany. But Clovis became a scholar of the France that once had been. He plowed through medieval manuscripts; he mastered archaic dialects; he reconstructed an imaginary world that was little more tangible than Perhapsy. But he enjoyed it, and in his enjoyment he managed to connect his fantasy with the real world. From the flintstone of his intellect he struck sparks of marketable excitement, and in this Clovis illustrates a skill critical to the creative process. There is no one lonelier than the artist whose work speaks to no one, and no life less appreciated than that of the scholar who, like Peter Penn, cannot make his ideas comprehensible to the world.

Clovis, in contrast, received the most prestigious scholarships for his graduate work, the warmest accolades of his instructors, and offers of tenure from great universities. He could say, "I have developed a sense of mission that I may contribute to this country a better appreciation of France's social and political values—not just of her literary and historical contributions." As a teacher, an author, and a scholar his career was an unqualified success.

His home life was not so smooth. His first wife was stricken with an encephalitis that distorted her personality and rendered her permanently bedridden and irrational. The marriage became progressively and mutually more painful. Neither of his parents had ever shown emotion, and Clovis too had learned early to keep his feelings private.

Stoically, he wrote to the Study of the "minor frustration caused by my wife's encephalitis . . . to discuss it further with my parents would only make them unhappy. . . . I sometimes feel desire to discuss these problems with some of my wife's women friends, but I have never done this because it would seem like complaining." He did, however, form close substitute relationships with women. After many years he also finally did obtain a divorce, although he felt profoundly guilty about it, and eventually remarried happily.

I first met Clovis in 1969 in the stacks of Yale's Sterling Library. He had a cramped, almost monastic cubicle filled with old books and musty manuscripts, but it was enlivened by a bright contemporary print. Clovis himself was a pleasant, good-looking forty-seven-year-old man. His somber gray suit was unexpectedly brightened by a flamboyant orange tie. He often looked away as he talked to me, and at first I experienced him as cold. But I soon realized that this was more self-discipline than emotional rigidity, and that he was a very private and self-contained individual whose emotional fires, though banked, gave off both light and warmth. The way he talked often moved me deeply. I learned that talking about people depressed him, but that the discussion of his work, French drama, would bring a smile to his face.

When Clovis told me of his father's death four years previously, his eyes filled with tears. He hastened to assure me, "At the time of my father's death, I had to suppress my feelings." When he talked of the tragedy of his first wife's illness, he became distant, acutely anxious, and distressed. Talking of his happier second marriage, he became warmer and maintained eye contact.

Clovis told me that before his first wife got sick, modern English-American theater had been a shared hobby for the two of them. Once she could no longer go to plays with him, he became passionately interested in early and arcane French dramatists. On the verge of a divorce that his conscience could not countenance, he chose to translate from medieval French "the romantic tragedy of the hopeless love of a married man and a prostitute, both of whom chose to commit suicide rather than to be false to their love." This self-denying professor, who in the light of day could not permit himself tears, found himself weeping alone in darkened French theaters at plays that were of only historical importance to most English-speaking people. "I cry along with the old ladies," he confessed, "and feel thrilled by the experience." But Clovis was no schizoid eccentric; he did not keep his source of comfort secret. At Yale he brought the early French theater vividly to the awareness of his students.

In 1972 Clovis's daughter was diagnosed with lupus erythematosus, a poorly understood disease that causes arthritis, irreversible kidney damage, and episodic emotional instability. He was affectively aware of his daughter's peril and the awful years of his first wife's illness were again before him. But he continued to embrace the one area of his life where he could always maintain control. He wrote that he felt "enthusiastic" about the future because of the scholarly writing on French drama that he had planned. He added, "The language of the material you read has an emotional, aesthetic satisfaction."

Had he been unable to communicate his secret world of medieval France to others, he might have become deeply depressed. But as long as he could empathically communicate the scholarship in which he sublimated his feelings, he could retain an almost palpable animal vigor and *joie de vivre.*

When I saw Clovis again in 1996, he was seventy-five, still happily married to his second wife, and close to his four children. And he had become—in a sense—a Guardian. But as Eriksonian made clear from the beginning, every developmental accomplishment has a shadow side. The shadow of Intimacy is Isolation; the shadow of Generativity is Stagnation. The shadow of Guardianship, which I have called Hoarding, is preservation for its own sake, and Clovis had gotten trapped there.

Not all traditions merit preservation, and very few merit preservation at any cost. The ones that *are* worthy have value beyond tradition itself. In Polonius, Shakespeare gives us a classic example of elderly rigidity, which is the potential vice of all grandparents.

Clovis still wanted to create, to research, to write. But alas, there was no time. He was now the curator of a great library collection; he was editing a history of medieval France; he was editing his wife's papers; he was certainly preserving the past for the future. But in these efforts he had exiled himself from the medieval world where his imagination could run free. To continue his own work, he felt, would be "self-interest and selfishness." When I asked him if he wouldn't rather write about early French theater than be a library curator, he lightened up briefly. Perhapsy lived on inside him. But he caught himself, and became again a "responsible" professor emeritus and grandfather.

Unlike the other Guardians we've seen in these pages, he had left his rich inner life behind him in the service of what felt to him like the obligations of late middle age. I found myself thinking of the eighty-year-old Robert Sears, a "Termite" (as the gifted Terman children called themselves) and a brilliant and creative professor of psychology. I remember watching him in the Terman Study archive, unpaid, sorting the old IBM punch cards—a thankless task if there ever was one—so that young whippersnappers like me could enjoy an orderly universe in which to make our reputations. He was cultivating his garden—as I cultivate mine in the Grant Study, long after retirement—and he did it with joy, not out of obligation.

Sublimation is a way of negotiating with the superego, of getting around guilt, of finding pleasure if not in the forbidden garden, then at least in the garden next door. It served Clovis very well for a long time, but his Calvinist conscience eventually got its own back. Clovis had never freed himself from his guilt over having divorced his first wife, and as new obligations piled up—the library, the papers, the editing—they fed his superego until it became strong enough to shut the door to his passionate medieval world. Like Newman, who was always part mystic and part engineer, Clovis was always part ascetic and part troubadour. In his case, the ascetic eventually won out. At least for the time being, Perhapsy was dead once more. People change; people don't.

Professors Clovis and Bright, however good they were at sublimation, were not examples of ideal flourishing any more than Beethoven was. Nor were they among the best outcomes in the Study. In fact, sublimation was only slightly more correlated with the success in the Decathlon of Flourishing than the neurotic defenses. Sublimation does not cure all ills. Its superiority lies in the fact that, unlike the neurotic defenses, it transforms agony into real pleasure—but a pleasure, unlike the more narcissistic defenses, that conforms to the needs of others. Beethoven's need to create magic in music was as involuntary as the clotting of his blood, but just as critical to his resilient survival. To replace valid dynamic defenses that can and do change with maturity in favor of the reliable Big Five would be as foolish as ignoring the girlfriends and best mates of the College sophomores and studying only their body build—a mistake that the Grant Study made once and will not encourage others to make again. Its demonstration that intangible coping styles are relevant to modern social science made the Grant Study worth every penny spent on it, and worth the five years of my life that I have continued to devote to it—like Professor Sears—with no compensation at all.

IF WISHES WERE HORSES . . .

I am often asked how I would design a lifetime study if I had the opportunity to build one from scratch, instead of just walking into the

Grant Study as I did and watching it grow. Let me close this history with my answer.

First, I would ask for an endowment to be run like an annuity, to last one hundred years. That would ensure that minimal continuity could be maintained even during lean years.

Second, in selecting my cohort I would not try for inclusive diversity, but I would include children of both sexes from no more than two contrasting ethnic groups. Sociology requires a large representative sample; biology needs small homogeneous samples. Historical change alters the rules of sociology; the rules of biology and ethology are more constant. In designing my ideal study I remain a biologist and an ethologist. And I would focus on blue-collar children, because (despite the reasoning of the early planners) the Inner City men provided even more surprises than the College men did.

Third, along with DNA and social security numbers for all the members of my cohort, I would collect the names, birth dates, and addresses of five relatives under sixty who do not live with them. The names and addresses of research subjects change with frustrating regularity, and this would spare us the huge investment in detective work that lost subjects demand.

Fourth, I would concentrate on finding and using the best measures available to assess the positive emotions (joy, compassion, trust, hope, and their like), and attachment.

Fifth, I would assess involuntary coping through a two-hour videotaped couples interview every twenty years in preference to the excerpted vignettes that I used. That way real-time rater reliability could be obtained, and a major source of bias (the excerpting) removed.

Sixth, consistent with safe x-ray exposure limits, I would collect neuroimaging data every five years. Like the men who started the Grant Study, I am not quite sure what I would be looking for—the brain changes that accompany the evolution of twenty-year-old narcissism into seventy-year-old empathy?—but the ambitious kids reviewing the data in 2050 certainly would be.

Finally, I would keep data collection simple enough so that after fifty years investigators would not be drowning in their own accumulations. The Grant Study was already struggling with this intrinsic problem of longitudinal studies all the way back in 1941, and that was only three years in! I deliberately kept data collection simple, and I take some credit for the long-term survival of the Study of Adult Development because of it. I would keep the Study cohort small enough that every member would remain a person to the Study, and not just a collection of numbers. Every member could maintain a warm alliance with the Study staff, and errors in data would be noticed. And yes, I would collect pencil-and-paper tests—belief systems need to be tested empirically.

CONCLUSION

At ten years old I entertained a Perhapsy-like wish for the greatest telescope in the world. When I was an assistant professor at Tufts Medical School and assumed responsibility for the Grant and Glueck Studies, I believed I had found it—the device that would let me see entire lifetimes at a single glance. In 1969, begging for money as usual, I paraphrased Archimedes to the Commissioner of Mental Health: "Give me the salary for a single research assistant, and I will move the world." He very graciously told me that he could not do that. But the Grant Foundation did, and I was off and running.

I have spent forty years peering through that telescope, and it has shown me world after world that I never dreamed of. Over those years I've developed convictions, and (I pride myself on this, too) exposed them to empirical scrutiny. Three big ones have stood the test of time, if not perfectly. One was that a warm childhood was a most important predictive factor and that a bad childhood was not. Another was the assertion that I made to *The Atlantic,* that the most important contributor to joy and success in adult life is love (or, in theoretical terms, *attachment*). My third great conviction was the identification of the involuntary adaptive "mechanisms of defense" as the second greatest contributor. Forty years of study have convinced me that in this I was right; mature defenses remain the *sine qua non* of warm relationships. Alas, however, they do not appear to be essential for sustained good health and successful physical aging—yet another favorite hypothesis washed up as I followed the men into old age.

This extraordinary telescope has brought great joy and meaning to my life. It allowed me to embark upon a quest that had haunted me from childhood, exploring questions that matter to me both personally and scientifically. It has provided me with wonderful companions on the way—not only my many colleagues over all these years, but also the Study members themselves. And I become more and more aware that the Study, and the work we've done with it, has encouraged other people to think about their own lives and the lives of others. Not statistically, perhaps, but with a little more curiosity and a little more interest and a little more kindness. And how can that hurt?

APPENDIXES

NOTES

ACKNOWLEDGMENTS

INDEX

APPENDIXES

THE INTERVIEW SCHEDULES

The Grant Study Interviews. Below are two examples of the semi-structured schedules of questions I used to guide my interviews with the men. Within reason, listed questions were always asked in the same order. The interviewer took longhand notes during the interview. Whenever a question elicited a problem area in a man's life, the interviewer probed for his particular coping style.

The Interview Schedule, age 45–55

I. **Work**

  a. What do you do? Any recent changes in responsibility?
  b. Ten years from now, where are you heading?
  c. What do you like and what do you dislike about your work?
  d. What for you is most difficult?
  e. What job would you have preferred?
  f. What are good and bad aspects about relations with your boss? With subordinates?
  g. How do you handle some of the problems that arise with these people?
  h. Looking back, how did you get into your present work?
  i. Were there people with whom you identified?
  j. What work do you do outside of your job—degree of responsibility?
  k. What plans for retirement?
  l. Ever unemployed for more than a month? Why?

m. What will you do the first week of retirement? Anticipated feelings?

II. Family

a. News of parents and siblings.

b. Describe each child, their problems and sources of concern to you.

c. How do you handle adolescence differently from your parents?

d. Any deaths: first response, second response, means of finally handling feelings.

e. This is the hardest question that I shall ask: Can you describe your wife?

f. Since nobody is perfect, what causes you concern about her?

g. Style of resolution of disagreements.

h. Has divorce ever been considered? Explain.

i. Quality of contact with parents and degree of pleasure derived.

j. Which of your parents wore the pants when you were younger?

III. Medical

a. How is your health overall?

b. How many days sick leave do you take a year?

c. When you get a cold, what do you do?

d. Specific medical conditions and means of coping with the disability.

e. Views and misconceptions about these conditions.

f. Injuries and hospitalizations since college.

g. Patterns of use of smoking and recollections about stopping.
h. Pattern of use of medicines and of alcohol.
i. Do you ever miss work due to emotional strain, fatigue, or emotional illness?
j. Effect of work on health and vice versa.
k. How easily do you get tired?
l. Effect of health on the rest of your life.

IV. **Psychological**
a. Biggest worries last year.
b. Dominant mood over past six months.
c. Some people have trouble going for help and advice: What do you do?
d. Can you talk about your oldest friends? What made them friends?
e. Who are the people (non-family) you would feel free to call on for help?
f. What social clubs do you belong to, and what is your pattern of entertaining?
g. How often do you get together with friends?
h. What do people criticize you for or find irritating about you?
i. What do they admire or find endearing?
j. What are your own satisfactions and dissatisfactions with yourself?
k. Ever seen a psychiatrist? Who? When? How long? What do you remember? What did you learn?
l. Persistent daydreams or concerns that you think about but don't tell others?
m. Effect of emotional stress?

n. Philosophy over rough spots?

o. Hobbies and use of leisure time? Athletics?

p. Vacations? How spent and with whom?

q. Any questions raised by my review of case record?

r. What questions do you have about the Study?

The Interview Schedule, age 65–80

I. Work

1. Why did you retire? When?

2. How was your job over the last two years? What did you miss most? Was there a retirement ceremony?

3. What did you do the first six weeks of retirement?

4. What takes the place of your job?

5. What is your most important activity now?

6. What is the best part of retirement?

7. The worst part?

8. The most difficult part?

9. How is your retirement financed?

10. Any job change if you were to do life over?

II. Social

1. What is it like being home for lunch?

2. How do you spend increased time with your wife? Problems?

3. How did your marriage last a quarter century?

4. What have you learned from your children?

5. Grandchildren?

6. Oldest friend?

7. Activities with others?

III. Psychological

1. Mood over last 6 months?
2. Worries last year?
3. Changes in religious beliefs?

IV. Health

1. Health since retirement?
2. Most annoying aspect of aging?
3. What physical activities have you given up?
4. Cold—what do you do?
5. Days of sick leave?
6. Hospital days total since 1970?
7. Medications?

THE INNER CITY MEN AND THE TERMAN WOMEN

In the late 1960s, another important longitudinal study became available to the Harvard Study of Adult Development as a foil for the homogeneous sample of privileged and intelligent white men that made up the College cohort. This study allowed us to compare the College men with a group of white men who were far less privileged, and (at least on IQ testing) less intelligent. A second study became available in the 1980s and offered a comparison with intellectually gifted, but not particularly privileged, women. These opportunities allowed us to draw conclusions about biology versus environment in some of our outcomes, and also, more specifically, about certain kinds of sociological influences. I refer to these two studies throughout the text in contexts where they shed light on, or in some cases expand, the Grant Study findings.

The Inner City Cohort (The Glueck Study of Juvenile Delinquency)

The second cohort joined us in 1969, through the generosity of Sheldon and Eleanor Glueck. Sheldon Glueck was a professor at Harvard Law School and his wife, Eleanor, a groundbreaking social worker there; both were world-famous criminologists. In the 1940s they began an intensive prospective study of 500 white male teenagers from inner-city Boston who had been remanded to reformatories—the Glueck Study of Juvenile Delinquency. They carefully matched them for IQ (their average Wechsler-Bellevue score, based on careful individual testing, was 95) and ethnicity to a control group of 500 boys who had no history of serious delinquency, but who came from the same high-crime neighborhoods, the same minority identifications, and the same impoverished urban classrooms as the youths who had ended up in reform schools. The Gluecks included no African Americans and no women, and they excluded any boys who by age fourteen had manifested any serious delinquency.

The initial phases of data collection for the Glueck Study were similar to the Grant Study's, with psychiatric interviews, anthropometric measurements, medical examinations, family histories, socioeconomic assessments, and even complete Rorschach tests. But the Gluecks stopped their follow-up in 1962, when their subjects were 32. Thanks to a grant from the National Institute on Alcohol Abuse and Alcoholism, I was able to integrate the two studies in 1970. The control cohort of Inner City men, renamed the *Glueck Study,* is now a part of the Harvard Study of Adult Development, enriching it by the contrast it provides with the socially privileged and high-IQ Harvard men. Like the Grant Study men, the Inner City men deserve great credit for their loyalty to the Study.

Since the integration, the College and Inner City cohorts have been followed in an identical manner, except for the years between 1962 and 1974, when the Inner City men were not followed.

Most of the Inner City men, who were born between 1925 and 1932, had early memories of discrimination and deprivation. But they too were beneficiaries of the G.I. Bill and of America's postwar economic boom. By the 1950s, they, children of once-devalued Irish and Italian immigrant parents, had become the voting majority and political masters of Boston. Their previous distinction as the most disparaged of city minorities had been passed on to recent African-American arrivals from the South. While only one in ten of their fathers belonged to the middle class, at age forty-seven half of the Inner City men had achieved middle-class status. In this achievement the Inner City men were a subsample of underclass "no hope" youth who became solidly middle class—certainly another reasonable definition of "success."

The Terman Women Cohort

In 1920, Lewis Terman, a renowned Stanford psychologist, started a study of roughly 1,500 grammar-school children. They were mostly 4th graders of a 1906–1911 birth cohort, and they represented all the children in Oakland, San Francisco, and Los Angeles with (carefully tested) IQs of over 140. They have been followed by questionnaires every five years since then, with little attrition, except by death, to this day. Of the Terman group, 672 are women, most of whom went to college.

In 1987, through the generosity of Stanford professors Robert Sears and Albert Hastorf, Caroline Vaillant and I spent a full year reviewing the records of the the 78- to 79-year-old women, and interviewing a representative surviving sample of 40. We used a similar interview to the one we used with the Grant Study men to facilitate comparison. This sample of women who were the Grant Study men's intellectual peers (or superiors) is the cohort that appears in this volume as the Terman women, and it allowed us to study some of the sociological effects of gender.

In the 1920's, California was still a young state. One Study member could recall watching the last sailing ships glide through the Golden Gate before the bridge was built. The population of Los Angeles was 500,000. The Terman women were descendants of pioneers. One woman's grandmother had saved her own life by killing an intruder with a tomahawk. One woman's father, a high school teacher, used routinely to disarm his students before class. Another father won a stagecoach line to Arizona in a poker game.

Twenty percent of the Terman women's fathers were in blue-collar occupations; thirty percent were in "the professions." Only one Terman woman's father was an unskilled laborer; he worked as a janitor at the University of California at Berkeley, so that his five bright children could all go to college for free.

The Terman women were physically and intellectually precocious as children. But their high intelligence—they had a mean IQ of 151—did not handicap them psychologically. On the contrary, their mental health was demonstrably better than their classmates'.

After following them until they were 75, Terman and his coworkers observed that the women showed significantly more humor, common sense, perseverance, leadership, and even popularity than their classmates. They were as likely as their classmates to marry, but their physical health was better. At age eighty, the mortality of the Terman women, like the College men's, has been only half the expected rate for white American women in their birth cohort. As in the Harvard sample, more than half of the Terman women have survived past eighty.

The career situation for these highly intelligent women was full of paradox, however. They grew up with mothers who did not yet have the right to vote. College tuition in California at the time was cheap enough ($25 to $50 a term at both Stanford and Berkeley) that a college degree was a realistic expectation for a bright woman. And the Depression, which began when they were twenty, and World War II, which began when they were thirty, put pressures on these women to enter the work force. But the jobs on offer were limited in scope, compensation, and opportunity. When asked what occupational opportunities World War II had opened for her, one Berkeley-educated woman replied dryly, "I finally learned to type."

Almost half of the Terman women held full-time jobs for most of their lives. Most had gone to college and many to graduate school. Nevertheless, their mean maximum annual income ($30,000 in 1989 dollars) was identical to that of the Inner City men, who had a mean IQ of 95 and an average of ten to eleven years of education. Wartime demands for Rosie the Riveter and her like might have been an economic boon for high school dropouts, but they were an economic millstone for the gifted Terman women.

For the Terman *men,* it is worth noting, the war really did create great opportunities. The G.I. Bill paid for their graduate schooling and allowed some of them to create the Los Alamos and Livermore laboratories and, ultimately, Silicon Valley. Some of the Terman men gave their brilliance to the Los Angeles entertainment industry. Lewis Terman's own gifted children, a boy and a girl, were both included in his study. Both graduated from Stanford and worked for the university for much of their lives. The son served as provost, and was mentor to many of the founders of Silicon Valley; the daughter was a secretary in one of the dormitories.

Thus, of our three samples, it was the college-educated middle-class Terman women, most of whose relatives had been in the United States for generations, who most clearly illustrated the negative effect of social bigotry upon development.

ASSESSMENT OF CHILDHOOD SCALES

1. *Child Temperament Scale (Age 0–10)*

1 = very shy, tics, phobias, bedwetting beyond age 8, dissocial, severe feeding problems, other noted problems.

3 = average.

5 = good-natured, normally social, an "easy child."

2. *Childhood Environmental Strengths Scale (Age 5–18)*

1. **Global Impression (rater's overall hunch)**

   1 = a negative, non-nurturing environment.

   3 = neither negative nor positive feeling about subject's childhood.

   5 = a positive, intact childhood; good relationships with parents, siblings, and others; environment seems conducive to developing self-esteem. A childhood that rater would have wanted.

2. **Relationship with Siblings**

   1 = severe rivalry, destructive relationship, sibling undermines child's self-esteem or no siblings.

   3 = no good information, not mentioned as good though not particularly bad.

   5 = close to at least one sibling.

3. Home Atmosphere

    1 = any noncongenial home, lack of family cohesiveness, parents not together, early maternal separation, known to many social agencies, many moves, financial hardship that impinged greatly on family life.

    3 = average home: doesn't stand out as good or bad; or lack of information.

    5 = warm, cohesive atmosphere, parents together, doing things as a family, sharing atmosphere, maternal and paternal presence, few moves, financial stability or special harmony in spite of difficulties.

4. Mother/Child Relationship

    1 = distant, hostile, blaming others (such as father, teachers) for wrong methods of upbringing, overly punitive, overprotective, expecting too much, mother absent, seductive, not en couraging feeling of self-worth in child.

    3 = mostly for lack of information or lack of distinct impression about mother.

    5 = nurturing, encouraging of autonomy, helping boy develop self esteem, warmth.

5. Father/Child Relationship

    1 = distant, hostile, overly punitive, expectations unrealistic or not what son wants for himself, paternal absence, negative or destructive relationship.

    3 = lack of information, no distinct impression about father.

    5 = warmth, encouraging of autonomy in child, helping to develop self-esteem, do things with son, discusses problems, interested in child.

ADULT ADJUSTMENT SCALES

I. Scale for Objective Mental Health from Age 30–50

1. Income over $20,000 in 1967 dollars

 1 = Above $20,000
 2 = Below

2. Steady promotion, 1967

 Examination of questionnaires at 5-year intervals from 1946–1967 reveals steady promotion or career progress
 1 = Yes
 2 = No steady promotion

3. Games, 1967

 Examination of 1951–1967 questionnaires and review of other data reveals games with non-family members (golf, bridge, tennis, etc.)
 1 = Yes
 2 = No games with others

4. Vacation, 1967

 Evidence from the 1957, 1964, and 1967 questionnaires that the subject took more than two weeks' vacation a year and had fun rather than just dutiful visits to relatives
 1 = Yes, takes them
 2 = No, ignores vacation

5. Enjoyment of job

 Evidence from 1946, 1951, 1954, 1960, 1964, and 1967 questionnaires that the subject enjoyed his job and was enthusiastic about it.
 1 = Unambiguous enjoyment
 2 = Enjoyment not clear
 3 = Definite lack of job enjoyment

6. Psychiatric visits, 1967

 Visits from college through 1967
 1 = Under 10 visits
 2 = 10 or more visits

7. Drug/alcohol use, 1967	Evidence either that (a) the subject used sleeping pills weekly for a year or tranquilizers and amphetamines daily for a month, or (b) for a period at least as long as a year (or for two points in time) the individual drank more than 8 ounces of alcohol a day or felt he had trouble with control or he, his family, and his friends thought he drank too much 1 = No to (a) and (b) 2 = Yes to (a) and/or (b)
8. Days' sick leave 1967	Based on questionnaires in 1944, 1946, and 1967 1 = Less than 5 days 2 = 5 or more days
9. Marital enjoyment, 1967	Averaging the husband's reports of his marriage in 1954 and 1967 with the wife's report of their marriage in 1967 1 = Good (marriage score of 4 or 5) 2 = Intermediate (marriage score of 6 or 7) 3 = Getting or considering a divorce (marriage score of 8+)

II. Scale for Objective Mental Health from Age 50–65

1. Career (3 questionnaires)	1 = Working full-time 2 = Significant reduction of work load 3 = Retired
2. Career success (3 questionnaires)	1 = Current (or pre-retirement) responsibilities/success as great or greater than 1970 2 = Demotions or reduced effectiveness (prior to retirement)
3. Career or retirement enjoyment (2 questionnaires)	1 = Meaningful, enjoyable 2 = Ambiguous 3 = Working only because he must or feels retirement demeaning/boring

4. Vacations
(2 questionnaires)

1 = 3+ weeks and fun
2 = Less than 3 weeks if working or un-
playful retirement

5. Psychiatrist use
(2 questionnaires)

1 = No visits
2 = 1–10 visits
3 = Psych hospitalization or 10+ visits

6. Tranquilizer use
(2 questionnaires)

1 = None
2 = One use to a month
3 = More than once a month's use

7. Days' sick leave (exclude irreversible
illness)
(2 questionnaires)

1 = Less than 5 days/year
2 = 5+ days

8. Marriage 1970–1984
(3 questionnaires)

1 = Clearly happy
2 = So-so
3 = Clearly unhappy or divorced

9. Games with others
(3 questionnaires)

1 = Regular social activities/sports
2 = Little or none

Total (low score is good)

9–14 = Score compatible with being classified mentally healthy
15–23 = Bottom quartile; excludes individual from being classified as mentally
healthy

III. Scale for Objective Mental Health from Age 65–80

1. Career or retirement enjoyment

1 = Still enjoying part-time work and/or
retirement
2 = Ambiguous or midrange
3 = Dissatisfied with retirement

2. Retirement success, age 65–80

1 = Still enjoying part-time work and/or
retirement
2 = Ambiguous or midrange
3 = Dissatisfied with retirement

3. Contact with younger relatives, age 65–80

1 = Meaningful, enjoyable family interaction
2 = Ambiguous or infrequent interaction with young relatives
3 = Avoids or shunned by kids, grandkids, nieces, nephews

4. Use of leisure time, age 65–80

1 = Varied, imaginative, and enjoyment of leisure time
2 = Some leisure activities with moderate enjoyment
3 = Bored, passive, unsatisfactory use of leisure time

5. Games with others, age 65–80

1 = Many regular social activities: bridge, lunches, golf
2 = Some social activities, but limited involvement
3 = Almost no social activities

6. Psychiatric use, age 65–80

1 = No visits for counseling
2 = 1–10 visits
3 = Psychiatric hospitalization or 10+ visits

7. Mood-altering drug use, age 65–80

1 = None
2 = 1–30 days use
3 = More than 1 month's use in a year

8. Marriage, age 65–80

1 = Clearly happy (until widowed)
2 = Never married, or so-so or fair if while married
3 = Clearly unhappily married, or divorced with no new intimate relationship

9. Rater's subjective impression, age 65–80

Rater's subjective impression after reviewing 6–7 questionnaires and other interview data in file
1 = Adjustment to aging is excellent
2 = Adjustment to aging is good or above average
3 = Ambiguous or average adjustment to aging
4 = Poor adjustment to aging
5 = Adjustment to aging worse than for most men

DOMINANT COLLEGE PERSONALITY TRAITS

Dominant College Personality Traits (N=251)

Trait*: (frequency), Definition	Important Correlates
Vital affect: (20%). *Expressive, forceful, spontaneous energy, animated*	
Sociable, friendly: (22%). *Naturally friendly, socially at ease, makes friends easily*	
Well integrated: (60%). *Steady, stable, dependable, trustworthy, surmounts problems that confront him*	Mature defenses—very significant Longevity—very significant Decathlon—very significant Eriksonian maturity—very significant Childhood strengths—significant No depression—significant Stable marriage—significant
Practical, organizing: (37%). *Practical not theoretical, organized not analytical, likes getting things done*	Conservative—very significant Decathlon—very significant Maturity of defense—very significant Eriksonian maturity—very significant No depression—very significant Stable marriage—significant
Humanistic: (16%). *Interested in people, wish to work with people*	
Pragmatic: (38%). *Practical, conforming, accept the mores of the times*	Conservative—very significant No depression—very significant Maturity of defenses—significant
Political: (17%). *Interested in government, social reform, public policy rather than people*	
Over-integrated, just so: (13%). *Neat, meticulous, rigid, depend on routine, systematic*	
Bland affect: (38%). *Not warm or positive mood, not rich or vital affect*	

Trait*: (frequency), Definition	Important Correlates
Self-driving: (14%). *Self-control, willpower, persevering, uneasy with leisure*	
Cultural: (22%). *Headed for artistic and literary or at least cultural careers*	Liberal—very significant
Verbalistic: (18%). *Facile, lucid, well-formulated and rich in their use of language*	
Inarticulate: (14%). *Inability to express themselves*	
Shy: (18%). *Embarrassed, reserved, awkward socially but like people*	
Physical science: (12%). *Mechanical, inductive, like lab work and things more than people*	
Sensitive affect: (17%). *Shy, subtle, aesthetic, poor adjustment to everyday realities*	Liberal—very significant
Creative and intuitive: (6%). *Original, literary, and artistic, spurn concrete forms of thought*	Liberal—very significant
Mood swings: (14%). *Strongly marked and/or fluctuations in moods*	
Inhibited: (19%). *Overly moral, indecisive on acting on desires*	
Ideational: (21%). *Theoretical, analytical, dislike routine, scholarly, prefer literature over science*	Liberal—very significant
Self-conscious, introspective: (25%). *More concerned with subjective feelings than others*	Liberal—very significant
Lack of purpose and values: (20%). *Drifting, unenthusiastic*	
Unstable autonomic functions: (14%). *Undue anxiety, tremulousness, blushing, sweating, palpitations, functional urinary or GI symptoms*	
Asocial: (10%). *Other people unimportant, prefer things and their own company*	

Trait*: (frequency), Definition	Important Correlates
Incompletely integrated: (15%). *Erratic, unreliable, undependable, little perseverance, poorly organized*	Decathlon—very significant
Psychopathic: (7%). *Confined to a small number of men who might be mentally ill*	

*The traits are arranged in the degree that they correlated with the Study's ABC adjustment. The "soundest" boys most commonly manifested Vital affect and Sociability.

STUDY BIBLIOGRAPHY

A. Books

1. Hooton EA: *Young Man, You Are Normal*. New York, G. P. Putnam's Sons, 1945.

2. Heath CW, et al.: *What People Are*. Cambridge, MA, Harvard University Press, 1945.

3. Monks John P: *College Men at War*. Boston, American Academy of Arts and Sciences, 1957.

4. Vaillant GE: *Adaptation to Life*. Boston, MA, Little, Brown, 1977 [reprinted with a new preface in 1995 by Harvard University Press, Cambridge, MA].

5. Vaillant GE: *Natural History of Alcoholism*. Cambridge, MA, Harvard University Press, 1983.

6. Vaillant GE: *Ego Mechanisms of Defense: A Guide for Clinicians and Researchers*. Washington, DC, American Psychiatric Press, 1992.

7. Vaillant GE: *The Wisdom of the Ego*. Cambridge, MA, Harvard University Press, 1993.

8. Vaillant GE: *Natural History of Alcoholism, Revisited*. Cambridge, MA, Harvard University Press, 1995.

9. Vaillant GE: *Aging Well*. Boston, Little, Brown, 2002 [also in Hebrew translation].

10. Vaillant GE: *Spiritual Evolution: A Scientific Defense of Faith*. New York, Doubleday Broadway, 2008.

11. Vaillant GE: *Triumphs of Experience: The Men of the Harvard Grant Study.* Cambridge, MA, Harvard University Press, 2012.

B. Papers

Clark Heath, Director

1. Johnson REL Brouha: Pulse rate, blood lactate and duration of effort in relation to ability to perform strenuous exercise. *Revue Canadienne de Biologie,* 1942, 1, 2, 171–178.

2. Davis, Pauline: Effect on the electroencephalogram of changing the blood sugar level. *Archives of Neurology and Psychiatry,* 1943, 49, 186–194.

3. Wells FL: A research focused upon the normal personality: A note. *Character and Personality,* 1944, 122, 299–301.

4. Wells FL: Mental factors in adjustment to higher education. *Journal of Consulting Psychology,* 1945, 9, 2, 67–86.

5. Savage, Beatrice M: Undergraduate ratings of courses in Harvard College. *Harvard Educational Review,* 1945, 15, 3, 168–172.

6. Seltzer CC: The relationship between the masculine component and personality. *American Journal of Physical Anthropology* New Series, 1945, 3, 33–47.

7. Bock AV: Selection of pre-medical students. *Bios,* 1945, 16, 199–209.

8. Seltzer CC: Chest circumference changes as a result of severe physical training. *American Journal of Physical Anthropology* New Series, 1946, 4, 3, 389–394.

9. Seltzer CC: Body disproportions and dominant personality traits. *Psychosomatic Medicine,* 1946, 8, 2.

10. Heath Clark W, and Lewise W Gregory: Problems of normal

college students and their families. *School and Society*, 1946, 63, 1638, 355–358.

11. Heath, Clark W, and Lewise W Gregory: What it takes to be an officer. *Infantry Journal*, March 1946.

12. Wells FL and WL Woods: Outstanding traits: In a selected college group, with some reference to career interests and war records. *Genetic Psychology Monographs*, 1946, 33, 127–249.

13. Wells FL: Verbal facility: Positive and negative associations. *The Journal of Psychology*, 1947, 23, 3–14.

14. Wells FL: Verbal excess over quantitation: Two case studies. *The Journal of Psychology*, 1947, 14, 4.

15. Wells FL: Personal history, handwriting and specific behavior. *The Journal of Psychology*, 1946, 23, 65–82.

16. Wells FL: Verbal facility, sensitive affect, psychometrics: Two "Palimpsest Personalities." *The Journal of Psychology*, 1947, 23, 179–191.

17. Wells FL: Inarticulate, without physical science motivations: Case studies V and VI. *The Journal of Psychology*, 1947, 24, 149–159.

18. Wells FL: Physical science motivations, inarticulate: Case studies VII and X. *The Journal of Psychology*, 1947, 24, 211–227.

19. Heath CW, Monks JP, and WL Woods: The nature of career selection in a group of Harvard undergraduates. *Harvard Education Review*, 1947, 17, 3, 190–197.

20. Heath CW: The hemoglobin of healthy college undergraduates and comparisons with various medical, social, physiologic and other factors. *Blood: The Journal of Hematology*, 1948.

21. Seltzer CC, FL Wells, and EB McTernan: A relationship be-

tween Sheldonian somatotype and psychotype. *Journal of Personality,* 1948, 16: 431–436.

22. Wells FL: Verbal aptitudes versus attitudes: Case Studies XI–XIV. *Journal of Genetic Psychology,* 1948, 72, 185–200.

23. Seltzer CC: Phenotype patterns of racial reference and outstanding personality traits. *Journal of Genetic Psychology,* 1948, 72, 221–245.

24. Wells FL: Quantitative-spatial aptitudes and motivations: Case Studies XV–XVIII. *Journal of Genetic Psychology,* 1948, 73, 119–140.

25. Wells FL: Clinical psychology in retrospect and prospect. *Journal of Psychology,* 1949, 27, 125–142.

26. Wells FL: Adjustment problems at upper extremes of test "intelligence": Cases XIX–XXVIII. *Journal of Genetic Psychology,* 1949, 74, 61–84.

27. Wells FL: Psychometric patterns in adjustment problems at upper extremes of test "intelligence": Cases XXIX–LVI Department of Hygiene Referrals. *Journal of Genetic Psychology,* 1950, 76, 3–37.

28. Wells FL: Some projective functions of simple geometrical figures: Cases LXXXV–XCV. *Journal of Genetic Psychology,* 1950, 77, 267–281.

29. Wells FL: Harvard National Scholars in a student frame of reference. *Journal of Genetic Psychology,* 1951, 79, 205–219.

30. Wells FL: Further notes on Rorschach and case history in Harvard National Scholars: Cases CXIII–CXXVI. *Journal of Genetic Psychology,* 1951, 79, 261–287.

31. Wells FL: Rorschach and Bernreuter procedures with Harvard

National Scholars in the Grant Study: Cases III, IX, X, XXVII, XXVIII, CII–CXII. *Journal of Genetic Psychology,* 1951, 79, 221–260.

32. Wells FL: The effects of need achievement on the content of TAT stories: A re-examination. *Journal of Abnormal and Social Psychology,* 1953, 48, 532–536.

Charles McArthur, Director

1. McArthur C: Long-term validity of the Strong Interest Test in two subcultures. *Journal of Applied Psychology,* 1954, 38, 346–353.

2. McArthur C: Personalities of public and private school boys. *Harvard Educational Review,* 1954, 24, 256 262.

3. McArthur C: Analyzing the clinical process. *Journal Counseling Psychology,* 1954, 1, 203–206.

4. McArthur C and LB Stevens: The validation of expressed interests as compared with inventoried interests: A fourteen-year follow up. *Journal of Applied Psychology,* 1955, 39, 184 189.

5. McArthur C: Predictive power of pattern analysis and of job scale analysis of the Strong. *Journal of Counseling Psychology,* 1955, 2, 205–206.

6. McArthur C: Personalities of first and second children. *Psychiatry,* 1956, 19, 47–54.

7. McArthur C: Upper class intelligence as the critical case for a theory of "middle-class bias." *Journal of Counseling Psychology,* 1957, 4, 23–30.

8. McArthur C: The psychology of smoking (with others). *Journal*

of Abnormal and Social Psychology, March 1958, 56, no. 2, 267–275.

9.   McArthur C: Sub-culture and personality during the college years. *Journal of Educational Sociology,* 1960, 33, 260–268.

10. Dill, David B: Fitness of Harvard men at twenty-two years. *Duke University Council on Gerontology, Proceedings 1961–65,* December 10, 1963, 226–239.

11. McArthur C: The validity of the Yale Strong Scales at Harvard. *Journal of Counseling Psychology,* 1965, 12, 35–38.

12. McArthur C: Career choice: It starts at home. *Think,* March–April 1966, 15–18.

13. Robinson S, Dill DB, Tsankoff SP, Wagner JA, and RD Robinson: Longitudinal studies of aging in 37 men. *Journal of Applied Physiology,* 1975, 38, 263–267.

George Vaillant, Director

1.   Vaillant GE, Brighton JR, MacArthur CC: Physicians' use of mood-altering drugs: A 20 year follow-up. *New England Journal of Medicine,* 1970, 282, 365–370.

2.   Vaillant GE: Theoretical hierarchy of adaptive ego mechanisms: A 30-year follow-up of 30 men selected for psychological health. (Helene and Felix Deutsch Prize, 1969) *Archives of General Psychiatry,* 1971, 24, 107–118.

3.   Vaillant GE: The evolution of adaptive and defensive behaviors during the adult life cycle. Abstracted in *Journal of the American Psychoanalytic Association,* 1971, 19, 110–115.

4.   Vaillant GE, MacArthur CC: A thirty-year follow-up of somatic symptoms under emotional stress. In *Life History Research*

in Psychopathology, vol 2. Edited by Roff M, Robins LN, Pollock M. Minneapolis, University of Minnesota Press, 1972.

5. Vaillant GE, Sobawale NC, MacArthur C: Some psychological vulnerabilities of physicians. *New England Journal of Medicine*, 1972, 287, 372–375.

6. Vaillant GE: Why men seek psychotherapy: I. Results of a survey of college graduates. *American Journal of Psychiatry*, 1972, 129, 645–651.

7. Vaillant GE, McArthur CC: The natural history of male psychological health: I. The adult life cycle from 18–50. *Seminars in Psychiatry*, 1972; 4:415–427.

8. Vaillant GE. Antecedents of healthy adult male adjustment. In *Life History Research in Psychopathology*, vol 3. Edited by Roff M, Ricks D. Minneapolis, University of Minnesota Press, 1973.

9. Vaillant GE, Vaillant CO: The relation of psychological adaptation to medical disability in middle aged men. In *Successful Aging: A Conference Report*. Edited by Pfeiffer E. Durham, NC, Center for the Study of Aging and Human Development, Duke University, 1974.

10. Vaillant GE: The natural history of male psychological health: II. Some antecedents of healthy adult adjustment. *Archives of General Psychiatry*, 1974, 31, 15–22.

11. Robinson S, Dill DB, Tsankoff SP, Wagner JA, and RD Robinson: Longitudinal studies of aging in 37 men. *Journal of Applied Physiology*, 1975, 38, 263–267.

12. Vaillant GE: The natural history of male psychological health: III. Empirical dimensions of mental health. *Archives of General Psychiatry*, 1975, 32, 420–426.

13. Vaillant GE: Natural history of male psychological health: V.

The relation of choice of ego mechanisms of defense to adult adjustment. *Archives of General Psychiatry,* 1976, 33, 535–545.

14. Vaillant GE: Natural history of male psychological health: IV. What kinds of men do not get psychosomatic illness. *Psychosomatic Medicine,* 1978, 40, 420–431.

15. Vaillant GE: Natural history of male psychological health: VI. Correlates of successful marriage and fatherhood. *American Journal of Psychiatry,* 1978, 135, 653–659.

16. Vaillant GE: Natural history of alcoholism: I. A preliminary report. In *Human Functioning in Longitudinal Perspective.* Edited by Sells SB, Crandall R, Roff M, Strauss JS, Pollin W. Baltimore, MD, Williams and Wilkins, 1979.

17. Vaillant GE: Health consequences of adaptation to life. *American Journal of Medicine,* 1979, 67, 732–734.

18. Vaillant GE: Natural history of male psychological health: Effects of mental health on physical health. *New England Journal of Medicine,* 1979, 301, 1249–1254.

19. Vaillant GE, Milofsky ES: Natural history of male psychological health: IX. Empirical evidence for Erikson's model of the life cycle. *American Journal of Psychiatry,* 1980, 137, 1348–1359.

20. Vaillant GE: Adolf Meyer was right: Dynamic psychiatry needs the life chart. Strecker Monograph Series, #16. *The Journal of the National Association of Private Psychiatric Hospitals,* 1980, 11, 4–14.

21. Vaillant GE: Natural history of male psychological health: VIII. Antecedents of alcoholism and "orality." *American Journal of Psychiatry,* 1980, 137, 181–186.

22. Vaillant GE: The doctor's dilemma. In *Alcoholism Treatment in Transition.* Edited by Grant M. London, Croom Helm, 1980.

23. Vaillant GE: Paths out of alcoholism. In *Evaluation of the Alcoholic: Implications for Research, Theory, and Treatment.* Edited by Meyer RE et al. Maryland, National Institute on Alcohol Abuse and Alcoholism, 1981.

24. Vaillant GE, Vaillant CO: Natural history of male psychological health: X. Work as a predictor of positive mental health. *American Journal of Psychiatry*, 1981, 138, 1433–1440.

25. Vaillant GE, Gale L, Milofsky ES: Natural history of male alcoholism: II. The relationship between different diagnostic dimensions. *Journal of Studies on Alcohol*, 1982, 43, 216–232.

26. McAdams DP, Vaillant GE: Intimacy, motivation and psychosocial adjustment. *Journal of Personality Assessment*, 1982, 46, 586–593.

27. Vaillant GE, Milofsky ES: The etiology of alcoholism: A prospective viewpoint. *American Psychologist*, 1982, 37, 494–503.

28. Vaillant GE, Milofsky E: The natural history of male alcoholism. IV. Paths to recovery. *Archives of General Psychiatry*, 1982, 39, 127–133.

29. Vaillant GE: Childhood environment and maturity of defense mechanisms. In *Human Development: An Interactional Perspective.* Edited by Magnusson D, Allen V. New York, Academic Press, 1983.

30. Vaillant GE, Clark W, Cyrus C, Milofsky ES, Kopp J, Wulsin V, Mogielnicki NP: Prospective study of alcoholism treatment: Eight-year follow-up. *American Journal of Medicine*, 1983, 75, 455–463.

31. Vaillant GE: Natural history of male alcoholism: V. Is alcoholism the cart or the horse to sociopathy? *British Journal of Addiction,* 1983, 78, 317–326.

32. Long JVF, Vaillant GE: Natural history of male psychological health: XI. Escape from the underclass. *American Journal of Psychiatry,* 1984, 141, 341–346.

33. Hartmann E, Milofsky E, Vaillant G, Oldfield M, Falke R, Ducey C: Vulnerability to schizophrenia: Prediction of adult schizophrenia using childhood information. *Archives of General Psychiatry,* 1984, 41, 1050–1056.

34. Beardslee WR, Vaillant GE: Prospective prediction of alcoholism and psychopathology. *Journal of Studies on Alcohol,* 1984, 45, 500–503.

35. Vaillant GE: The Study of Adult Development at Harvard Medical School. In *Handbook of Longitudinal Research,* vol. 2. Edited by Mednick SA, Harway M, Finello KM. New York, Praeger, 1984.

36. Vaillant GE: An empirically derived hierarchy of adaptive mechanisms and its usefulness as a potential diagnostic axis. *Acta Psychiatrica Scandinavica,* 1985; Supplementum no. 319, 71, 171–180.

37. Drake RE, Vaillant GE: A validity study of Axis II of DSM-III. *American Journal of Psychiatry,* 1985, 142, 553–558.

38. Snarey J, Vaillant GE: How lower- and working-class youth become middle-class adults: The association between ego defense mechanisms and upward social mobility. *Child Development,* 1985, 56, 899–910.

39. Vaillant GE, Bond M, Vaillant CO: An empirically validated hi-

erarchy of defense mechanisms. *Archives of General Psychiatry,* 1986, 43, 786–794.

40. Vaillant GE: Cultural factors in the etiology of alcoholism: A prospective study, in *Alcohol and Culture: Comparative Perspectives from Europe and America,* Annals of the New York Academy of Sciences, vol. 472. Edited by Babor TF. New York, The New York Academy of Sciences, 1986.

41. McCullough L, Vaillant C, Vaillant GE: Toward reliability in identifying ego defenses through verbal behavior. In *Empirical Studies of Ego Mechanisms of Defense.* Edited by Vaillant GE. Washington, DC, American Psychiatric Press, 1986.

42. Vaillant GE, Vaillant CO: A cross-validation of two empirical studies of defenses. In *Empirical Studies of Ego Mechanisms of Defense.* Edited by Vaillant GE. Washington, DC, American Psychiatric Press, 1986.

43. Beardslee WR, Son L, Vaillant GE: Exposure to parental alcoholism during childhood and outcome in adulthood: A prospective longitudinal study. *British Journal of Psychiatry,* 1986; 149:584–591.

44. Vaillant GE: Time: An important dimension of psychiatric epidemiology. In *Psychiatric Epidemiology. Progress and Prospects.* Edited by Cooper B. London, Croom Helm, 1987.

45. Felsman JK, Vaillant GE: Resilient children as adults: A forty year study. In *The Invulnerable Child.* Edited by Anthony EJ, Cohler BJ. New York, Guilford Press, 1987.

46. Vaillant GE: An empirically derived hierarchy of adaptive mechanisms and its usefulness as a potential diagnostic axis. In *Diagnosis and Classification in Psychiatry: A Critical Appraisal of*

DSM-III. Edited by Tischler GL. New York, Cambridge University Press, 1987.

47. Phillips KA, Vaillant GE, Schnurr P: Some physiologic antecedents of adult mental health. *American Journal of Psychiatry,* 1987; 144:1009–1013.

48. Vaillant C, Milofsky E, Richards R, Vaillant GE: A social casework contribution to understanding alcoholism. *Health and Social Work,* 1987; 12:169–176.

49. Vaillant GE: A developmental view of old and new perspectives of personality disorders. *Journal of Personality Disorders,* 1987; 1:146–156.

50. Vaillant GE, McCullough L: The Washington University Sentence Completion Test compared with other measures of adult ego development. *American Journal of Psychiatry,* 1987; 144:1189–1194.

51. Snarey J, Son L, Kuehne VS, Hauser S, Vaillant G: The role of parenting in men's psychosocial development: A longitudinal study of early adulthood infertility and midlife generativity. *Developmental Psychology,* 1987; 23:593–603.

52. Vaillant GE: Some differential effects of genes and environment on alcoholism. In *Alcoholism: Origins and Outcome.* Edited by Rose RM, Barrett JE. New York, Raven Press, 1988.

53. Drake RE, Vaillant GE: Predicting alcoholism and personality disorder in a 33-year longitudinal study of children of alcoholics. *British Journal of Addiction,* 1988; 83:799–807.

54. Drake RE, Vaillant GE: Introduction: Longitudinal views of personality disorder. *Journal of Personality Disorders,* 1988; 2:44–48.

55. Drake RE, Adler DA, Vaillant GE: Antecedents of personality disorders in a community sample of men. *Journal of Personality Disorders,* 1988; 2:60–68.

56. Vaillant GE, Schnurr P: What is a case? A 45-year study of psychiatric impairment within a college sample selected for mental health. *Archives of General Psychiatry,* 1988; 45:313–319.

57. Peterson C, Seligman MEP, Vaillant GE: Pessimistic explanatory style is a risk factor for physical illness: A thirty-five year longitudinal study. *Journal of Personality and Social Psychology,* 1988; 55:23–27.

58. Vaillant GE: What can long-term follow-up teach us about relapse and prevention of relapse in addiction. *British Journal of Addiction,* 1988; 83:1147–1157.

59. Vaillant GE: Attachment, loss and rediscovery. *Hillside Journal of Clinical Psychiatry,* 1988; 10:148–164.

60. Vaillant GE: The evolution of defense mechanisms during the middle years. In *The Middle Years: New Psychoanalytic Perspectives.* Edited by Oldham JM, Liebert RS. New Haven, Yale University Press, 1989.

61. Vaillant GE, Vaillant CO: Natural history of male psychological health: XII. A 45-year study of successful aging at age 65. *American Journal of Psychiatry,* 1990; 147:31–37.

62. Schnurr PP, Vaillant CO, Vaillant GE: Predicting exercise in late midlife from young adult personality characteristics. *International Journal of Aging and Human Development,* 1990; 30:153–161.

63. Vaillant GE: Repression in college men followed for half a century. In *Repression and Dissociation: Implications for Personality*

Theory, Psychopathology, and Health. Edited by Singer JL. Chicago, University of Chicago Press, 1990.

64. Vaillant GE: Avoiding negative life outcomes: Evidence from a forty-five year study. In *Successful Aging: Perspectives from the Behavioral Sciences.* Edited by Baltes PB, Baltes MM. Cambridge, Cambridge University Press, 1990.

65. Vaillant GE, Vaillant CO: Determinants and consequences of creativity in a cohort of gifted women. *Psychology of Women Quarterly,* 1990; 14:607–616.

66. Vaillant GE: Prospective evidence for the effects of environment upon alcoholism. In *Alcoholism and the Family: The 4th International Symposium of the Psychiatric Research Institute of Tokyo.* Edited by Saitoh S, Steinglass P, Schuckit MA. New York, Brunner-Mazel, 1992.

67. Vaillant GE, Schnurr PP, Baron JA, Gerber PD: A prospective study of the effects of cigarette smoking and alcohol abuse on mortality. *Journal of General Internal Medicine,* 1991; 6:299–304.

68. Vaillant GE: The association of ancestral longevity with successful aging. *Journal of Gerontology,* 1991; 46:292–298.

69. Vaillant GE, Milofsky ES: The etiology of alcoholism: A prospective viewpoint. In *Society, Culture, and Drinking Patterns Reexamined.* Edited by Pittman, DJ, White, HR. New Brunswick, NJ, Rutgers Center of Alcohol Studies, 1991.

70. Vaillant GE: Physician, cherish thyself: The hazards of self-prescribing. *Journal of the American Medical Association,* 1992; 267:2373–2374.

71. Vaillant GE, Roston D, McHugo GJ: An intriguing association

between ancestral mortality and male affective disorder. *Archives of General Psychiatry*, 1992; 49:709–715.

72. Vaillant CO, Vaillant GE: Is the U-curve of marital satisfaction an illusion? A 40-year study of marriage. *Journal of Marriage and the Family*, 1993; 55:230–239.

73. Vaillant GE, Koury SH: Late midlife development. In *The Course of Life*. Edited by Pollock GH, Greenspan SI. Madison, CT, International Universities Press, 1993.

74. Vaillant GE: Is alcoholism more often the cause or the result of depression? *Harvard Review of Psychiatry*, 1993; 1:94–99.

75. Vaillant GE: Successful aging and psychosocial well-being. In *Older Men's Lives*. Edited by Thompson EH. Thousand Oaks, California, Sage Publications, 1994.

76. Vaillant GE: Ego mechanisms of defense and personality psychopathology. *Journal of Abnormal Psychology*, 1994; 103:44–50.

77. Vaillant GE: La créativité chez les hommes et les femmes "ordinaires." *Nervuure*, 1994; 7:25–27.

78. Lee KA, Vaillant GE, Torrey WC, Elder GH: A 50-year prospective study of the psychological sequelae of World War II combat. *American Journal of Psychiatry*, 1995; 152:516–522.

79. Vaillant GE: Addictions over the life course. In *Psychotherapy, Psychological Treatments and the Addictions*. Edited by Edwards G, Dare C. Cambridge, England, Cambridge University Press, 1996.

80. Vaillant GE: A long-term follow-up of male alcohol abuse. *Archives of General Psychiatry*, 1996; 53:243–249.

81. Vaillant GE, Orav J, Meyer SE, McCullough-Vaillant L, Ros-

ton D: Late life consequences of affective spectrum disorder. *International Psychogeriatrics,* 1996; 8:1–20.

82. Cui X, Vaillant GE: The antecedents and consequences of negative life events in adulthood: A longitudinal study. *American Journal of Psychiatry,* 1996; 152:21–26.

83. Vaillant GE, Gerber PD: Natural history of male psychological health: XIII. Who develops high blood pressure and who responds to treatment. *American Journal of Psychiatry,* 1996; 153:24–29.

84. Cui X, Vaillant GE: Does depression generate negative life events? *Journal of Nervous and Mental Disease,* 1997; 185:145–150.

85. Vaillant GE: The natural history of alcoholism and its relationship to liver transplantation. *Liver Transplantation and Surgery,* 1997; 3:304–310.

86. Vaillant GE: Poverty and paternalism: A psychiatric viewpoint. In *The New Paternalism: Supervisory Approaches to Poverty,* 279–304. Edited by L. Meade. Washington, D.C., Brookings, 1998.

87. Vaillant GE: Natural history of male psychological health: XIV. Relationship of mood disorder vulnerability to physical health. *American Journal of Psychiatry,* 1998; 155:184–191.

88. Soldz S, Vaillant GE: A 50-year longitudinal study of defense use among inner city men: A validation of the DSM-IV defense axis. *Journal of Nervous and Mental Disease,* 1998; 186:104–III.

89. Vaillant GE, Meyer SE, Mukamal K, Soldz S: Are social supports in late midlife a cause or a result of successful physical aging? *Psychological Medicine,* 1998; 28:1159–1168.

90. Wulsin LR, Vaillant GE, Wells VE: A systematic review of the mortality of depression. *Psychosomatic Medicine,* 1999; 61:6–17.

91. Vaillant GE: Lessons learned from living. *Scientific American Presents,* 1999; 10:32–37.

92. Soldz S, Vaillant GE: The big five personality traits and the life course: A 45-year longitudinal study. *Journal of Research in Personality,* 1999; 33:208–232 [1999 Best Article].

93. Laub JH, Vaillant GE: Delinquency and mortality: A 50-year follow-up study of 1000 delinquent and nondelinquent boys. *American Journal of Psychiatry,* 2000; 157:96–102.

94. Vaillant GE: Adaptive mental mechanisms: Their role in a positive psychology. *American Psychologist,* 2000; 55:89–98.

95. Vaillant GE: Prevention of alcoholism: Reflections of a naturalist. In *Childhood Onset of Adult Psychopathology.* Ed. by JL Rapoport. Washington, DC, American Psychiatric Press, 2000.

96. Vaillant GE, Davis JT: Social/emotional intelligence and midlife resilience in schoolboys with low tested intelligence. *American Journal of Orthopsychiatry,* 2000; 70:215–222.

97. Vaillant GE: If addiction is involuntary, how can punishment help? In *Drug Addiction and Drug Policy.* Edited by PB Heymann and WN Brownsberger. Cambridge, MA, Harvard University Press, 2001, pp. 144–167.

98. Cui X and Vaillant GE: Stressful life events and late adulthood adaptation. In *Aging in Good Health.* Edited by SE Levkoff, YK Cheej, and S Noguchi. New York, Springer, 2001.

99. Vaillant GE and K Mukamal: Successful aging. *American Journal of Psychiatry,* 2001, 158.839–847.

100. Vaillant GE: The value of a hierarchy of defenses. In *Entwicklung und Risiko.* Edited by G. Roper, G. Noam, and C. von Hagen. Berlin, Kohlhammer, 2001.

101. Vaillant GE, Vaillant CO: The Study of Adult Development. In *Landmark Studies in the Twentieth Century.* Edited by Anne Colby and Erin Phelps. New York, Russell Sage Foundation, 2002.

102. Vaillant GE: Healthy aging among inner-city men. *International Psychogeriatrics,* 2002; 13:425–437.

103. Vaillant GE: Natural history of addiction and pathways to recovery. In *Principles of Addiction Medicine,* 3rd edition. Edited by AW Graham et al., Chevy Chase, MD, American Society of Addiction Medicine, 2003.

104. Vaillant GE: Mental health. *American Journal of Psychiatry,* 2003; 160:1373–1384.

105. Vaillant GE: A 60-year follow-up of male alcoholism. *Addiction,* 2003; 98:1043–1051.

106. Isaacowitz DM, Vaillant GE, Seligman MEP: Strength and satisfaction across the adult lifespan. *International Journal of Aging and Human Development,* 2003; 27:181–201.

107. Vaillant GE: Mental health. In *Comprehensive Textbook of Psychiatry,* 8th edition. Edited by BJ Sadock and VA Sadock. Philadelphia, Williams and Wilkins, 2004.

108. Vaillant GE: Positive aging. In *Positive Psychology in Practice.* Edited by Linley PA and Joseph S. Hoboken, NJ: John Wiley & Sons, 2004.

Robert Waldinger, Director

1. Vaillant GE, Batalden MB, Orav J, Roston D, Barrett JE: Evidence for an X-linked personality trait related to affective illness. *Australian and New Zealand Journal of Psychiatry,* 2005; 39:730–735.

2. Vaillant GE: Alcoholics Anonymous: Cult or cure. *Australian and New Zealand Journal of Psychiatry*, 2005; 39: 431–436.

3. Vaillant GE, DiRago AC, Mukamal K: Natural history of male psychological health XV: Retirement satisfaction. *American Journal of Psychiatry*, 2006; 163:682–688.

4. Vaillant GE, DiRago AC: Satisfaction with retirement in men's lives. In *Charting a New Life: The Crown of Life Dynamics of the Early Retirement Period*. Edited by J James and P Wink. New York: Springer, 2006.

5. Waldinger RJ, Vaillant GE, Orav EJ: Childhood sibling relationships as predictors of major depression in adulthood: A thirty-year prospective study. *American Journal of Psychiatry*, 2006; 164:949–954.

6. Vaillant GE: Generativity— A form of unconditional love. In *Altruism and Health*. Edited by S. Post. New York, Oxford University Press, 2007.

7. DiRago AC, Vaillant GE (2007). Resilience in Inner City youth: Childhood predictors of occupational status across the lifespan. *Journal of Youth and Adolescence, 36:61–70.*

8. Vaillant GE, Templeton J, Ardelt M, Meyer S: Natural history of male mental health: Health and religious involvement. *Social Science and Medicine, 2008, 66:221–231.*

9. Koenig L and Vaillant GE: A prospective study of church attendance and health over the lifespan. *Health Psychology, 28,* 2009:117–124.

10. Frosch ZA, Dierker LC, Rose JS, and Waldinger RJ: Smoking trajectories, health, and mortality across the adult lifespan. *Addictive Behaviors, 2009, 34:701–704.*

11. Beardslee WR, Vaillant GE: Adult development. In *Psychiatry,*

2nd ed. Edited by Tasman A, Kay J, Lieberman J. New York, Saunders, 2008.

12. DiRago A, Vaillant GE: Mature defense mechanisms. In *The Encyclopedia of Positive Psychology*. Edited by SJ Lopez. 2:600–605. Malden, MA, Wiley-Blackwell, 2009.

13. Grenfield EA, Vaillant GE, Marks NF: Do formal religious participation and spiritual perceptions have independent linkages with diverse dimensions of psychological well-being? *Journal of Health and Social Behavior,* 2009; 50:196–212.

14. Vaillant GE, Vaillant CO: Normality and mental health. In *Comprehensive Textbook of Psychiatry,* 9th edition. Edited by BJ Sadock, VA Sadock, and P. Ruiz. Philadelphia, Williams and Wilkins, 2009.

15. Martin-Joy J and George E. Vaillant: Recognizing and promoting resilience. In *Successful Cognitive and Emotional Aging.* Edited by C Depp and DV Jeste. Arlington VA: American Psychiatric Press, 2010.

16. Katie A McLaughlin, Laura D Kubzansky, Erin C Dunn, Robert Waldinger, George Vaillant, and Karestan C Koenen: Childhood social environment, emotional reactivity to stress, and mood and anxiety disorders across the life course. *Depression and Anxiety,* 2010; 27: 1087–1094 (2010).

17. Waldinger RA and MS Schulz: Facing the music or burying our heads in the sand? Adaptive emotion regulation in midlife and late life. *Research in Human Development,* 2010; 7:292–306.

18. Waldinger RJ, Schulz MS: What's love got to do with it? Social connections, perceived health stressors, and daily mood in married octogenarians. *Psychology and Aging,* 2010, 25: 422–431.

19. Pergakis MB, Hasan NS, Heller NR, Waldinger RJ: Octoge-

narian reports of lifetime spiritual experiences: Types of experience and early life predictors. *Journal of Religion, Spirituality, and Aging,* 2010; 22: 220–238.

20. Waldinger RJ, Schulz MS: Linking hearts and minds in couple interactions: Intentions, attributions and overriding sentiments. *Journal of Family Psychology,* 2006, 20:494–504.

21. Waldinger RJ, Kensinger EA, Schulz MS: Neural activity, neural connectivity, and the processing of emotionally-valenced information in older adults: Links with life satisfaction *Cognitive, Affective & Behaviorial Neuroscience,* 2011, 11:426–436.

22. Liu L, Cohen S, Schulz MS, Waldinger RJ: Sources of somatization: Exploring the roles of insecurity in relationships and styles of anger experience and expression. *Social Science & Medicine,* 2011, 73:1436–1443.

23. Cohen S, Schulz MS, Weiss E, Waldinger RJ: "A" for effort: The individual and dyadic contributions of empathic accuracy and empathic effort to relationship satisfaction. *Journal of Family Psychology,* 2012, 26:236–245.

NOTES

1. MATURATION MAKES LIARS OF US ALL

(Epigraph) Popular adage attributed to Heraclitus.

1. Gail Sheehy, *Passages* (New York: Dutton, 1976); Daniel J. Levinson, *Seasons of a Man's Life* (New York: Knopf, 1978).

2. Sheldon Glueck and Eleanor Glueck, *Unraveling Juvenile Delinquency* (New York: Commonwealth Fund, 1950).

3. Lewis M. Terman and Melita H. Oden, *The Gifted Child Grows Up: Genetic Studies of Genius*, vol. 4 (Stanford, CA: Stanford University Press, 1947); Melita H. Oden, "The Fulfillment of Promise: Forty-Year Follow-up of the Terman Gifted Group," *Genetic Psychology Monographs* 77 (1968): 3–93; Carole K. Holahan and Robert R. Sears, *The Gifted Group in Maturity* (Stanford, CA: Stanford University Press, 1995); Howard S. Friedman and Leslie R. Martin, *The Longevity Project* (New York: Hudson Street Press, 2011).

4. George E. Vaillant, *The Natural History of Alcoholism Revisited* (Cambridge: Harvard University Press, 1995); George E. Vaillant, *Aging Well* (Boston: Little, Brown and Company, 2002).

5. Jack Block, *Lives through Time* (Berkeley: Bancroft Books, 1971); Dorothy H. Eichorn, John A. Clausen, Norma Haan, et al., *Present and Past in Middle Life* (New York: Academic Press, 1981); Glen H. Elder, Jr., *Children of the Great Depression* (Chicago: University of Chicago Press, 1984); Erik H. Erikson, Joan M. Erikson, and Helen Q. Kivnick, *Vital Involvement in Old Age* (New York: W. W. Norton, 1986); John A. Clausen, *American Lives* (New York: Free Press, 1993).

6. Thomas R. Dawber, *The Framingham Study* (Cambridge, MA: Harvard University Press, 1980); Susan E. Hankinson, Graham A. Colditz, JoAnn E. Manson, et al., *Healthy Women Healthy Lives: A Guide to Preventing Disease from the Landmark Nurses' Health Study* (New York: Free Press, 2002); Edward Giovanucci, Mier J. Stampfer, Graham A. Colditz, et al., "Multivitamin Use, Folate, and Colon Cancer in Women in the Nurses' Health Study," *Annals of Internal Medicine*, 129 (1998): 517–524; Michel Lucas, Fariba Mirzaei, An Pan, et al., "Coffee, Caffeine, and Risk of Depression among Women," *Archives of Internal Medicine* (2011), 171(17): 1571–1578.

7. George E. Vaillant, *Adaptation to Life* (Boston, MA: Little, Brown & Co., 1977).

8. Robert M. Hauser, William H. Sewell, John A. Logan, et al., 1992. "The Wisconsin Longitudinal Study: Adults as Parents and Children at Age 50," *IASSIST Quarterly* 16:23–38; William H. Sewell, Robert M. Hauser, Kristen W. Springer, et al. "As We Age: A Review of the Wisconsin Longitudinal Study, 1957–2001," *Research in Social Stratification and Mobility*, 20 (2004): 3–111, ed. Kevin T. Leicht; Wisconsin Longitudinal Study: "The Class of 1957 at Age 65; A First Look: A Letter to Wisconsin's High School Cass of 1957 and Their Families" (Madison, Wisconsin: privately printed by the Wisconsin Longitudinal Study 2006).

9. Vaillant, *Adaptation to Life.*

10. Erik H. Erikson, *Childhood and Society* (New York: Norton, 1951).

11. Jane Loevinger, *Ego Development* (San Francisco: Jossey-Bass, 1976).

2. THE PROOF OF THE PUDDING

(Epigraph) P. D. Scott, review of Lee N. Robins, "Deviant Children Grown-Up," *British Journal of Psychiatry* 113 (1967): 929–930.

1. Joshua W. Shenk, "What Makes Us Happy," *The Atlantic,* June 2009, 36–53.

2. George E. Vaillant, "A Rewarding Life," *The Australian Financial Review,* Aug. 21, 2009, 1–2.

3. Earnest Hooton, *Young Man, You Are Normal* (New York: G. P. Putnam & Sons, 1945), 103.

4. Hooton, *Young Man, You Are Normal.*

5. Virgil, *Eclogues,* 10:69, 37 B.C.

3. A SHORT HISTORY OF THE GRANT STUDY

(Epigraph) Benjamin C. Bradlee, *A Good Life* (New York: Simon and Schuster, 1995), 15–16.

1. *Official Register of Harvard University,* vol. 35, 404–405.

2. Ibid.

3. Dan H. Fann, "Grant Study Analyzes 'Normal' Individuals," *Harvard Crimson,* May 13, 1942.

4. Adolf Meyer, "The Life Chart," in *Contributions to Medical and Biological Research* (New York: Paul B. Hoeber, 1919), 102.

5. Walter B. Cannon, *The Wisdom of the Body* (New York: Norton, 1932).

6. George E. Vaillant, *The Wisdom of the Ego* (Cambridge: Harvard University Press, 1993).

7. *Harvard University Gazette,* November 19, 1998.

8. Clark W. Heath, *What People Are* (Cambridge: Harvard University Press, 1945).

9. Ibid.

10. Frederic L. Wells and William L. Woods, "Outstanding Traits," *Genetic Psychological Monographs 33* (1946): 127–249.

11. Harry F. Harlow, "The Nature of Love," *American Psychologist* 13 (1958): 678.

12. Ernst Kretschmer, *Physique and Character* (Abington, England: Routledge, 1931).

13. Carl C. Seltzer, Frederic L. Wells, and E. B. McTernan, "A Relationship between Sheldonian Somatotype and Psychotype," *Journal of Personality* 16 (1948): 431–436.

14. William H. Sheldon, *Atlas of Men: A Guide for Somatotyping the Adult Male at All Ages* (New York: Gramercy Publishing Company, 1954).

15. Heath, *What People Are.*

16. Earnest Hooton, *Young Man, You Are Normal* (New York: G. P. Putnam's Sons, 1945), 102, 207.

17. Ibid., 86.

18. Ibid.

19. Ibid., 103.

20. Ibid., 82.

21. Heath, *What People Are;* Wells and Woods, "Outstanding Traits."

22. John P. Monks, *College Men at War* (Boston: American Academy of Arts and Sciences, 1957).

23. Wells and Woods, "Outstanding Traits."

24. William Grant to Arlie Bock, August 7, 1944, in folder "Correspondence," Grant Study Archives.

25. Donald W. Hastings, "Follow-Up Results In Psychiatric Illness," *American Journal of Psychiatry* 114 (1958): 1057–1066.

26. Kimberly A. Lee, George E. Vaillant, William C. Torrey, et al., "A 50-year Prospective Study of the Psychological Sequelae of World War II Combat," *American Journal of Psychiatry* 152 (1995): 516–522.

27. Caroline O. Vaillant and George E. Vaillant, "Is the U-curve of Marital Satisfaction an Illusion? A 40-year Study of Marriage," *Journal of Marriage and the Family* 55 (1993): 230–239.

28. Dan P. McAdams and George E. Vaillant, "Intimacy, Motivation and Psychosocial Adjustment," *Journal of Personality Assessment* 46 (1982): 586–593.

29. Grant Study Archives.

30. Charles McArthur, "Long-Term Validity of the Strong Interest Test in Two Subcultures," *Journal of Applied Psychology* 38 (1954): 346–353.

31. George E. Vaillant, Jane R. Brighton, and Charles McArthur, "Physicians' Use of Mood-Altering Drugs: A 20-Year Follow-up Report," *New England Journal of Medicine* 282 (1970): 365–370.

32. George E. Vaillant, *Natural History of Alcoholism* (Cambridge: Harvard University Press, 1983).

33. Robert W. White, *Lives in Progress* (New York: Holt, Rinehart and Winston, 1972).

4. HOW CHILDHOOD AND ADOLESCENCE AFFECT OLD AGE

(Epigraph) William Wordsworth, "Ode: Intimations of Immortality."

1. Joseph Conrad, *Victory* (New York: Doubleday, 1915), 383.

2. Michael G. Marmot, G. Davey Smith, Stephen Stansfield, et al., "Health Inequalities among British Civil Servants: The Whitehall II Study," *Lancet* 337 (1991): 1387–1393.

3. George E. Vaillant, *Natural History of Alcoholism Revisited* (Cambridge: Harvard University Press, 1995); George E. Vaillant, *Aging Well* (New York: Little, Brown, 2002).

4. Vaillant, *Natural History of Alcoholism Revisited*.

5. Ibid.

6. Erik H. Erikson, *Childhood and Society,* 2nd ed. (New York: W. W. Norton & Co., 1963); George E. Vaillant and Eva Milofsky, "Natural History of Male Psychological Health, IX: Empirical Evidence for Erikson's Model of the Life-cycle," *American Journal of Psychiatry* 137 (1980): 1348–1359.

7. George E. Vaillant, "Natural History of Male Psychological Health II: Some Antecedents of Healthy Adult Adjustment," *Archives of General Psychiatry* 31 (1974): 15–22.

8. Vaillant, "Natural History of Male Psychological Health II."

9. George E. Vaillant, "Why Men Seek Psychotherapy: I. Results of a Survey of College Graduates," *American Journal of Psychiatry* 129 (1972): 645–651.

10. Erikson, *Childhood and Society.*

11. Aaron Lazare, Gerald L. Klerman, and David J. Armor, "Oral, Obsessive, and Hysterical Personality Patterns: An Investigation of Psychoanalytic Concepts by Means of Factor Analysis," *Archives of General Psychiatry,* 14(6) (1966): 624–630.

12. George E. Vaillant and Caroline O. Vaillant, "Natural History of Male Psychological Health, X: Work as a Positive Predictor of Mental Health," *American Journal of Psychiatry* 138 (1981): 1433–1440.

13. George E. Vaillant, *Spiritual Evolution: A Scientific Defense of Faith* (New York: Doubleday Broadway, 2008).

14. George E. Vaillant, *Adaptation to Life* (Boston: Little, Brown, 1977), 297.

15. Paul T. Costa, Jr., and Robert R. McCrae, *The NEO Personality Inventory Manual* (Odessa, FL: Psychological Assessment Resources, 1985).

16. Robert R. McCrae and Paul T. Costa, Jr., *Emerging Lives, Enduring Dispositions: Personality in Adulthood* (Boston: Little, Brown, 1984).

17. Brent W. Roberts and Daniel Mroczek, "Personality Trait Change in Adulthood," *Current Directions in Psychological Science* 17 (2008): 31–35.

18. Stephen Soldz and George E. Vaillant, "The Big Five Personality Traits and the Life Course: A 45-Year Longitudinal Study," *Journal of Research in Personality* 33 (1999): 208–232.

19. William G. Iacono and Matthew McGue, "Minnesota Twin Family Study," *Twin Research* 5 (2002): 482–487; Thomas J. Bouchard, Jr., David T. Lykken,

Matthew McGue, et al., "Intrinsic and Extrinsic Religiousness: Genetic and Environmental Influences and Personality Correlates, *Twin Research 2* (1999): 88–98.

20. Clark W. Heath, *What People Are* (Cambridge: Harvard University Press, 1945).

21. Leo Tolstoy, *Childhood, Boyhood and Youth* (New York: Scribner's, 1904), 109.

22. Michael Rutter, *Maternal Deprivation Reassessed,* 2nd ed. (Harmondsworth, England: Penguin Books, 1981).

23. Earnest Hooton, *Young Man, You Are Normal* (New York: Putnam, 1945).

24. Vaillant, *Natural History of Alcoholism Revisited.*

25. Melita H. Oden, "The Fulfillment of Promise: 40-Year Follow-up of the Terman Gifted Group," *Genetic Psychology Monographs* 77 (1968): 3–93; Howard S. Friedman and Leslie R. Martin, *The Longevity Project* (New York: Hudson Street Press, 2011).

26. Sheldon Glueck and Eleanor Glueck, *Delinquents and Nondelinquents in Perspective* (Cambridge: Harvard University Press, 1968); William McCord and Joan McCord, *Origins of Crime* (New York: Columbia University Press, 1959).

5. MATURATION

(Epigraph) William James, "The Laws of Habit," *Popular Science Monthly* 30 (1887): 433–451, quote p. 447.

1. Robert V. Kail and John C. Cavanaugh, *Human Development,* 6th ed. (Belmont, CA: Wadsworth, 2010), 488.

2. Erik H. Erikson, *Identity: Youth and Crisis* (New York: Norton, 1968), 136.

3. Lester Luborsky, "Clinicians' Judgments of Mental Health," *Archives of General Psychiatry* 7 (1962): 407–417.

4. American Psychiatric Association, *Diagnostic and Statistical Manual of Mental Disorders,* 4th ed. (Washington, D.C., 1994); George E. Vaillant, "The Natural History of Male Psychological Health, III: Empirical Dimensions of Mental Health," *Archives of General Psychiatry* 32 (1975): 420–426.

5. George E. Vaillant and Eva Milofsky, "Natural History of Male Psychological Health, IX: Empirical Evidence for Erikson's Model of the Lifecycle," *American Journal of Psychiatry* 137 (1980): 1348–1359.

6. Erik H. Erikson, *Childhood and Society*, 2nd ed. (New York: W. W. Norton & Co., 1963), 272.

7. George E. Vaillant, "Natural History of Male Psychological Health, V: The Relation of Choice of Ego Mechanisms of Defense to Adult Adjustment," *Archives of General Psychiatry* 33 (1976): 535–545; George E. Vaillant, *The Wisdom of the Ego* (Cambridge: Harvard University Press, 1993).

8. Vaillant and Milofsky, "Natural History of Male Psychological Health, IX: Empirical Evidence for Erikson's Model of the Lifecycle."

9. George E. Vaillant, *Spiritual Evolution: A Scientific Defense of Faith* (New York: Doubleday Broadway, 2008); Antonio Damasio, *Looking for Spinoza* (New York: Houghton Mifflin Harcourt, 2003).

10. Paul I. Yakovlev and André R. Lecours, "The Myeogenetic Cycles of Regional Maturation of the Brain," *Regional Development of the Brain in Early Life,* ed. Alexandre Minkowski (Oxford: Blackwell Scientific Publications, 1967); Francine M. Bencs, Mary Turtle, Yusuf Khan, et al., "Myelinization of a Key Relay in the Hippocampal Formation Occurs in the Human Brain During Childhood, Adolescence and Adulthood," *Archives of General Psychiatry* 51 (1994): 477–484.

11. Monika Ardelt and George E. Vaillant, "What Affects the Growth and Loss of Wisdom Throughout the Life Course? Evidence from Three Case Studies," *Gerontologist* 48 (2008): 353.

12. Carol Gilligan, personal communication, 1990.

13. Erikson, *Childhood and Society.*

14. Ravenna Helson, Constance Jones, and Virginia S. Y. Kwan, "Personality Change over 40 Years of Adulthood: Hierarchical Linear Modeling Analyses of Two Longitudinal Samples," *Journal of Personality and Social Psychology* 83 (2002): 752–766; Constance Jones and Harvey Peskin, "Psychological Health from the Teens to the 80s: Multiple Developmental Trajectories," *Journal of Adult Development* 17 (2010): 20–32.

15. Jack Block, *Lives through Time* (Berkeley: Bancroft Books, 1971); Glen H. Elder, Jr., *Children of the Great Depression* (Chicago: University of Chicago Press, 1974); Dan P. McAdams and Jennifer L. Pals, "A New Big Five," *American Psychologist,* 61 (2006): 204–217; Robert R. McCrae and Paul T. Costa, Jr., "The Stability of

Personality," *Current Directions in Psychological Sciences* 3 (1994): 173–175; Robert W. White, *Lives in Progress,* 3rd ed. (NewYork: Henry Holt & Co., 1972); Daniel J. Levinson, Charlotte M. Darrow, Edward B. Klein, et al., "The Psychosocial Development of Men in Early Adulthood and the Mid-Life Transition," *Life History Research in Psychopathology,* vol. 3, ed. David F. Ricks, Alexander Thomas, and Merrill Roff (Minneapolis: Minnesota University Press, 1974), 243–258; George E. Vaillant and Charles C. McArthur, "The Natural History of Male Psychological Health, I: The Adult Life Cycle from 18–50," *Seminars in Psychiatry* 4 (1972): 415–427.

16. Robert J. Havighurst, *Developmental Tasks and Education* (NewYork: David Mc-Kay, 1972).

17. Vaillant, *The Wisdom of the Ego.*

18. Vaillant and McArthur, "The Natural History of Male Psychological Health, I: The Adult Life Cycle from 18–50."

19. Monika Ardelt and George E. Vaillant, "The Presence and Absence of Wisdom in Everyday Life: Evidence from Two Longitudinal Case Studies," Gerontological Society of America Annual Meetings, Atlanta, GA, November 2009.

20. Eleanor H. Porter, *Pollyanna* (Boston: Page Co., 1914).

21. Martin E. P. Seligman, *Flourish* (NewYork: Free Press, 2011).

22. Laura Carstensen, Derek M. Isaacowitz, and Susan T. Charles, "Taking Time Seriously: A Theory of Socioemotional Selectivity," *American Psychologist,* 54 (1999): 165–181; Susan T. Charles, Chandra A. Reynolds, and Margaret Gatz, "Age Related Differences and Changes in Positive and Negative Affect over 23 Years," *Journal of Personality and Social Psychology* 80 (2001): 136–151.

23. Jane Loevinger, *Ego Development* (San Francisco: Jossey-Bass, 1976); Francine M. Benes, Mary Turtle, Yusuf Khan, et al., "Myelinization of a Key Relay in the Hippocampal Formation Occurs in the Human Brain During Childhood, Adolescence and Adulthood," *Archives of General Psychiatry* 51(1994): 477–484.

24. Benes et al., "Myelinization of a Key Relay."

25. Laura L. Carstensen, *A Long Bright Future* (NewYork: Crown, 2009); Laura L. Carstensen, Monisha Pasupathi, Ulrich Mayr, et al., "Emotion Experience in Everyday Life Across the Adult Life Span," *Journal of Personality and Social Psychology* 79 (2000):644–655; Charles, Reynolds, and Gatz, "Age-Related Differences."

26. Laura L. Cartensen, "Social and Emotional Patterns in Adulthood: Support for Socioemotional Selectivity Theory," *Psychology and Aging* 7 (1992): 331–338.

27. Vaillant and Milofsky, "Empirical Evidence for Erikson's Model of the Lifecycle."

28. George E. Vaillant, Janice Templeton, Monika Ardelt, et al., "Natural History of Male Mental Health: Health and Religious Involvement," *Social Science and Medicine* 66 (2008): 221–231.

29. Paul B. Baltes and Jacqui Smith, "Toward a Psychology of Wisdom and Its Ontogenesis," *Wisdom: Its Nature, Origins and Development*, ed. Robert J. Sternberg (Cambridge, England: Cambridge University Press, 1990): 87–120.

30. Jane Loevinger, *Ego Development* (San Francisco, CA: Jossey-Bass, 1976).

31. Baltes and Smith, "Toward a Psychology of Wisdom."

32. Monika Ardelt, "Empirical Assessment of a Three-Dimensional Wisdom Scale," *Research on Aging* 25 (2003): 275–324.

33. Aaron A. Lazare, Gerald L Klerman, and David J. Armor, "Oral Obsessive and Hysterical Patterns," *Archives of General Psychiatry* 14 (1966): 624–630; Derek M. Isaacowitz, George E. Vaillant, and Martin E. P. Seligman, "Strengths and Satisfaction across the Adult Lifespan," *International Journal of Aging and Human Development* 57 (2003): 181–201.

34. Monika Ardelt, "Empirical Assessment of a Three-Dimensional Wisdom Scale."

6. MARRIAGE

1. G. E. Vaillant, *Adaptation to Life* (Boston: Little, Brown, 1977), 320.

2. Robert J. Waldinger and Marc S. Schulz, "Facing the Music or Burying Our Heads in the Sand? Adaptive Emotion Regulation in Midlife and Latelife," *Research in Human Development* 7 (2010): 292–306.

3. George E. Vaillant and Caroline O. Vaillant, "Is the U-Curve of Marital Satisfaction an Illusion?" *Journal of Marriage and the Family* 55 (1993): 230–239.

4. Vaillant, *Adaptation to Life.*

5. Lewis M. Terman, *Psychological Factors in Marital Happiness* (New York: McGraw-Hill, 1938); John M. Gottman, *What Predicts Divorce?* (Hillsdale, NJ: Lawrence Erlbaum, 1994).

6. George E. Vaillant, *The Natural History of Alcoholism Revisited* (Cambridge: Harvard University Press, 1995).

7. Mary W. Hicks and Marilyn Platt, "Marital Happiness and Stability: A Review of Research in the 60s," *Journal of Marriage and the Family* 32 (1970): 553–574; Sylvia Weishaus and Dorothy Field, "A Half-Century of Marriage: Continuity or Change?" *Journal of Marriage and the Family* 50 (1988): 763–774.

8. Henri Troyat, *Tolstoy* (Garden City, N.Y.: Doubleday, 1967).

9. Mary Ainsworth, Mary C. Blehar, Everett Waters, et al., *Patterns of Attachment* (Hillsdale N.J.: Erlbaum, 1978).

10. Emmy E. Werner and Ruth S. Smith, *Vulnerable but Invincible: A Longitudinal Study of Resilient Children and Youth* (New York: McGraw-Hill, 1982).

11. George E. Vaillant, *Wisdom of the Ego* (Cambridge: Harvard University Press, 1993).

12. U.S. National Center of Health Statistics, "Divorces and Divorce Rates, U.S.," *Vital and Health Statistics,* series 21 (1978), no. 29.

13. Alan Booth, David R. Johnson, Lynn K. White, et al., "Divorce and Marital Instability over the Life Course," *Journal of Family Issues* 7 (1986): 421–442.

14. Laura L. Cartensen, "Social and Emotional Patterns in Adulthood: Support for Socioemotional Selectivity Theory," *Psychology and Aging* 7 (1992): 331–338.

15. Jan Hoffman, "Embracing Divorce as an Apple-Pie Institution," *Orange County Life* (April 29, 1989).

7. LIVING TO NINETY

(Epigraph) Michael Bury and Anthea Holme, *Life after Ninety* (London: Routledge, 1991), 43.

1. Ian J. Deary, Lawrence J. Whalley, and John M. Starr, *A Lifetime of Intelligence: Follow-up Studies of the Scottish Mental Health Surveys of 1932 and 1947* (Washington, D.C.: American Psychological Association, 2009); Hilary Lapsley, personal communication, 2011.

2. Oliver Wendell Holmes, *The Complete Poetical Works of Oliver Wendell Holmes* (Boston: Houghton Mifflin, 1908), 147.

3. James F. Fries, "Aging, Natural Death and the Compression of Morbidity," *New*

England Journal of Medicine 303 (1980): 130–135; Thomas T. Perls and Margery H. Silver, *Living to 100* (New York: Basic Books, 1999), 130.

4. *U.S.A. Today,* March 17, 2011, 3A.

5. Keiko A. Taga, Howard S. Friedman, and Leslie R. Martin, "Early Personality Traits as Predictors of Mortality Risk Following Conjugal Bereavement," *Journal of Personality* 77 (2009): 669–690.

6. Nathan W. Shock, *Normal Human Aging* (Washington, DC: US Government Printing Office, 1984).

7. James F. Fries and Lawrence M. Crapo, *Vitality and Aging* (San Francisco: W. H. Freeman, 1981).

8. David A. Drachman, "Do We Have Brain to Spare?" *Neurology* 64 (2005): 2004–2005.

9. Dan G. Blazer, Dana C. Hughes, and Linda K. George, "The Epidemiology of Depression in an Elderly Community Population," *Gerontologist* 27 (1987): 281–287; Paul B. Baltes and Karl U. Mayer, eds., *The Berlin Aging Study* (Cambridge, England: Cambridge University Press, 1999).

10. Laura L. Carstensen, "Social and Emotional Patterns in Adulthood: Support for Socio-Emotional Selectivity Theory," *Psychology and Aging* 7 (1992): 331–338.

11. John W. Rowe and Robert L. Kahn, *Successful Aging* (New York: Dell, 1999).

12. Jason Brandt, Miriam Spencer, and Marshall Folstein, "The Telephone Interview for Cognitive Status," *Neuropsychiatry, Neuropsychology and Behavioral Neurology* 1 (1988): 111–117; B. D. Carpenter, M. E. Straus, and A. M. Ball, "Telephone Assessment of Memory in the Elderly," *Journal of Clinical Geropsychology* 1 (1995): 107–117.

13. Rowe and Kahn, *Successful Aging;* Bury and Holme, *Life after Ninety.*

14. David Snowdon, *Aging with Grace: The Nun Study and the Science of Old Age* (London: Harper Collins, 2001).

15. Sigmund Freud, "Dostoevsky and Parricide (1928)," *Standard Edition,* 21:177–196 (London: Hogarth Press, 1961), 177.

16. E. M. Forster: *Howards End* (New York: Edward Arnold, 1973), 183–184.

17. Francine Benes, "Human Brain Growth Spans Decades." *American Journal of Psychiatry* 155 (1998): 1489.

18. Baltes and Mayer, eds., *The Berlin Aging Study.*

19. Rowe and Kahn, *Successful Aging.*

20. Warner Schaie, "The Course of Adult Intellectual Development," *American Psychologist* 49 (1994): 304–313.

21. Bury and Holme, *Life after Ninety,* 59.

22. George E. Vaillant, "Natural History of Male Psychological Health: Effects of Mental Health on Physical Health," *New England Journal of Medicine* 301 (1979): 1249–1254.

23. Howard S. Friedman and Leslie R. Martin, *The Longevity Project* (New York: Hudson Street Press, 2011).

24. G. E. Vaillant, *Aging Well* (New York: Little, Brown, 2002).

25. Ibid.

26. Paola Sebastiani, Nadia Solovieff, Andrew T. DeWan, et al., "Genetic Signatures of Exceptional Longevity in Humans," *Science Press,* www.sciencemag.org, July 1, 2010.

27. Brandt, Spencer, and Folstein, "Telephone Interview."

28. Vaillant, *Aging Well.*

29. Ibid.

30. Rowe and Kahn, *Successful Aging,* 28; Birgit Ljungquist, Stig Berg, Jan Lanke, et al., "The Effect of Genetic Factors for Longevity: A Comparison of Identical and Fraternal Twins in the Swedish Twin Registry," *Journal of Gerontology* 53A (1998): M441–M446.

31. Hans Selye, *The Stress of Life* (New York: McGraw-Hill, 1956); Franz Alexander, Thomas M. French, and George Pollock, eds., *Psychosomatic Specificity: Experimental Study and Results,* vol. 1 (Chicago: Chicago University Press, 1968); Helen F. Dunbar, *Psychosomatic Diagnosis* (New York: Hoefer, 1943).

32. Edwin F. Gildea, "Special Features of Personality Which Are Common to Certain Psychosomatic Disorders," *Psychosomatic Medicine* 11 (1949): 273–281.

33. Felix Deutsch, *The Mysterious Leap from the Mind to the Body* (Madison, CT: International Universities Press, 1973).

34. George E. Vaillant, "Why Men Seek Psychotherapy, I: Results of a Survey of College Graduates," *American Journal of Psychiatry* 129 (1972): 645–651.

35. George E. Vaillant, "Natural History of Male Psychological Health, IV: What Kinds of Men Do Not Get Psychosomatic Illness," *Psychosomatic Medicine* 40 (1978): 420–431.

36. George E. Vaillant, "Natural History of Male Psychological Health, II: Some Antecedents of Healthy Adult Adjustment," *Archives of General Psychiatry* 31 (1974): 15–22.

37. Harlan M. Krumholz, Teresa E. Seeman, Susan S. Merrill, et al., "Lack of Association Between Cholesterol and Coronary Heart Disease Mortality and Morbidity and All-Cause Mortality in Persons Older than 70 Years," *Journal of the American Medical Association* 272 (1994). 1335.

38. Clark W. Heath, *What People Are* (Cambridge, MA: Harvard University Press, 1945).

39. Joseph R. DiFranza and Mary P. Guerrera, "Alcoholism and Smoking," *Journal of Studies on Alcohol and Drugs* 51 (1990): 130–135; George E. Vaillant, Paula P. Schnurr, John A. Baron, et al., "A Prospective Study of the Effects of Cigarette Smoking and Alcohol Abuse on Mortality," *Journal of General Internal Medicine* 6 (1991): 299–304.

40. Paula P. Schnurr, Caroline O. Vaillant, and George E. Vaillant, "Predicting Exercise in Late Midlife from Young Adult Personality Characteristics," *International Journal of Aging and Human Development* 30 (1990). 153–161, H. Taylor, *Harvard Sports Code,* rev. ed. (Cambridge, MA: Harvard University Press, 1979).

41. James S. House, Karl R. Landis, and Debra Umberson: "Social Relationships and Health," *Science* 248 (1998): 540–545.

42. George E. Vaillant, Stephanie E. Meyer, Kenneth J. Mukamal, et al., "Are Social Supports in Late Midlife a Cause or a Result of Successful Physical Aging?" *Psychological Medicine* 28 (1998): 1159–1168.

8. RESILIENCE AND UNCONSCIOUS COPING

(Epigraph) Sigmund Freud, "Analysis Terminable and Interminable (1937)," *Standard Edition,* vol. 23, ed. J. Strachey (London: Hogarth Press, 1964), 237.

1. Claude Bernard, *An Introduction to the Study of Experimental Medicine* (1865) (New York: Macmillan, 1927), 188.

2. Adolf Meyer, *The Collected Papers of Adolf Meyer,* vol. 4: *Mental Hygiene* (1908), ed. Eunice E. Winters (Baltimore: Johns Hopkins University Press, 1950–1952).

3. Sigmund Freud, "The Neuro-Psychoses of Defense (1894)," in *Standard Edition* vol. 3, ed. J. Strachey (London: Hogarth Press, 1964), 45–61; Sigmund Freud, "Inhibitions, Symptoms, and Anxiety (1926)," in *Standard Edition,* vol. 20, ed. J. Strachey (London: Hogarth Press, 1964), 87–157.

4. George E. Vaillant, "Theoretical Hierarchy of Adaptive Ego Mechanisms," *Archives of General Psychiatry* 24 (1971): 107–118.

5. George E. Vaillant, *Adaptation to Life* (Boston: Little, Brown, 1977); George E. Vaillant, *The Wisdom of the Ego* (Cambridge: Harvard University Press, 1993); George E. Vaillant, *Ego Mechanisms of Defense: A Guide for Clinicians and Researchers* (Washington, D.C.: American Psychiatric Publishing, 1994).

6. Phebe Cramer, *The Development of Defense Mechanisms* (New York: Springer Verlag, 1991); Andrew E. Skodol and John C. Perry, "Should an Axis for Defense Mechanisms Be Included in DSM-IV?" *Comprehensive Psychiatry* 34 (1993): 108–119.

7. American Psychiatric Association, *Diagnostic and Statistical Manual of Mental Disorders,* 4th ed. (Washington, D.C.: American Psychiatric Publishing, 1992), 751–753.

8. Vaillant, *Ego Mechanisms of Defense.*

9. Dan P. McAdams, *The Redemptive Self* (New York: Oxford University Press, 2006).

10. George E. Vaillant, "Adaptive Mental Mechanisms: Their Role in a Positive Psychology," *American Psychologist* 55 (2000): 89–98.

11. George E. Vaillant and Caroline O. Vaillant, "Natural History of Male Psychological Health, XII: A Forty-Five Year Study of Successful Aging at Age 65," *American Journal of Psychiatry* 147 (1990): 31–37.

12. Stephen G. Post, ed., *Altruism and Health* (New York: Oxford, 2007).

13. George E. Vaillant and Leigh McCullough, "The Washington University Sentence Completion Test Compared with Other Measures of Adult Ego Development," *American Journal of Psychiatry* 144 (1987): 1189–1194.

14. George E. Vaillant, "Involuntary Coping Mechanisms: A Psychodynamic Perspective," *Dialogues in Clinical Neuroscience* 13 (2011): 366–370.

9. ALCOHOLISM

(Epigraph) *Alcoholics Anonymous* (New York: Alcoholics Anonymous World Services, 2001), 58–59.

1. George E. Vaillant, *The Natural History of Alcoholism Revisited* (Cambridge: Harvard University Press, 1995).

2. Thomas R. Dawber, *The Framingham Study* (Cambridge: Harvard University Press, 1980); Lisa F. Berkman and S. Leonard Syme, "Social Networks, Host Resistance, and Mortality: A Nine-Year Follow-Up of Alameda County Residents," *American Journal of Epidemiology* 109 (1979): 186–201.

3. Vaillant, *The Natural History of Alcoholism Revisited.*

4. Sheldon Glueck and Eleanor Glueck, *Unraveling Juvenile Delinquency* (New York: Commonwealth Fund, 1950).

5. American Psychiatric Association, *Diagnostic and Statistical Manual of Mental Disorders,* 3rd ed. (Washington, DC: American Psychiatric Association, 1980).

6. Vaillant, *The Natural History of Alcoholism Revisited;* Melvin L. Seltzer, "The Michigan Alcoholism Screening Test: The Quest for a New Diagnostic Instrument," *American Journal of Psychiatry* 127 (1971): 1653–1658.

7. Vaillant, *The Natural History of Alcoholism Revisited,* 161.

8. William Hogarth, *Engravings* (Mineola, NY: Dover Publications, 1973).

9. E. Morton Jellinek, *The Disease Concept of Alcoholism* (New Haven: Hillhouse Press, 1960).

10. *Alcoholics Anonymous* (New York: Alcoholics Anonymous World Services, 2001).

11. J. Michael Polich, David J. Armor, and Harriet B. Braiker, *The Course of Alcoholism* (New York: Wiley, 1981); Jim Orford and Griffith Edwards, *Alcoholism* (Oxford: Oxford University Press, 1977); Don Cahalan and Robin Room, *Problem Drinkers among American Men* (New Brunswick, NJ: Rutgers Center for Alcohol Studies, 1974).

12. Vaillant, *The Natural History of Alcoholism Revisited,* 202.

13. Karl A. Menninger, *Man against Himself* (New York: Harcourt Brace, 1938), 177.

14. Robert P. Knight, "The Dynamics and Treatment of Chronic Alcohol Addiction," *Bulletin of the Menninger Clinic* 1 (1937), 234.

15. Paul Schilder, "The Psychogenesis of Alcoholism," *Quarterly Journal of Studies on Alcohol* 2 (1940): 277–292.

16. Jellinek, *The Disease Concept of Alcoholism,* 153.

17. Michael L. Selzer, "Alcoholism and Alcoholic Psychoses," *Comprehensive Textbook of Psychiatry,* 3rd ed., ed. Harold I. Kaplan, Alfred M. Freedman, and Benjamin J. Sadock (Baltimore: Williams and Wilkins, 1980), 1629.

18. Ernest Simmel, "Alcoholism and Addiction," *Psychoanalytic Quarterly* 17 (1948): 6–31; Howard T. Blane, *The Personality of the Alcoholic: Guises of Dependency* (New York: Harper and Row, 1968); George E. Vaillant, "The Natural History of Male Psychological Health, VIII: Antecedents of Alcoholism and 'Orality,'" *American Journal of Psychiatry* 17 (1980): 181–186; George Winokur, Paula J. Clayton, and Theodore Reich, *Manic Depressive Illness* (St. Louis: C.V. Mosby, 1969); Lee N. Robins, *Deviant Children Grown Up: A Sociological and Psychiatric Study of Sociopathic Personality* (Baltimore: Williams and Wilkins, 1966); Vaillant, *The Natural History of Alcoholism Revisited.*

19. George E. Vaillant, "Is Alcoholism More Often the Cause or the Result of Depression?" *Harvard Review of Psychiatry* 1 (1993): 94–99; George E. Vaillant, "A Long-Term Follow-Up of Male Alcohol Abuse," *Archives of General Psychiatry* 53 (1996): 243–249.

20. Vaillant, *The Natural History of Alcoholism Revisited.*

21. George E. Vaillant, "Natural History of Male Alcoholism, V: Is Alcoholism the Cart or the Horse to Sociopathy?" *British Journal of Addiction* (1983), 78: 317–326.

22. Vaillant, *The Natural History of Alcoholism Revisited.*

23. Marc Alan Schuckit, "Advances to Understanding the Vulnerability to Alcoholism," *Addictive States,* ed. Charles P. O'Brien and Jerome H. Jaffe (New York: Raven Press, 1992), 192.

24. Lee N. Robins and Darrel A. Regier, *Psychiatric Disorders in America* (New York: The Free Press, 1991).

25. Vaillant, *The Natural History of Alcoholism Revisited,* chapter 4, Table 4.1a.

26. Ibid.

27. Keith Humphreys, "Alcohol & Drug Abuse: A Research-Based Analysis of the Moderation Management Controversy," *Psychiatric Services* 54 (2003): 621–622.

28. George E. Vaillant, *The Natural History of Alcoholism* (Cambridge, MA: Harvard University Press, 1983), 229.

29. Vaillant, *The Natural History of Alcoholism Revisited,* 282.

30. Ibid., 181; Vaillant, "A Long-Term Follow-Up of Male Alcohol Abuse"; George E. Vaillant, "What Can Long-Term Follow-Up Teach Us about Relapse and Prevention of Relapse in Addiction?" *British Journal of Addiction* 83 (1988): 1147–1157.

31. Robb Stall and Patrick Biernacki, "Spontaneous Remission from the Problematic Use of Substances: An Inductive Model Derived from a Comparative Analysis of the Alcohol, Opiate, Tobacco and Food/Obesity Literatures," *International Journal of the Addictions* (1986): 1–23.

32. Vaillant, *The Natural History of Alcoholism Revisited.*

10. SURPRISING FINDINGS

1. Peter Townsend, Nick Davidson, and Margaret Whitehead, eds., *Inequalities in Health: The Black Report and the Health Divide* (Harmondsworth, England: Penguin, 1988).

2. Marcia Angell, "Privilege and Health—What Is the Connection?" *New England Journal of Medicine* 329 (1993): 126–127.

3. Michael G. Marmot, Hans Bosma, Harry Hemingway, et al., "Contribution of Job Control and Other Risk Factors and Social Variations of Coronary Heart Disease Incidence," *Lancet* 337 (1991): 1387–1393.

4. George E. Vaillant, *Natural History of Alcoholism Revisited* (Cambridge: Harvard University Press, 1995).

5. Howard S. Friedman and Leslie R. Martin, *The Longevity Project* (New York: Hudson Street Press, 2011).

6. David W. Baker, Michael S. Wolf, Joseph Feinglass, et al., "Health Literacy and Mortality among Elderly Persons," *Archives of Internal Medicine* 167 (2007): 1503–1509.

7. John P. Monks, *College Men at War* (Boston: American Academy of Arts and Science, 1951).

8. Paul T. Costa, Jr., and Robert R. McCrae, *The NEO Personality Inventory Manual* (Odessa, FL: Psychological Assessment Resources Inc., 1985).

9. Kimberly A. Lee, George E. Vaillant, William C. Torrey, et al., "A 50-Year Prospective Study of the Psychological Sequelae of World War II Combat," *American Journal of Psychiatry* 152 (1995): 516–522.

10. Mary L. Pendery, Irving M. Malzmann, and L. Jolyon West, "Controlled Drinking by Alcoholics? New Findings and a Revolution of a Major Affirmation Study," *Science* 217 (1982): 169–175; Vaillant, *Natural History of Alcoholism Revisited.*

11. Frederick L. Wells and William L. Woods, "Outstanding Traits," *Genetic Psychology Monographs* 33 (1945): 127–249; Stephen Soldz and George E. Vaillant, "The Big Five Personality Traits and the Life Course: A 45-Year Longitudinal Study," *Journal of Research in Personality,* 33 (1999): 208–232.

12. Paul Heelas and Linda Woodhead, *The Spiritual Revolution: Why Religion Is Giving Way to Spirituality* (London: Blackwell, 2005); Richard Dawkins, "Is Science a Religion?" *The Humanist* 26 (1997).

13. George Gallup, Jr., and D. Michael Lindsay, *Surveying the Religious Landscape: Trends in US Beliefs* (Harrisburg, PA: Morehouse, 1999); Harold G. Koenig, Judy C. Hays, David B. Larson, et al., "Does Religious Attendance Prolong Survival? A Six-Year Follow-Up Study of 3,968 Older Adults," *Journal of Gerontology, Series A, Biological Sciences & Medical Sciences,* 54A (1999): M370–M376; Harold G. Koenig, Michael E. McCullough, and David B. Larson, *Handbook of Religion and Health* (New York: Oxford University Press, 2001).

14. Richard P. Sloan, Emilia Bagiella, and Tia Powell, "Religion, Spirituality, Medicine," *The Lancet* 353 (1999): 664–667.

15. George E. Vaillant, Janice Templeton, Monika Ardelt, et al., "The Natural History of Male Mental Health: Health and Religious Involvement," *Social Science & Medicine* 66 (2008): 221–231.

16. August B. Hollingshead and Frederick C. Redlich, *Social Class and Mental Illness* (New York: Wiley, 1958); George E. Vaillant, "Natural History of Male Psychological Health, II: Some Antecedents of Healthy Adult Adjustment," *Ar-*

chives of General Psychiatry 31 (1974): 15–22; Clark W. Heath, *What People Are* (Cambridge: Harvard University Press, 1945);Vaillant, *Natural History of Alcoholism Revisited.*

17. George E.Vaillant, Stephanie E. Meyer, Kenneth J. Mukamal, et al., "Are Social Supports in Late Midlife a Cause or a Result of Successful Physical Aging?" *Psychological Medicine* 28, 69 (1998): 1159–1168; George E.Vaillant, "Natural History of Male Psychological Health, XIV: Relationship of Mood Disorder Vulnerability to Physical Health," *American Journal of Psychiatry* 55 (1998a): 184–191; Xing-Jia Cui and George E.Vaillant, "Antecedents and Consequences of Negative Life Events in Adulthood: A Longitudinal Study," *American Journal of Psychiatry* 152 (1996): 21–26.

18. Aaron Lazare, Gerald L. Klerman, and David J. Armor, "Oral, Obsessive and Hysterical Personality Patterns," *Archives of General Psychiatry* 14 (1966): 624–630; Derek M. Isaacowitz, George E. Vaillant, and Martin E. P. Seligman, "Strength and Satisfaction Across the Adult Lifespan," *International Journal of Aging and Human Development* 27 (2003): 181–201.

19. Vaillant, "Health and Religious Involvement."

20. Koenig, McCullough, and Larson, *Handbook of Religion and Health.*

21. Sloan, Bagiella, and Powell, "Religion, Spirituality, Medicine."

22. Marc A. Musick, James S. House, and David R. Williams, "Attendance at Religious Services and Mortality in a National Sample," *Journal of Health and Social Behavior* 45 (2004), 204.

23. Vaillant, *Natural History of Alcoholism Revisited.*

24. Vaillant, "Health and Religious Involvement."

25. R. Bayliss, C. Clarke, A. G. H. Whitfield, "Problems in Comparative Longevity," *Journal of Royal College of Physicians* (London) 21 (1987): 134–139.

26. George E.Vaillant, Diane Roston, and Gregory J. McHugo, "An Intriguing Association Between Ancestral Mortality and Male Affective Disorder," *Archives of General Psychiatry* 49 (1992): 709–715.

27. Aaron J. Rosanoff, Leva H. Handy, and Isabel R. Plesset, "The Etiology of Manic-Depressive Syndromes with Special Reference to Their Occurrence in Twins," *American Journal of Psychiatry* 91 (1935): 725–740.

28. Elliot S. Gershon, "Genetics," *Manic-Depressive Illness,* ed. Frederick K. Good-

win and Kay R. Jamison (New York: Oxford University Press, 1990), 373–401; Stephen V. Faraone, William S. Kremen, and Ming T. Tsuang, "Genetic Transmission of Major Affective Disorders: Quantitative Models and Linkage Analyses," *Psychological Bulletin* 108 (1990): 109–127; Julien Mendlewicz, Serge Sevy, Huguette Brocas, et al., "Polymorphic DNA Marker on X-Chromosome and Manic Depression," *Lancet* 1 (1987): 1230–1232.

11. SUMMING UP

1. Robert R. McCrae and Paul T. Costa, Jr., "The Stability of Personality: Observations and Evaluations," *Current Directions in Psychological Sciences* 3 (1994): 173–175.

2. K. Warner Schaie, *Longitudinal Studies of Adult Psychological Development* (New York: Guilford Press, 1983).

3. Erik H. Erikson, *Childhood and Society* (New York: Norton, 1951).

4. Paul T. Costa, Jr., and Robert R. McCrae, *The NEO Personality Inventory Manual* (Odessa, FL: Psychological Assessment Resources, 1985); Robert R. McCrae and Paul T. Costa, Jr., *Emerging Lives, Enduring Dispositions* (Boston: Little, Brown, 1984).

5. Stephen Soldz and George E. Vaillant, "The Big Five Personality Traits and the Life Course: A 45-Year Longitudinal Study," *Journal of Research in Personality* 33 (1999): 208–232.

6. George E. Vaillant and Caroline O. Vaillant, "Natural History of Male Psychological Health, X: Work as a Predictor of Positive Mental Health," *American Journal of Psychiatry* 138 (1981): 1433–1440; George E. Vaillant, *Natural History of Alcoholism Revisited* (Cambridge: Harvard University Press, 1995).

7. John A. Clausen, *American Lives* (Berkeley: University of California Press, 1995).

8. Howard S. Friedman and Leslie R. Martin, *The Longevity Project* (New York: Hudson Street Press, 2011).

9. George E. Vaillant, *Natural History of Alcoholism* (Cambridge: Harvard University Press, 1983).

10. Vaillant, *Natural History of Alcoholism Revisited*.

11. Sigmund Freud, "The Neuropsychoses of Defense (1894)," in *Standard Edition,* vol. 3, ed. J. Strachey (London: Hogarth Press, 1962), 41–61; George E. Vaillant, "The Historical Origins and Future Potential of Sigmund Freud's Concept of the Mechanisms of Defense," *International Review of Psychoanalysis* 19 (1992): 35–50.

12. Robert V. Kail and John C. Cavanaugh, *Human Development,* 6th ed. (Belmont, CA: Wadsworth, 2010).

13. Kail and Cavanaugh, *Human Development;* William C. Crain, *Theories of Development* (New York: Prentice Hall, 2011).

14. Joshua W. Shenk, "What Makes Us Happy," *Atlantic* 303 (2009): 36–53.

ACKNOWLEDGMENTS

The title page of this book is misleading, for it suggests a single authorship. In truth, the book represents a vast collaborative effort that has continued for seventy-five years. The effort began in the late 1930s as two separate studies: a study of juvenile delinquency by Sheldon and Eleanor Glueck at Harvard Law School, and the Grant Study by Clark Heath and Arlie Bock at the Harvard University Health Services. In 1972 the studies were brought together under the auspices of the Harvard University Health Services as the Study of Adult Development. I have written this book as one member of a very large team, on which I have had the privilege of playing for forty-six of its seventy-five years. Many others can also claim authorship.

I am most deeply indebted to the several hundred erstwhile college sophomores and Boston schoolboys who are the Study's subjects. Since 1940, they have generously shared their time, their lives, and their experiences. This work is also indebted to the two independent teams of researchers who conceived, sustained, funded, and guided this longitudinal research for the first thirty years of the Study. For the 456 Inner City schoolboys this meant Sheldon Glueck and Eleanor Glueck and their team at the Harvard Law School. For the 268 Grant Study College men this meant William T. Grant, Arlie Bock, Clark Heath, Lewise Davies, Charles McArthur, and their team at the Harvard University Health Services.

In the forty years since the Grant and the Glueck studies were consolidated into the Study of Adult Development, many individuals have played critical roles and have been formally acknowledged in Study books. One research associate who has not been properly acknowledged is John Martin-Joy, who for almost two decades has obtained reliable evidence (as yet unpublished) that after age fifty, defenses in healthy men continue to mature until age seventy-five.

For this book I wish to acknowledge the very special contributions of five extraordinary women and one extraordinary man who in different ways played major roles in bringing this seventy-five-year study to fruition. Eva Milofsky for ten years was the keystone to the Study's staff and conducted almost half of the Grant Study midlife interviews. Maren Batalden conducted many of the retirement interviews, and her prose is the most beautiful in the book. Caroline Vaillant labored for twenty years in countless ways on the Study, giving wise advice and displaying an uncanny knack for finding "lost" subjects. For the last twenty years Robin Western has been responsible for this being a seventy-five-year, not a fifty-five-year study. Her tact and grace in contacting the men and their doctors have kept attrition to an absolute minimum, and her orderly mind has kept the vast Study archives and her boss in praiseworthy order. For more than fifteen years the internist Ken Mukamal was the extremely thorough blind rater of the men's physical exams. Finally, I want to acknowledge the extraordinary contribution of Eve Golden, who spent hour after hour above and beyond the call of editorial duty transforming my disorderly academic prose into lucid English. Only my own greediness keeps me from acknowledging Eve as the book's coauthor.

I hope that the many research associates who have helped me to create this book but whose names do not appear will understand that the limiting factor is space and not gratitude. I am also grateful to the thousands of physicians who generously supported our efforts without billing their medical examinations of the men to the Study.

Over the years, four broad-minded department chairmen, Paul Myerson, John Mack, Miles Shore, and Jonathan Borus, helped to ensure that I had time for this Study. As directors of the Harvard University Health Services, Dana Farnsworth, then Warren Wacker, and recently David Rosenthal have played indispensible roles as hosts and advisers.

This book is indebted to funding from the William T. Grant Foundation, The John T. Templeton Foundation, the Harvard Neuro-discovery Center, the Fidelity Foundation, the National Institute on Alcohol Abuse and Alcoholism, the National Institute of Mental Health, and the National Institute on Aging. God bless their philan-thropy.

Finally, anyone who has ever written an academic text knows that it is senior editors and critics smarter than oneself who make such books possible. My heartfelt thanks go to Robert Waldinger (Study Director), Elizabeth Knoll (Executive Editor at Harvard Press), Bob Drake, Dan McAdams, and my wife, Diane Highum, M D.

INDEX

200; behavior during interviews, 174;
questionnaires sent to, 193

Woods, William, 62, 74, 101, 142, 355;
limitations of methodology, 130; personality profiling scheme of, 77–78, 82,
130

Word association test, 18, 73

Wordsworth, William, 108

Work, 374–375, 377; enjoyment of job,
386; hours worked in old age, 167; lack
of pleasure in, 173–174; work problems,
142

World War I, 61, 68

World War II, 67, 68, 166, 281, 333; censorship during, 83; combat exposure in,
247; "combat fatigue" in, 101–102, 319;
creative process in veteran of, 363;

Grant Study men in armed services,
82; identification of potential officers,
79–80; Manhattan Project, 57; military
rank in, 43, 45, 172, 251, 323; officer selection in, 35; Terman women and, 382,
383

Wright, Thomas, 62

W. T. Grant Foundation, 107

X-chromosome-linked disorders, 348–
349

Yakovlev, Paul, 147

Young, Algernon, 180–186, 214

Young Man, You Are Normal (Hooton), 77,
140